Fundamentals of Pattern Recognition and Machine Learning

Ulisses Braga-Neto

Fundamentals of Pattern Recognition and Machine Learning

 Springer

Ulisses Braga-Neto
Department of Electrical
and Computer Engineering
Texas A&M University
College Station, TX, USA

ISBN 978-3-030-27658-4   ISBN 978-3-030-27656-0 (eBook)
https://doi.org/10.1007/978-3-030-27656-0

This Springer imprint is published by the registered company Springer Nature Switzerland AG
The registered company address is: Gewerbestrasse 11, 6330 Cham, Switzerland

To Flávia

Preface

> "Only the educated are free."
> –Epictetus.

The field of pattern recognition and machine learning has a long and distinguished history. In particular, there are many excellent textbooks on the topic, so the question of why a new textbook is desirable must be confronted. The goal of this book is to be a concise introduction, which combines theory and practice and is suitable to the classroom. It includes updates on recent methods and examples of applications based on the python programming language. The book does not attempt an encyclopedic treatment of pattern recognition and machine learning, which has become impossible in any case, due to how much the field has grown. A stringent selection of material is mandatory for a concise textbook, and the choice of topics made here, while dictated to a certain extent by my own experience and preferences, is believed to equip the reader with the core knowledge one must obtain to be proficient in this field. Calculus and probability at the undergraduate level are the minimum prerequisites for the book. The appendices contain short reviews of probability at the graduate level and other mathematical tools that are needed in the text.

This book has grown out of lecture notes for graduate classes on pattern recognition, bioinformatics, and materials informatics that I have taught for over a decade at Texas A&M University. The book is intended, with the proper selection of topics (as detailed below), for a one or two-semester introductory course in pattern recognition or machine learning at the graduate or advanced undergraduate level. Although the book is designed for the classroom, it can also be used effectively for self-study.

The book does not shy away from theory, since an appreciation of it is important for an education in pattern recognition and machine learning. The field is replete with classical theorems, such as the Cover-Hart Theorem, Stone's Theorem and its corollaries, the Vapnik-Chervonenkis Theorem, and several others, which are covered in this book. Nevertheless, an effort is made in the book to strike a balance between theory and practice. In particular, examples with datasets from applications in Bioinformatics and Materials Informatics are used throughout the book to illustrate the theory. These datasets are also used in end-of-chapter coding assignments based on python. All plots in the text were generated using python scripts, which can be downloaded from the book website. The reader is encouraged to experiment with these scripts and use them in the coding assignments. The book website also contains datasets from Bioinformatics and Materials Informatics applications, which are used in the plots and coding assignments. It has been my experience in the classroom that the understanding of the subject by students is increased significantly once they engage in assignments involving coding and data from real-world applications.

The book is organized as follows. Chapter 1 is a general introduction to motivate the topic. Chapters 2–8 concern classification. Chapters 2 and 3 on optimal and general sample-based classification are the foundational chapters on classification. Chapters 4-6 examine the three main categories of classification rules: parametric, nonparametric, and function-approximation, while Chapters 7 and 8 concern error estimation and model selection for classification. Chapter 9 on dimensionality reduction still deals with classification, but also includes material on unsupervised methods. Finally, Chapters 10 and 11 deal with clustering and regression. There is flexibility for the instructor or reader to pick topics from these chapters and use them in a different order. In particular, the "Additional Topics" sections at the end of most chapters cover miscellaneous topics, and can be included or not, without affecting continuity. In addition, for the convenience of instructors and readers, sections that contain material of a more technical nature are marked with a star. These sections could be skipped at a first reading.

The Exercises section at the end of most chapters contain problems of varying difficulty; some of them are straightforward applications of the concepts discussed in the chapter, while others introduce new concepts and extensions of the theory, some of which may be worth discussing in class. Python Assignment sections at the end of most chapters ask the reader to use python and scikit-learn to implement methods discussed in the chapter and apply them to synthetic and real data sets from Bioinformatics and Materials Informatics applications.

Based on the my experience teaching the material, I suggest that the book could be used in the classroom as follows:

1. A one-semester course focusing on classification, covering Chapters 2-9, while including the majority of the starred and additional topics sections.

2. An applications-oriented one-semester course, skipping most or all starred and additional topics sections in Chapters 2-8, covering Chapters 9-11, and emphasizing the coding assignments.

3. A two-semester sequence covering the entire book, including most or all the starred and additional topics sections.

This book is indebted to several of its predecessors. First, the classical text by Duda and Hart (1973, updated with Stork in 2001), which has been a standard reference in the area for many decades. In addition, the book by Devroye, Györfi and Lugosi (1996), which remains the gold standard in nonparametric pattern recognition. Other sources that were influential to this text are the books by McLachlan (1992), Bishop (2006), Webb (2002), and James et al. (2013).

I would like to thank all my current and past collaborators, who have helped shape my understanding of this field. Likewise, I thank all my students, both those whose research I have supervised and those who have attended my lectures, who have contributed ideas and corrections to the text. I would like to thank Ed Dougherty, Louise Strong, John Goutsias, Ascendino Dias e Silva, Roberto Lotufo, Junior Barrera, and Severino Toscano, from whom I have learned much. I thank Ed Dougherty, Don Geman, Al Hero, and Gábor Lugosi for the comments and encouragement received while writing this book. I am grateful to Caio Davi, who drew several of the figures. I appreciate very much the expert assistance provided by Paul Drougas at Springer, during difficult times in New York City. Finally, I would like to thank my wife Flávia and my children Maria Clara and Ulisses, for their patience and support during the writing of this book.

Ulisses Braga-Neto
College Station, TX
July 2020

Contents

Chapter 1

Introduction

"The discipline of the scholar is a consecration
to the pursuit of the truth."
–Norbert Wiener, *I am a Mathematician*, 1956.

After a brief description of the pattern recognition and machine learning areas, this chapter sets down basic mathematical concepts and notation used throughout the book. It introduces the key notions of prediction and prediction error for supervised learning. Classification and regression are introduced as the main representatives of supervised learning, while PCA and clustering are mentioned as examples of unsupervised learning. Classical complexity trade-offs and components of supervised learning are discussed. The chapter includes examples of application of classification to Bioinformatics and Materials Informatics problems.

1.1 Pattern Recognition and Machine Learning

A *pattern* is the opposite of *randomness*, which is closely related to the notions of *uniformity* and *independence*. For example, the outcomes of an unloaded die are "random," and so is the sequence of digits in the decimal expansion of π, because the frequency distribution of outcomes is uniform. Had that not been the case, the series of outcomes would have *revealed a pattern*, in the form of "clumps" in their frequency distribution. Pattern recognition in this sense is the domain of *unsupervised learning*. On the other hand, there is "randomness" between two events if they are independent. For example, musical preference is independent of the occurrence of heart disease, but food preference is not: there is a pattern of association between a high-fat diet and heart disease. Pattern recognition in this sense is the domain of *supervised learning*.

© Springer Nature Switzerland AG 2020
U. Braga-Neto, *Fundamentals of Pattern Recognition and Machine Learning*,
https://doi.org/10.1007/978-3-030-27656-0_1

Furthermore, the wakeful human mind is constantly acquiring sensorial information from the environment, in the form of vision, hearing, smell, touch, and taste signals. The human mind is the best learning system there is to process these kind of data, in the sense that no computer as yet can consistently outperform a rested and motivated person in recognizing images, sounds, smells, and so on. Machine learning applications in fields such as computer vision, robotics, speech recognition and natural language processing, generally have as their goal to emulate and approach human performance as closely as possible.

Pattern recognition became a significant engineering field during the American space program in the 1960's. Initially, it was closely associated with the analysis of digital images transmitted by deep space vehicles and probes. From this beginning in image analysis, pattern recognition has expanded today to a very broad spectrum of applications in imaging, signal processing, and more. On the other hand, machine learning originated mainly in the neuroscience and computer science areas. This is a field that has achieved great popularity in recent years. Pattern recognition and machine learning have substantial overlap with each other, and a common mathematical setting. In this book, we treat these two subjects as complementary parts of a whole. Other identifiable areas closely related to this topic are artificial intelligence, data science, discriminant analysis, and uncertainty quantification.

1.2   Basic Mathematical Setting

In supervised learning, information about the problem is summarized into a vector of measurements $\mathbf{X} \in R^d$, also known as a *feature vector*, and a *target* $Y \in R$ to be predicted. The relationship between the feature vector $\mathbf{X}$ and the target Y is, in practice, rarely deterministic, i.e., in real applications it is seldom the case that there is a function f such that $Y = f(\mathbf{X})$. Instead, the relationship between $\mathbf{X}$ and Y is specified by a joint *feature-target distribution* $P_{\mathbf{X},Y}$. See Figure 1.1 for an illustration. This state of uncertainty is due mainly to the presence of: 1) hidden or latent factors, that is, factors on which Y depends but that are not available to be observed or measured; 2) measurement noise in the values of the predictor $\mathbf{X}$ itself.

1.3   Prediction

The objective in supervised learning is to *predict* Y given $\mathbf{X}$. A *prediction rule* produces a predictor $\psi : R^d \to R$, such that $\psi(\mathbf{X})$ predicts Y. Notice that the predictor itself is not random; this is the case since in practice one is interested in definite predictions (however, a few examples of random ψ are considered briefly in the book). Design of the predictor ψ uses information about the joint

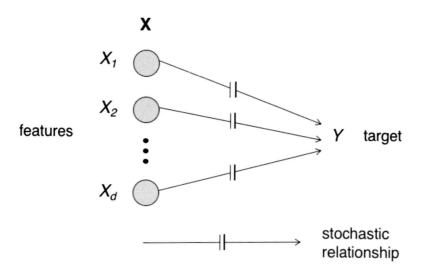

Figure 1.1: Stochastic relationship between the features and target in supervised learning.

feature-target distribution $P_{\mathbf{X},Y}$, which can be:

- Direct knowledge about $P_{\mathbf{X},Y}$.

- Indirect knowledge about $P_{\mathbf{X},Y}$ through an independent and identically distributed (i.i.d.) *sample* $S_n = \{(\mathbf{X}_1, Y_1), \ldots, (\mathbf{X}_n, Y_n)\}$ from $P_{\mathbf{X},Y}$; this is often called the *training data*. (However, as seen in Section 3.5.2, the i.i.d. assumption does not hold, even approximately, in certain problems.)

Any predictor design method employs a combination of these two sources of information. In the extreme case where complete knowledge of $P_{\mathbf{X},Y}$ is available, an *optimal predictor* $\psi^*(\mathbf{X})$ can be obtained in principle, and data are not needed (optimal predictors are discussed in Chapter 2 and Chapter 11). In the other direction, if no knowledge about $P_{\mathbf{X},Y}$ is available, then a *data-driven* prediction rule must rely solely on S_n. Surprisingly, there are conditions under which certain data-driven predictors can approach the optimal predictor as $n \to \infty$, regardless of what the unknown $P_{\mathbf{X},Y}$ is; however, the convergence rate must be arbitrarily slow in the worst case. Therefore, for finite n, which is the practical case, having some knowledge about $P_{\mathbf{X},Y}$ is necessary *to guarantee* good performance; this is known as a *no-free-lunch theorem* (such convergence issues are discussed in Chapter 3).

1.4   Prediction Error

Validity of a predictor, including its optimality, is defined with respect to a prespecified *loss function* $\ell : R \times R \to R$ that measures the "distance" between the predicted value $\psi(\mathbf{X})$ and the target Y. For example, common loss functions are the *quadratic loss* $\ell(\psi(\mathbf{X}), Y) = (Y - \psi(\mathbf{X}))^2$, the *absolute difference loss* $\ell(\psi(\mathbf{X}), Y) = |Y - \psi(\mathbf{X})|$, and the *misclassification loss*

$$\ell(\psi(\mathbf{X}), Y) = I_{Y \neq \psi(\mathbf{X})} = \begin{cases} 1, & Y \neq \psi(\mathbf{X}), \\ 0, & Y = \psi(\mathbf{X}), \end{cases} \tag{1.1}$$

where $I_A \in \{0, 1\}$ is an *indicator variable*, such that $I_A = 1$ if and only if A is true.

As we are dealing with a stochastic model, the loss of a predictor must be averaged over the random variables $\mathbf{X}$ and Y. Accordingly, the *expected loss* of ψ, or its *prediction error*, is defined by

$$L[\psi] = E[\ell(Y, \psi(\mathbf{X}))] . \tag{1.2}$$

An optimal predictor ψ^* minimizes $L[\psi]$ over all $\psi \in \mathcal{P}$, where $\mathcal{P}$ is the class of all predictors under consideration.

1.5   Supervised vs. Unsupervised Learning

In supervised learning, the target Y is always defined and available. There are two main types of supervised learning problems:

- In *classification*, $Y \in \{0, 1, \dots, c - 1\}$, where c is the number of *classes*. Variable Y is called a *label* to emphasize that it has no numeric meaning but simply codes for the different categories. In *binary classification*, $c = 2$ and there are only two classes. For example, the binary classes "healthy" and "diseased" can be coded into the numbers 0 and 1, respectively. A predictor in this case is called a *classifier*. The basic loss criterion for classification is the misclassification loss in (1.1), in which case (1.2) yields the *classification error rate*

$$\varepsilon[\psi] = E[I_{Y \neq \psi(\mathbf{X})}] = P(Y \neq \psi(\mathbf{X})) , \tag{1.3}$$

  i.e., simply the probability of an erroneous classification. In the case of binary classification, $I_{Y \neq \psi(\mathbf{X})} = |Y - \psi(\mathbf{X})| = (Y - \psi(\mathbf{X}))^2$, so that the quadratic, absolute difference, and misclassification losses all yield the classification error in (1.3). Classification is the major concern of the book, being covered extensively in Chapters 2–9.

- In *regression*, Y represents a numerical quantity, which could be continuous or discrete. A common loss function used in regression is the quadratic loss, in which case the prediction error in (1.2) is called the *mean-square error*. In real-valued regression, $Y \in R$ is a continuously-varying real number. For example, the lifetime of a device is a positive real number. Regression methods are discussed in detail in Chapter 11.

On other other hand, in *unsupervised learning*, Y is unavailable and only the distribution of $\mathbf{X}$ specifies the problem. Therefore, there is no prediction and no prediction error, and it is not straightforward to define a criterion of performance. Unsupervised learning methods are mainly concerned in detecting structure in the distribution of $\mathbf{X}$. Examples include dimensionality reduction methods such as Principal Component Analysis (PCA), to be discussed in Chapter 9, and clustering, which is the topic of Chapter 10.

If the target Y is available for only a subpopulation of the feature vector $\mathbf{X}$, we have a hybrid case, called *semi-supervised learning*, which can be semi-supervised classification or regression, according to the nature of Y. The main question in semi-supervised learning is when and how the set of points with missing Y can increase the accuracy of classification or regression.

In addition to supervised and unsupervised learning, another area that is often associated with machine learning is known as *reinforcement learning*. This is a somewhat different problem, however, since it concerns decision making in continuous interaction with an environment, where the objective is to minimize a cost (or maximize a reward) over the long run.

1.6 Complexity Trade-Offs

Complexity trade-offs involving sample size, dimensionality, computational complexity, interpretability, and more, are a characteristic feature of supervised learning methods. The resolution of these trade-offs often involve difficult choices.

A key complexity trade-offis known as the *curse of dimensionality* or the *peaking phenomenon*: for a fixed sample size, the expected classification error initially improves with increasing number of features, but eventually starts to increase again. This is a consequence of the large size of high dimensional spaces, which require correspondingly large training sample sizes in order to design good classifiers. The peaking phenomenon is illustrated in Figure 1.2(a), which displays the expected accuracy in a discrete classification problem for various training sample sizes as a function of the number of predictors. The plot is based on exact formulas for the expected classification accuracy derived by G. Hughes in a classic paper (this was the first paper to demonstrate the peaking phenomenon, which is thus also known as *Hughes Phenomenon*). One can observe that accuracy

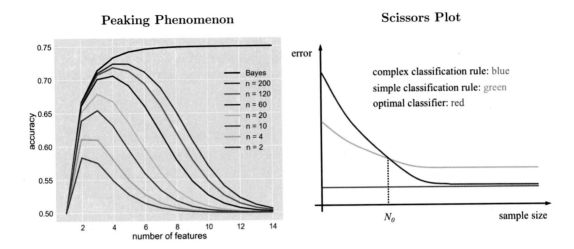

Figure 1.2: Two basic complexity trade-offs in supervised learning. Left: the peaking phenomenon (plot generated by c01_hughes.py). Right: the scissors plot.

increases and then decreases as the number of predictors increases. The error increases eventually due to *overfitting*, which occurs when the sample size is too small (compared to the dimensionality of the problem) for the classification rule to learn the classifier properly. Therefore, the optimal number of features moves to the right (i.e., accuracy "peaks later") with increasing sample size. Notice that the expected accuracy in all cases decreases to the no-information value of 0.5, except for the optimal classification error, which corresponds in this case to an infinite training sample size. The optimal error can never decrease as more features are added (the optimal classifier and classification error are studied in detail in Chapter 2).

Another complexity trade-off in supervised learning can be seen in the so-called *scissors plot*, displayed in Figure 1.2(b), which displays the expected errors of two classification rules and the error of the optimal classifier. The complex classification rule in this example is *consistent*, i.e., its expected error converges to the optimal error as sample size increases, whereas the simple classification rule is not consistent. Perhaps the complex classification rule in this example is *universally consistent*, i.e., consistent under any feature-label distribution (see Chapter 3). However, in this example, the simple classification rule performs better under small sample sizes, by virtue of requiring less data. There is a problem-dependent critical sample size N_0, under which one is in the "small-sample" region, and should use the simpler, non-consistent classification rule. Here, as elsewhere in the book, "small sample" means a small number of training points in comparison to the dimensionality or complexity of the problem. The powerful *Vapnik-Chervonenkis theory* provides distribution-free results relating classification performance to the ratio between sample size and complexity (see Chapter 8).

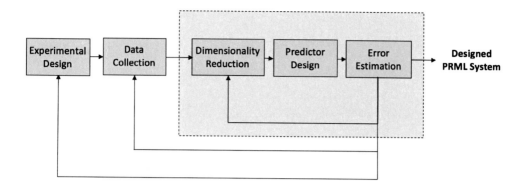

Figure 1.3: The design cycle in supervised learning.

1.7 The Design Cycle

The typical design cycle in supervised learning problems is depicted in Figure 1.3. The process starts with the key step of *experimental design*, which includes framing the question that needs to be addressed, identifying the populations and features of interest, determining the appropriate sample sizes and sampling mechanism (e.g., whether the populations are to be sampled jointly or separately). Next, there is the step of data collection itself, followed by the three main steps: *dimensionality reduction*, which seeks to extract the important discriminatory information from the data, *predictor design*, which might be a classifier or regressor, and *error estimation*, where the accuracy of the constructed predictor is evaluated. If the estimated accuracy is not desirable, the process may cycle back over the dimensionality reduction and predictor design steps. In case no good predictor can be found, it is possible that the data collection process was flawed, e.g., due to the presence of bad sensors. Finally, it is possible that the experiment was badly designed, in which case the process has to restart again from the beginning. In this book, we focus on the three key steps of dimensionality reduction, predictor design, and error estimation. We however also examine briefly experimental design issues, such as the effects of sample size and sampling mechanism on prediction accuracy (see Chapter 3).

1.8 Application Examples

In this section, we illustrate concepts described previously by means of two classification examples from real-world problems in Bioinformatics and Materials Informatics. This section also serves as an introduction to these application areas, which are employed in examples and coding assignments throughout the book.

1.8.1   Bioinformatics

A very important problem in modern medicine is classification of disease using the activity of genes as predictors. All cells in a given organism contain the same complement of genes (DNA), also known as a **genome**, but different cell types are associated with different levels of activation of the genes in the genome. A gene could be silent, or it could be active. When active, the genetic code is transcribed into messenger RNA (mRNA), which in turn is translated into a protein, according to the "fundamental dogma of molecular biology":

$$\text{DNA} \;\rightarrow\; \text{mRNA} \;\rightarrow\; \text{Protein}.$$

A gene is more or less expressed according to how much mRNA is transcribed, and therefore how much protein is produced. This can be measured on a genome-wide scale by means of high-throughput gene-expression experiments, e.g., by hybridization of the mRNA to *DNA microarrays*, or by direct sequencing and counting of the mRNA molecules, also known as *RNA-seq* technology. In this context, specific mRNA sequences are also called *transcripts*. Most transcripts can be mapped to unique genes.

Example 1.1. This example concerns the prediction of the clinical outcome of dengue fever. Dengue is a viral disease, which is transmitted by mosquitos in the genus *Aedes* and is endemic in tropical regions. Its clinical outcome can be categorized into "classical" dengue fever (DF), which is a debilitating disease but typically nonlethal, and dengue hemorrhagic fever (DHF), which has a significant mortality rate. It is important during outbreaks to be able to determine which variety is present in patients, in order to avoid overwhelming medical services with unnecessary hospitalizations. Unfortunately, this discrimination can only be made by clinical methods in the second week after the onset of fever. Nascimento et al. [2009] hypothesized that pattern recognition and machine learning methods could be applied to gene expression data from cells of the immune system to predict the development of DHF much earlier than available clinical methods. To investigate this hypothesis, they employed a data set with the expression of 1981 transcripts in 26 patients on the early days of fever, who were later diagnosed with dengue fever, dengue hemorrhagic fever, or nonspecific fever (see Section A8.2). Figure 1.4 displays microarray data for 40 transcripts of the DF and DHF patients from the original data set. These transcripts were obtained by univariate *filter feature selection*, which is discussed in detail in Chapter 9. The particular method used here was to rank the transcripts according to discriminatory power between the DF and DHF classes, as measured by the absolute value of the test statistic of a two-sample t-test, and keep the top 40 transcripts. The data matrix corresponding to the 40 transcripts measured on the DF and DHF

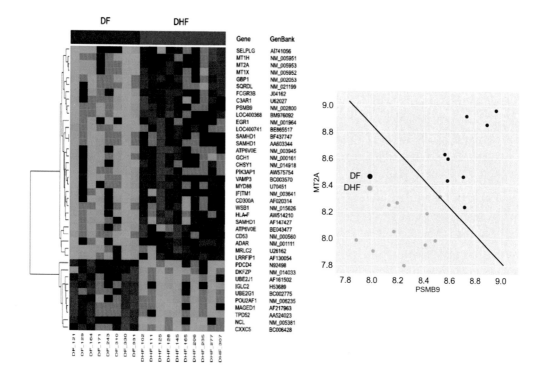

Figure 1.4: Dengue fever prognosis example. Left: heatmap for the gene expression microarray data matrix. Red and green code for high and low expression values, respectively. Rows are organized by hierarchical clustering, with dendrogram displayed on the left of the heatmap. Right: bivariate LDA classifier using two chosen transcripts (plot generated by c01_bioex.py). Left panel from [Nascimento et al., 2009], reprinted under a creative commons attribution (CC BY) license.

patients is displayed in the *heatmap* on the left of Figure 1.4. The heatmap uses a color scale to represent expression level, whereby one goes from bright red to bright green as gene expression goes from high to low (with respect to the average expression). Each column of the data matrix is the gene expression profile for the corresponding patient. The rows of the matrix correspond to the expression of each gene, and are organized here according to similarity using *hierarchical clustering*. This produces the *dendrogram*, seen on the left of the data matrix. Hierarchical clustering, which is discussed in detail in Chapter 10, allows us to see clear patterns in the heatmap. In particular, we can observe that the expression of the transcripts at the top, from SELPLG to LRRFIP1, is predominantly low in the DF patients, while expression of these genes is mixed, but mostly high, in the DHF patients. The reverse is observed in the bottom transcripts, from PDCD4 to CXXC5. Hence, simple *univariate classifiers*, consisting of a single transcript, could be used to discriminate the two classes.

However, using *multivariate classifiers* based on multiple variables makes it is possible to discriminate the two classes with higher accuracy. Indeed, accurate univariate classifiers do not exist in most problems, which requires pattern recognition and machine learning methods to be inherently multivariate. We illustrate the multivariate approach with a two-dimensional classifier based on two of the 40 transcripts: PSMB9 and MTA2. The two-dimensional space corresponding to the expression of these two variables is the *feature space*. Each point in this space represents a patient in the sample data, which are used to train a *Linear Discriminant Analysis* (LDA) classifier (see Chapter 4). The *decision boundary* of this classifier is a line in two dimensions (and a hyperplane in higher dimensions). The feature space, sample points, and linear decision boundary are plotted on the right side of Figure 1.4. Since the transcripts come from the top of the data matrix, high expression of both, corresponding to the upper right area in the plot, is predictive of DF (good prognosis). On the other hand, low expression of these transcripts, corresponding to the lower left area, is predictive of DHF (poor prognosis). This *interpretability* property of linear classifiers is a big advantage in many applications: it leads to simple and testable scientific hypotheses about the phenomena at hand (which is not the case if complex nonlinear classification rules are used). It also offers the opportunity for validation using prior domain knowledge; in this particular example, it was previously known that PSMB9 participates in key cellular viral defense mechanisms, so that its lower expression is compatible with an increased exposure to the severe form of the disease. The general problem of selecting the appropriate variables for discrimination is called *feature selection*, which is a dimensionality reduction technique, to be discussed in detail in Chapter 9.

The problem of error estimation, mentioned previously, concerns how to use sample data to estimate the error of a predictor, such as the one in Figure 1.4. Ideally, one would have available a large amount of independent *testing data*, which is never used in training the classifier, and would compute a *test-set error estimate* as the total number of disagreements between the label predicted by the classifier and the actual label of each testing point divided by the total number of testing points. This estimator is guaranteed to be *unbiased*, but its variance depends on the testing sample size, so that the estimate will be accurate only if there is a large amount of labeled testing data, which is often an unrealistic requirement in real applications. The classifier on the right of Figure 1.4 used the entire data set for training, so there is no testing data available. An alternative is to test the classifier on the training data itself. We can see in Figure 1.4 that there are 18 training points and the classifier makes one error. Hence, the training-set error, also known as the *apparent error* or the *resubstitution error estimator*, comes out to $1/18 \approx 5.55\%$. This seems good, but one must be careful, because under small sample sizes, as is the case here, the resubstitution error estimator can display substantial *optimistic bias*, i.e., be significantly smaller on average than the true error, due to overfitting. Error estimation for classification is discussed in detail in Chapter 7. ◇

1.8.2 Materials Informatics

Pattern recognition and machine learning are typically used in Materials Science to establish *Quantitative Structure Property Relationships* (QSPR), i.e., predictive models that link the structure or composition of a material to its macroscopic properties, such as strength, ductility, malleability, and so on, with the ultimate goal of accelerating the discovery of new materials.

Example 1.2. This example concerns experimentally recorded values of atomic composition and stacking fault energy (SFE) in austenitic stainless steel specimens, obtained by Yonezawa et al. [2013] (See Section A8.4 for details about this data set.) The stacking fault energy is a microscopic property related to the resistance of austenitic steels. The purpose of the experiment is to develop a model to classify a steel sample as high-SFE or low-SFE based only on the atomic composition; high-SFE steels are less likely to fracture under strain and may be desirable in certain applications. The data set contains 17 features corresponding to the atomic composition (percentage weight of each atomic element) of 473 steel specimens. We face the problem that the data matrix contains many zero values, which are typically measurements that fell below the sensitivity of the experiment, and are therefore unreliable. These constitute *missing values* (this can occur for other reasons as well, such as a faulty or incomplete experiment). One option to address this issue is to apply *data imputation*. For example, a simple imputation method is to fill in missing values with an average of neighboring values. Given a large sample size and abundance of features, a simpler, and possibly safer, option is to discard measurements containing zero/missing values. Here, we discard all features that do not have at least 60% nonzero values across all sample points, and then remove any remaining sample points that contain zero values. The remaining training points are categorized into high-SFE steels (SFE$\geq$45) vs. low-SFE steels (SFE$\leq$ 35), with points of intermediate SFE value being dropped. This results in a reduced data set containing 123 specimens and 7 features.

Univariate histograms[1] of the two classes for three of the atomic features are plotted at the top of Figure 1.5. Also displayed are kernel estimates of the corresponding probability densities (see Chapter 5 for a discussion of kernel density estimators). From the left to the right, we can see that there is substantial overlap between the class histograms and density estimates for the first feature (Cr, chromium). Intuitively, this should not be a good feature for discriminating between low and high stacking fault energy steels. The next feature (Fe, iron) seems much more promising for classification, given the small overlap between the density estimates. In fact, one could draw a rough decision boundary, according to which steels with an iron percent weight less than about 67.5% is predicted to be a high-SFE material; otherwise, it is predicted to be a low-SFE material (in other words, iron content and stacking fault energy are *negatively correlated*, according to this data set). The next feature (Ni, Nickel) appears to be even more predictive, as the overlap between the

---

[1]Histograms are normalized counts of instances that fall inside intervals (also called "bins") that partition the domain. A histogram is a rough approximation of the *probability density* associated with a numerical measurement.

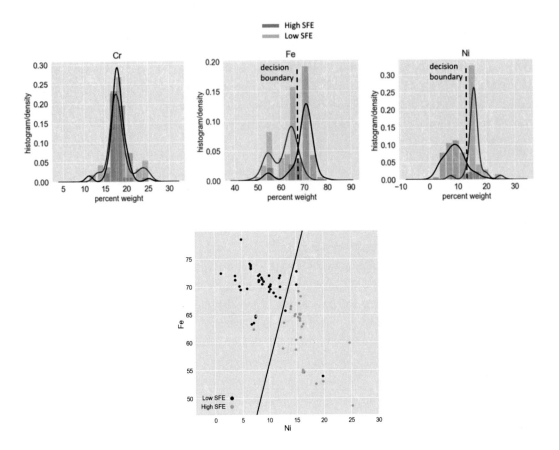

Figure 1.5: Materials Structure-Property Relationships. Top: univariate class-specific histograms, probability density estimates, and decision boundaries for a few of the features in the data set. Bottom: bivariate LDA classifier (plots generated by c01_matex.py).

density estimates is very small. Here, nickel content is positively correlated with SFE. Once again, we can draw a decision boundary, according to which a nickel content of roughly more than 13% seems to be predictive of a high-SFE material. These density estimates are approximations of the *class-conditional densities*, to be defined in Chapter 2, where it is shown that the classification error is determined by the overlap between the class-conditional densities weighted by the corresponding prior probabilities. As already mentioned in Example 1.1, one need not restrict oneself to classifying with a single feature; the two good features Ni and Fe can be combined to obtain a bivariate linear classifier, again an LDA classifier, displayed at the bottom of Figure 1.5. Both the univariate and bivariate classifiers in Figure 1.5 are linear, and therefore interpretable: it is easy to conclude from them that a steel with a smaller content of iron and a larger content of nickel likely exhibits high stacking fault energy. ◇

We caution that the bivariate linear classifier in the previous example was put together by selecting two good-looking univariate features; this would be an example of filter feature selection, already mentioned in Example 1.2. However, theoretical and empirical considerations indicate that this could be a bad idea (e.g., for a theoretical result, see "Toussaint's Counter-Example" in Chapter 9). The issue is that looking at individual features misses the synergistic multivariate effects that occur when features are joined. So bad features in isolation could in theory produce a good classifier when put together; and the opposite could also happen, good features in isolation can join to produce a bad classifier (this is not a common occurrence, but it is possible nevertheless).

1.9 Bibliographical Notes

The standard reference in pattern recognition, since its first edition in 1973, has been Duda et al. [2001]. The classical references on parametric and nonparametric methods in pattern recognition, respectively, are the treatises by McLachlan [1992] and Devroye et al. [1996]. Other references include the books by Hastie et al. [2001], Webb [2002], James et al. [2013], Bishop [2006], and Murphy [2012b]. The latter two references take a mostly Bayesian perspective of the field.

The area of statistical classification was inaugurated by Fisher [1936], where the famous *Iris data set* was introduced. Regression is a much older problem in statistics. The basic method of fitting a regression line is the least-squares method, invented by Gauss more than two centuries ago. The name "regression" was coined by Sir Francis Galton, an English Victorian statistician, in the late 1800's [Galton, 1886], where he observed that children of short or tall parents tend to be taller or shorter than them, respectively (thus, "regression towards the mean"). Duda et al. [2001] attribute the beginnings of unsupervised learning and clustering to the work of K. Pearson on mixtures of Gaussians in 1894. The classical paper by Hughes [1968] demonstrated the peaking phenomenon analytically for the first time.

A popular introduction to the field of reinforcement learning is the book by Sutton and Barto [1998]. A recent edited volume on semi-supervised learning [Chapelle et al., 2010] provides a comprehensive overview of this topic.

A standard reference on molecular biology, including the "fundamental dogma," is Alberts et al. [2002]. The original papers on DNA microarray technology are Schena et al. [1995]; Lockhart et al. [1996], while RNA-seq technology is covered in Marguerat and Bahler [2010]. A detailed description of the classifier depicted on the right side of Figure 1.4 can be found in Braga-Neto [2007]. A good general reference on Bioinformatics is Kohane et al. [2003]. An up-to-date reference on Materials Informatics is the recent edited book by Rajan [2013].

Chapter 2

Optimal Classification

> "But although all our knowledge begins with experience,
> it does not follow that it all arises from experience."
> – Immanuel Kant, *Critique of Pure Reason*, 1781.

As discussed in Chapter 1, the main objective of supervised learning is to predict a target variable Y given the information in a vector of measurements or features $\mathbf{X}$. In classification, Y has no numerical meaning, but codes for a number of different *classes*. If complete knowledge about the joint *feature-label distribution* $P_{\mathbf{X},Y}$ is available, then an optimal classifier can be obtained, in principle, and no training data are needed. Classification becomes in this case a purely probabilistic problem, not a statistical one. In this chapter, we consider at length the probabilistic problem of optimal classification, including the important Gaussian case. We assume that $Y \in \{0, 1\}$; extensions to the multiple-label case $Y \in \{0, 1, \ldots, c - 1\}$, for $c > 2$, are considered in the Exercises section. The reason is that the multiple-label case introduces extra complexities that may obscure the main issues about classification, which become readily apparent in the binary case.

2.1   Classification without Features

We begin with the very simple case where there are no measurements or features to base classification on. In this case, a predictor $\hat{Y}$ of the binary label $Y \in \{0, 1\}$ must be constant, returning always 0 or 1 for all instances to be classified. In such a case, it seems to natural to make the following decision:

$$\hat{Y} = \begin{cases} 1, & P(Y = 1) > P(Y = 0), \\ 0, & P(Y = 0) \geq P(Y = 1). \end{cases} \tag{2.1}$$

© Springer Nature Switzerland AG 2020
U. Braga-Neto, *Fundamentals of Pattern Recognition and Machine Learning*,
https://doi.org/10.1007/978-3-030-27656-0_2

i.e., assign the label of the most common class. The probabilities $P(Y = 1)$ and $P(Y = 0)$ are called the class *prior probabilities* or *prevalences*. (Since $P(Y = 0) + P(Y = 1) = 1$, only one of the prior probabilities need be specified.) Notice that $\hat{Y} = 1$ if and only if $P(Y = 1) > 1/2$.

Notice also that this is equivalent to assigning the label that is closest to the mean $E[Y] = P(Y = 1)$.

The predictor $\hat{Y}$ is indeed optimal, in the sense that its classification error rate

$$\varepsilon[\hat{Y}] = P(\hat{Y} \neq Y) = \min\{P(Y = 1), P(Y = 0)\} \tag{2.2}$$

is minimum over all constant predictors, as can be readily verified. Note that if either of the prevalences $P(Y = 1)$ or $P(Y = 0)$ is small, i.e., one of the two classes is unlikely to be observed, then $\varepsilon[\hat{Y}]$ is small, and this simple classifier without features actually has a small classification error. However, consider testing for a rare disease: it clearly will not do to call all patients healthy without any examination (we return to this topic in Section 2.4 below).

2.2   Classification with Features

There is something odd about classifying without features: one calls the same label for all instances being considered. Luckily, in practice, one always has access to a feature vector $\mathbf{X} = (X_1, \ldots, X_d) \in R^d$ to help classification; each X_i is a *feature*, for $i = 1, \ldots, d$, and R^d is called the *feature space*.

We assume for definiteness that $\mathbf{X}$ is a continuous feature vector in each class, which is an important case in practice. Formally, this means there are two nonnegative functions $p(\mathbf{x} \mid Y = 0)$ and $p(\mathbf{x} \mid Y = 1)$ on R^d, called the *class-conditional densities*, such that:

$$\begin{aligned}
P(\mathbf{X} \in E, Y = 0) &= \int_E P(Y = 0) p(\mathbf{x} \mid Y = 0) \, d\mathbf{x}, \\
P(\mathbf{X} \in E, Y = 1) &= \int_E P(Y = 1) p(\mathbf{x} \mid Y = 1) \, d\mathbf{x},
\end{aligned} \tag{2.3}$$

for any *Borel set* $E \subseteq R^d$, i.e., a set to which a probability can be assigned.[1] With $E = R^d$, the left-hand sides yield the prior probabilities, which also shows that the class-conditional densities integrate to 1. Furthermore, by adding the two equations in (2.3), it follows that $p(\mathbf{x}) = P(Y = 0)p(\mathbf{x} \mid Y = 0) + P(Y = 1)p(\mathbf{x} \mid Y = 1)$. Hence, the feature-label distribution $P_{\mathbf{X},Y}$ is completely specified by the class-conditional densities $p(\mathbf{x} \mid Y = 0)$ and $p(\mathbf{x} \mid Y = 1)$ weighted by the prior probabilities $P(Y = 0)$ and $P(Y = 1)$, respectively. The case of discrete $P_{\mathbf{X},Y}$ is considered briefly in Exercise 2.1 as well as in Chapter 3. See Figure 2.1 for a univariate example.

---

[1]Section 2.6.3 and Appendix A1 contain the full technical details.

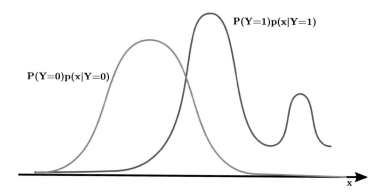

Figure 2.1: Weighted class-conditional densities in a univariate classification problem.

Assuming that $p(\mathbf{x}) > 0$,[2] define the *posterior probabilities* of Y given $\mathbf{X} = \mathbf{x}$ by

$$
\begin{aligned}
P(Y = 0 \mid \mathbf{X} = \mathbf{x}) &= \frac{P(Y = 0)p(\mathbf{x} \mid Y = 0)}{p(\mathbf{x})} = \frac{P(Y = 0)p(\mathbf{x} \mid Y = 0)}{P(Y = 0)p(\mathbf{x} \mid Y = 0) + P(Y = 1)p(\mathbf{x} \mid Y = 1)}, \\
P(Y = 1 \mid \mathbf{X} = \mathbf{x}) &= \frac{P(Y = 1)p(\mathbf{x} \mid Y = 1)}{p(\mathbf{x})} = \frac{P(Y = 1)p(\mathbf{x} \mid Y = 1)}{P(Y = 0)p(\mathbf{x} \mid Y = 0) + P(Y = 1)p(\mathbf{x} \mid Y = 1)}.
\end{aligned}
\tag{2.4}
$$

Posterior probabilities are not probability densities (e.g., they do not integrate to 1) but are simply probabilities. In particular, their values are always between 0 and 1, and

$$
P(Y = 0 \mid \mathbf{X} = \mathbf{x}) + P(Y = 1 \mid \mathbf{X} = \mathbf{x}) = 1,
\tag{2.5}
$$

for all $\mathbf{x} \in R^d$, so that only one of the posterior probabilities need be specified. We pick one arbitrarily and define the *posterior-probability function*

$$
\eta(\mathbf{x}) = E[Y \mid \mathbf{X} = \mathbf{x}] = P(Y = 1 \mid \mathbf{X} = \mathbf{x}), \quad \mathbf{x} \in R^d,
\tag{2.6}
$$

which plays an important role in the sequel. See Figure 2.2 for a univariate example.

The goal of classification is to predict Y accurately using a *classifier*, i.e. a $\{0, 1\}$-valued function of $\mathbf{X}$. Formally, a classifier is defined as a *Borel-measurable function* $\psi : R^d \to \{0, 1\}$, i.e. a very general function that still allows probabilities, such as the classification error, to be computed (see Appendix A1). A classifier partitions the feature space R^d into two sets: the 0-decision region $\{\mathbf{x} \in R^d \mid \psi(\mathbf{x}) = 0\}$ and the 1-decision region $\{\mathbf{x} \in R^d \mid \psi(\mathbf{x}) = 1\}$. The boundary between these two regions is called the *decision boundary*. Figure 2.3 depicts a univariate classifier with decision boundary x_0.

---

[2]We can assume an effective feature space $S = \{\mathbf{x} \in R^d \mid p(\mathbf{x}) > 0\}$ — regions of the feature space with probability zero can be ignored.

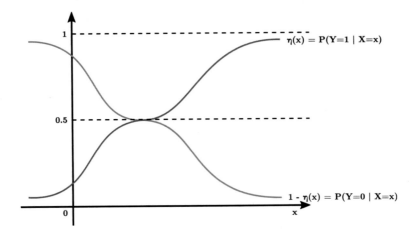

Figure 2.2: Posterior probabilities in a univariate classification problem.

The *error* of a classifier ψ is the probability of misclassification:

$$\varepsilon[\psi] \;=\; P(\psi(\mathbf{X}) \neq Y)\,. \tag{2.7}$$

Measurability of ψ guarantees that this probability is well defined. The *class-specific errors* of ψ are defined as

$$\varepsilon^0[\psi] \;=\; P(\psi(\mathbf{X}) = 1 \mid Y = 0) \;=\; \int_{\{\mathbf{x}\mid\psi(\mathbf{x})=1\}} p(\mathbf{x} \mid Y = 0)\,d\mathbf{x}\,,$$

$$\varepsilon^1[\psi] \;=\; P(\psi(\mathbf{X}) = 0 \mid Y = 1) \;=\; \int_{\{\mathbf{x}\mid\psi(\mathbf{x})=0\}} p(\mathbf{x} \mid Y = 1)\,d\mathbf{x}\,. \tag{2.8}$$

These are the errors committed in each class separately. In some contexts, $\varepsilon^0[\psi]$ and $\varepsilon^1[\psi]$ are known as the classifier *false positive* and *false negative* error rates, respectively. In addition, $1 - \varepsilon^1[\psi]$ and $1 - \varepsilon^0[\psi]$ are sometimes called the classifier *sensitivity* and *specificity*, respectively.

Notice that

$$\varepsilon[\psi] \;=\; P(\psi(\mathbf{X}) \neq Y) \;=\; P(\psi(\mathbf{X}) = 1, Y = 0) + P(\psi(\mathbf{X}) = 0, Y = 1)$$

$$=\; P(\psi(\mathbf{X}) = 1 \mid Y = 0)P(Y = 0) + P(\psi(\mathbf{X}) = 0 \mid Y = 1)P(Y = 1)$$

$$=\; P(Y = 0)\,\varepsilon^0[\psi] + P(Y = 1)\,\varepsilon^1[\psi] \tag{2.9}$$

$$=\; \int_{\{\mathbf{x}\mid\psi(\mathbf{x})=1\}} P(Y = 0)p(\mathbf{x} \mid Y = 0)\,d\mathbf{x} + \int_{\{\mathbf{x}\mid\psi(\mathbf{x})=0\}} P(Y = 1)p(\mathbf{x} \mid Y = 1)\,d\mathbf{x}\,.$$

With $p = P(Y = 1)$, we have $\varepsilon[\psi] = (1 - p)\varepsilon^0[\psi] + p\varepsilon^1[\psi]$, i.e., a linear combination of the class-specific error rates, with weights given by the corresponding prevalences. Also, $\varepsilon[\psi]$ is the sum of the integrals of the weighted densities over the opposite decision regions. These integrals are the shaded areas in Figure 2.3.

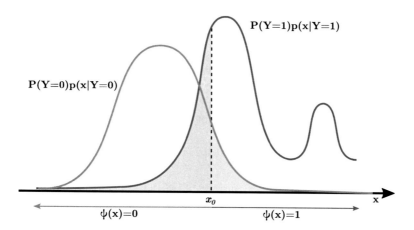

Figure 2.3: Univariate classifier, with indicated decision boundary x_0 and corresponding decision regions. The blue and orange shaded areas are equal to $P(Y = 1)\varepsilon^1[\psi]$ and $P(Y = 0)\varepsilon^0[\psi]$, respectively. The classification error is the sum of the two shaded areas.

The *conditional error* of a classifier ψ is defined as

$$\varepsilon[\psi \mid \mathbf{X} = \mathbf{x}] = P(\psi(\mathbf{X}) \neq Y \mid \mathbf{X} = \mathbf{x}). \qquad (2.10)$$

This can be interpreted as the "error at each point $\mathbf{x}$ of the feature space." Using the "Law of Total Probability" for random variables (A.53), we can express the classification error as the "average" conditional classification error over the feature space:

$$\varepsilon[\psi] = E[\varepsilon[\psi \mid \mathbf{X} = \mathbf{x}]] = \int_{\mathbf{x} \in R^d} \varepsilon[\psi \mid \mathbf{X} = \mathbf{x}] \, p(\mathbf{x}) \, d\mathbf{x}. \qquad (2.11)$$

Therefore, knowing the error at each point $\mathbf{x} \in R^d$ of the feature space plus its "weight" $p(\mathbf{x})$ is enough to determine the overall classification error.

In addition, given a classifier ψ, the conditional classification error $\varepsilon[\psi \mid \mathbf{X} = \mathbf{x}]$ is determined by the posterior-probability function $\eta(\mathbf{x})$, as follows

$$
\begin{aligned}
\varepsilon[\psi \mid \mathbf{X} = \mathbf{x}] &= P(\psi(\mathbf{X}) = 0, Y = 1 \mid \mathbf{X} = \mathbf{x}) + P(\psi(\mathbf{X}) = 1, Y = 0 \mid \mathbf{X} = \mathbf{x}) \\
&= I_{\psi(\mathbf{x})=0} \, P(Y = 1 \mid \mathbf{X} = \mathbf{x}) + I_{\psi(\mathbf{x})=1} P(Y = 0 \mid \mathbf{X} = \mathbf{x}) \\
&= I_{\psi(\mathbf{x})=0} \, \eta(\mathbf{x}) + I_{\psi(\mathbf{x})=1}(1 - \eta(\mathbf{x})) =
\begin{cases}
\eta(\mathbf{x}), & \text{if } \psi(\mathbf{x}) = 0, \\
1 - \eta(\mathbf{x}), & \text{if } \psi(\mathbf{x}) = 1.
\end{cases}
\end{aligned}
\qquad (2.12)
$$

2.3   The Bayes Classifier

The primary criterion of performance in classification is the error rate in (2.7). Therefore, we would like to find an optimal classifier that minimizes it.

A *Bayes classifier* is defined as a classifier that minimizes the classification error in (2.7),

$$\psi^* = \arg\min_{\psi \in \mathcal{C}} P(\psi(\mathbf{X}) \neq Y), \tag{2.13}$$

over the set $\mathcal{C}$ of all classifiers. In other words, a Bayes classifier is an optimal minimum-error classifier. Since $Y \in \{0, 1\}$, we have

$$\varepsilon[\psi] = P(\psi(\mathbf{X}) \neq Y) = E[|\psi(\mathbf{X}) - Y|] = E[|\psi(\mathbf{X}) - Y|^2]. \tag{2.14}$$

Therefore, a Bayes classifier is also the MMSE classifier and the minimum absolute deviation (MAD) classifier.

There may be more than one solution to (2.13), i.e., there may be more than one Bayes classifier; in fact, there may be an infinite number of them, as we show below. Knowledge of the feature-label distribution $P_{\mathbf{X},Y}$ must be enough, in principle, to obtain a Bayes classifier. The next theorem shows that, in fact, only knowledge of the posterior-probability function $\eta(\mathbf{x}) = P(Y = 1 \mid \mathbf{X} = \mathbf{x})$ is needed.

Theorem 2.1. *(Bayes classifier.) The classifier*

$$\psi^*(\mathbf{x}) = \arg\max_{i} P(Y = i \mid \mathbf{X} = \mathbf{x}) = \begin{cases} 1, & \eta(\mathbf{x}) > \frac{1}{2}, \\ 0, & \text{otherwise,} \end{cases} \tag{2.15}$$

for $\mathbf{x} \in R^d$, satisfies (2.13).

Proof. (Fun with indicator variables.) We show that $\varepsilon[\psi] \geq \varepsilon^*[\psi]$, for all $\psi \in \mathcal{C}$. From (2.11), it is enough to show that

$$\varepsilon[\psi \mid \mathbf{X} = \mathbf{x}] \geq \varepsilon[\psi^* \mid \mathbf{X} = \mathbf{x}], \quad \text{for all } \mathbf{x} \in R^d. \tag{2.16}$$

Using (2.12), we can write, for any $\mathbf{x} \in R^d$,

$$\varepsilon[\psi \mid \mathbf{X} = \mathbf{x}] - \varepsilon[\psi^* \mid \mathbf{X} = \mathbf{x}] = \eta(\mathbf{x})(I_{\psi(\mathbf{x})=0} - I_{\psi^*(\mathbf{x})=0}) + (1 - \eta(\mathbf{x}))(I_{\psi(\mathbf{x})=1} - I_{\psi^*(\mathbf{x})=1}). \tag{2.17}$$

Now, by exhausting all possibilities for $\psi(\mathbf{x})$ and $\psi^*(\mathbf{x})$ (there are four total cases), we can see that

$$I_{\psi(\mathbf{x})=0} - I_{\psi^*(\mathbf{x})=0} = -(I_{\psi(\mathbf{x})=1} - I_{\psi^*(\mathbf{x})=1}). \tag{2.18}$$

Substituting this back into (2.17), we get

$$\varepsilon[\psi \mid \mathbf{X} = \mathbf{x}] - \varepsilon[\psi^* \mid \mathbf{X} = \mathbf{x}] = (2\eta(\mathbf{x}) - 1)(I_{\psi(\mathbf{x})=0} - I_{\psi^*(\mathbf{x})=0}). \tag{2.19}$$

Now, there are only two possibilities: either $\eta(\mathbf{x}) > 1/2$ or $\eta(\mathbf{x}) \le 1/2$. In the first case, the terms in parentheses in the right-hand side of (2.19) are both nonnegative, while in the second case, they are both nonpositive. In either case, the product is nonnegative, establishing (2.16). $\diamond$

The optimal decision boundary is the set $\{\mathbf{x} \in R^d \mid \eta(\mathbf{x}) = 1/2\}$. Notice that this does not have to be a thin boundary (e.g., a set of measure zero), though it often is. In the example in Figure 2.2, the decision boundary is the single point where the two posterior functions intersect, so it is a set of measure zero, in this case.

In Theorem 2.1, the decision boundary was assigned to class 0, though the proof allows it to be assigned to class 1, or even be split between the classes. All of these classifiers are therefore optimal classifiers, so that there may be an infinite number of them. Note also that the Bayes classifier is defined locally (pointwise) to minimize $\varepsilon[\psi \mid \mathbf{X} = \mathbf{x}]$ at each value $\mathbf{x} \in R^d$, which also minimizes $\varepsilon[\psi]$ globally. Hence, the Bayes classifier is both the local and global optimal predictor.

Example 2.1.[3] Educational experts have built a model to predict whether an incoming freshman will pass or fail their introductory calculus class, based on the number of hours/day spent studying the lectures (S) and doing the homework (H). Using a binary random variable Y to code for pass/fail, the model is given by:

$$Y = \begin{cases} 1 \text{ (pass)}, & \text{if } S + H + N > 5, \\ 0 \text{ (fail)}, & \text{otherwise}, \end{cases} \tag{2.20}$$

for $S, H, N \ge 0$, where N is an unobservable variable corresponding to factors such as motivation, focus, discipline, etc., here translated to the equivalent in hours/day. The variable N acts as a noise term and models the uncertainty of the model. The variables S, H, and N are modeled as independent and exponential with parameter $\lambda = 1$ (the exponential distribution being a common model for nonnegative continuous-valued variables). We compute the optimal predictor of whether a given student will pass the class based on the observed values of S and H. Notice that this is a problem of optimal classification, and the optimal predictor is the Bayes classifier. From Thm. 2.1, we need to find the posterior-probability function $\eta(\mathbf{x}) = P(Y = 1 \mid \mathbf{X} = \mathbf{x})$, where $\mathbf{X} = (S, H)$,

---

[3]Examples 2.1 and 2.2 are adapted from the example in [Devroye et al., 1996, Section 2.3].

and then apply (2.15). Using (2.20), we have

$$
\begin{aligned}
\eta(s,h) &= P(Y = 1 \mid S = s, H = h) = P(S + H + N > 5 \mid S = s, H = h) \\
&= P(N > 5 - (s+h) \mid S = s, H = h) \\
&= P(N > 5 - (s+h)) =
\begin{cases}
e^{s+h-5}, & \text{if } s+h < 5, \\
1, & \text{otherwise,}
\end{cases}
\end{aligned}
\tag{2.21}
$$

for $s, h \geq 0$. Here we used the fact that the upper tail of an exponential random variable X with parameter λ is $P(X > x) = e^{-\lambda x}$, if $x \geq 0$ (being trivially equal to 1 if $x < 0$). In the next-to-last inequality, we also used the independence of N from S, H. The optimal decision boundary $D = \{\mathbf{x} \in R^d \mid \eta(\mathbf{x}) = 1/2\}$ is thus determined by:

$$
\eta(s,h) = 1/2 \Rightarrow e^{s+h-5} = 1/2 \Rightarrow s+h = 5 - \ln 2 \approx 4.31 .
\tag{2.22}
$$

The optimal decision boundary is therefore a line. Notice that $\eta(s,h) > 1/2$ iff $s + h > 5 - \ln 2$. The optimal classifier is thus

$$
\psi^*(s,h) =
\begin{cases}
1 \ (\text{pass}), & \text{if } s+h > 5 - \ln 2, \\
0 \ (\text{fail}, & \text{otherwise,}
\end{cases}
\tag{2.23}
$$

for $s, h \geq 0$. In other words, if the given student spends around 4.31 hours/day, at a minimum, studying the lectures or doing the homework, we optimally predict that the student will pass the class. Comparing this to (2.20), we note that the term $\ln 2$ allows for the uncertainty associated with the lack of information about N. If there were no noise ($N = 0$), then the optimal decision boundary would be at $s + h = 5$ — the boundary moves due to the uncertainty introduced by N. Notice that this is *quantified* uncertainty, as the complete probability structure of the problem is known. See Figure 2.4 for an illustration. $\diamond$

We close this section with additional ways to obtain a Bayes classifier. First, note that (2.15) is equivalent to

$$
\psi^*(\mathbf{x}) =
\begin{cases}
1, & \eta(\mathbf{x}) > 1 - \eta(\mathbf{x}), \\
0, & \text{otherwise,}
\end{cases}
\tag{2.24}
$$

for $\mathbf{x} \in R^d$. It follows from (2.4) and (2.24) that

$$
\psi^*(\mathbf{x}) =
\begin{cases}
1, & P(Y = 1)p(\mathbf{x} \mid Y = 1) > P(Y = 0)p(\mathbf{x} \mid Y = 0), \\
0, & \text{otherwise,}
\end{cases}
\tag{2.25}
$$

for $\mathbf{x} \in R^d$. Hence, a Bayes classifier can be determined by comparing, at each point in the feature space, the weighted class-conditional densities. The optimal decision boundary is the loci of points

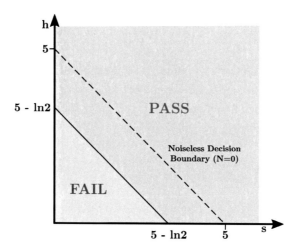

Figure 2.4: Optimal classifier in Example 2.1.

where these functions meet; see Figure 2.5 in the next section for an illustration in a simple univariate case. Notice that increasing $P(Y = 1)$ or $P(Y = 0)$ would have the effect of pushing the optimal decision boundary *away* from the corresponding class center. In the case $P(Y = 1) = P(Y = 0) = \frac{1}{2}$, a Bayes classifier can be determined by comparing the unweighted class-conditional densities to each other directly.

In addition, a simple manipulation of (2.25) allows us to write a Bayes classifier in the following form:

$$\psi^*(\mathbf{x}) = \begin{cases} 1, & D^*(\mathbf{x}) > k^*, \\ 0, & \text{otherwise}, \end{cases} \qquad (2.26)$$

where the *optimal discriminant* $D^* : R^d \to R$ is given by

$$D^*(\mathbf{x}) = \ln \frac{p(\mathbf{x} \mid Y = 1)}{p(\mathbf{x} \mid Y = 0)}, \qquad (2.27)$$

for $\mathbf{x} \in R^d$, with optimal threshold

$$k^* = \ln \frac{P(Y = 0)}{P(Y = 1)}. \qquad (2.28)$$

If $p(\mathbf{x} \mid Y = 1) = 0$ or $p(\mathbf{x} \mid Y = 0) = 0$, we define $D^*(\mathbf{x})$ to be $-\infty$ or ∞, respectively. The case where both $p(\mathbf{x} \mid Y = 1)$ and $p(\mathbf{x} \mid Y = 0)$ are zero can be ignored, since in that case $p(\mathbf{x}) = 0$.

The optimal discriminant D^* is also known in the statistics literature as the *log-likelihood function*. If $P(Y = 0) = P(Y = 1)$ (equally-likely classes), then $k^* = 0$, and the decision boundary is implicitly determined by the simple equation $D^*(\mathbf{x}) = 0$. In many cases, it is more convenient to work with discriminants than directly with class-conditional densities or posterior-probability functions. Examples of discriminants are given in Section 2.5.

2.4   The Bayes Error

Given a Bayes classifier ψ^*, the error $\varepsilon^* = \varepsilon[\psi^*]$ is a fundamental quantity in supervised learning, known as the *Bayes error*. Of course, all Bayes classifiers share the same Bayes error, which is unique. The Bayes error is the lower bound on the classification error that can be achieved in a given problem. It should be sufficiently small if there is to be any hope of designing a good classifier from data.

Theorem 2.2. *(Bayes error.) In the two-class problem,*

$$\varepsilon^* = E[\min\{\eta(\mathbf{X}), 1 - \eta(\mathbf{X})\}]. \tag{2.29}$$

Furthermore, the maximum value ε^* can take is 0.5.

Proof. It follows from (2.11), (2.12), and (2.24) that

$$\varepsilon^* = \int (I_{\eta(\mathbf{X}) \leq 1-\eta(\mathbf{X})}\, \eta(\mathbf{X}) + I_{\eta(\mathbf{X}) > 1-\eta(\mathbf{X})}(1 - \eta(\mathbf{X})))p(\mathbf{x})\, d\mathbf{x} = E[\min\{\eta(\mathbf{X}), 1 - \eta(\mathbf{X})\}]. \tag{2.30}$$

Now, applying the identity

$$\min\{a, 1 - a\} = \frac{1}{2} - \frac{1}{2}|2a - 1|, \quad 0 \leq a \leq 1, \tag{2.31}$$

to (2.30), we can write

$$\varepsilon^* = \frac{1}{2} - \frac{1}{2}E[|2\eta(\mathbf{X}) - 1|], \tag{2.32}$$

from which it follows that $\varepsilon^* \leq \frac{1}{2}$. $\diamond$

Therefore, the maximum optimal error in a two-class problem is $\frac{1}{2}$, not 1. This can be understood intuitively by reasoning that the long-run error achieved by repeatedly flipping an unbiased coin to make a binary decision is 50%.. As the classification error lower bound, the Bayes error cannot exceed that. In addition, from (2.4) in the proof of the previous theorem, we can see that

$$\varepsilon^* = \frac{1}{2} \Leftrightarrow E[|2\eta(\mathbf{X}) - 1|] = 0 \Leftrightarrow \eta(\mathbf{X}) = \frac{1}{2} \text{ with probability } 1. \tag{2.33}$$

(Since expectation is not affected by what happens over regions of probability zero.) In other words, the maximum Bayes error rate is achieved when $\eta(\mathbf{X}) = 1 - \eta(\mathbf{X}) = \frac{1}{2}$ with probability 1, i.e., there is no separation between the posterior-probability functions over all regions of the feature space with positive probability (refer to Figure 2.2). This means that there is total confusion, and no discrimination is possible between $Y = 0$ and $Y = 1$ using $\mathbf{X}$ as feature vector. In this case, the best one can do is indeed equivalent to flipping a coin. A problem where the Bayes error is 0.5

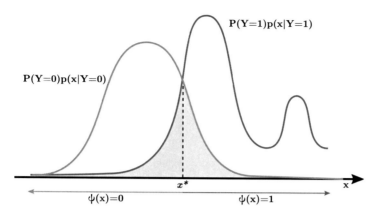

Figure 2.5: Class-conditional weighted by prior probabilities in a univariate problem. The optimal decision boundary is the point x^*. The Bayes error is the sum of the orange and blue shaded areas.

(or very close to it) is hopeless: no classification method, no matter how sophisticated, can achieve good discrimination. In this case, a different feature vector $\mathbf{X}$ must be sought to predict Y.

Application of (2.9) and (2.25) yields:

$$
\begin{aligned}
\varepsilon^* &= P(Y=0)\,\varepsilon^0[\psi^*] + P(Y=1)\,\varepsilon^1[\psi^*] \\
&= \int_{\{\mathbf{x}|P(Y=1)p(\mathbf{x}|Y=1)>P(Y=0)p(\mathbf{x}|Y=0)\}} P(Y=0)p(\mathbf{x}\mid Y=0)\,d\mathbf{x} + \int_{\{\mathbf{x}|P(Y=1)p(\mathbf{x}|Y=1)\leq P(Y=0)p(\mathbf{x}|Y=0)\}} P(Y=1)p(\mathbf{x}\mid Y=1)\,d\mathbf{x}.
\end{aligned}
\tag{2.34}
$$

See Figure 2.5 to see an illustration in a simple univariate problem. The orange and blue shaded regions are equal to $P(Y=1)\varepsilon^1[\psi^*]$ and $P(Y=0)\varepsilon^0[\psi^*]$, respectively, and the Bayes error is the sum of the two areas. We can see that the Bayes error has to do with the amount of "overlap" between the (weighted) class-conditional densities. Comparing Figures 2.3 and 2.5 reveals that shifting the decision boundary increases the classification error, so that x^* is indeed optimal.

An important consequence of Thm 2.2 follows from an application of Jensen's inequality (A.66):

$$
\begin{aligned}
\varepsilon^* &= E[\min\{\eta(\mathbf{X}), 1-\eta(\mathbf{X})\}] \\
&\leq \min\{E[\eta(\mathbf{X})], 1-E[\eta(\mathbf{X})]\} = \min\{P(Y=1), P(Y=0)\},
\end{aligned}
\tag{2.35}
$$

where we used the facts that the function $f(u) = \min\{u, 1-u\}$ is concave and that

$$
E[\eta(\mathbf{X})] = \int_{R^d} P(Y=1 \mid \mathbf{X}=\mathbf{x})p(\mathbf{x})\,d\mathbf{x} = P(Y=1).
\tag{2.36}
$$

It follows from (2.35) that the error of the optimal classifier with any feature vector $\mathbf{X}$ is bounded from above by the error of the optimal classifier without features in (2.2). (As far as optimal prediction is concerned, having any information is at least as good as having no information.) It also follows from (2.35) that if one of the two classes is unlikely to be observed, the Bayes error is small no matter what. Caution must be exercised here; for example, as pointed out in Section 2.1, in the case of a rare disease, a small error rate means little. The solution in this case would be to look at the class-specific error rates $\varepsilon^0[\psi^*]$ and $\varepsilon^1[\psi^*]$ — one would like both to be small. This observation is consistent with (2.34): $\varepsilon^* = P(Y = 0)\,\varepsilon^0[\psi^*] + P(Y = 1)\,\varepsilon^1[\psi^*]$, so that the Bayes error will be small even if $\varepsilon^0[\psi^*]$ or $\varepsilon^1[\psi^*]$ are large, as long as $P(Y = 0)$ or $P(Y = 1)$ is small, respectively.

Incidentally, the "optimal" class-specific error rates $\varepsilon^0[\psi^*]$ and $\varepsilon^1[\psi^*]$ are not optimal in the same sense that ε^* is optimal: for a given classifier ψ, it may occur that $\varepsilon^0[\psi] < \varepsilon^0[\psi^*]$ or $\varepsilon^1[\psi] < \varepsilon^1[\psi^*]$; but *both* cannot occur, as can be seen by comparing (2.9) and (2.34). (It is a general fact that the decision boundary of a classifier can be shifted to make one class-specific error rate as small as desired, while making the other one large.)

Next, we consider the effect on the Bayes error of a transformation applied to the feature vector. This is a common occurrence in practice: for example, preprocessing or normalization may be applied to the data, or dimensionality reduction (to be considered at length in Chapter 9) may be employed, in which a transformation "projects" the data from a high-dimensional space to a lower-dimensional one.

Theorem 2.3. *Let $\mathbf{X} \in R^p$ be the original feature vector, $t : R^p \to R^d$ be a (Borel-measurable) transformation between feature spaces, and $\mathbf{X}' = t(\mathbf{X}) \in R^d$ be the transformed feature vector. Let $\varepsilon^*(\mathbf{X}, Y)$ and $\varepsilon^*(\mathbf{X}', Y)$ be the Bayes errors corresponding to the original and transformed problems. Then*

$$\varepsilon^*(\mathbf{X}', Y) \geq \varepsilon^*(\mathbf{X}, Y)\,, \tag{2.37}$$

with equality if t is invertible.

Proof. First we show that $\eta(\mathbf{X}') = E[\eta(\mathbf{X}) \mid t(\mathbf{X})]$ (note that, strictly speaking, we should write $\eta'(\mathbf{X}')$ since η' is in general a different function than η, but we will ignore that, as it creates no confusion). Using Bayes Theorem, we have:

$$\begin{aligned}
\eta(\mathbf{x}') &= P(Y = 1 \mid \mathbf{X}' = \mathbf{x}') = P(Y = 1 \mid t(\mathbf{X}) = \mathbf{x}') \\
&= \frac{P(Y = 1)P(t(\mathbf{X}) = \mathbf{x}' \mid Y = 1)}{P(t(\mathbf{X}) = \mathbf{x}')} \\
&= \frac{1}{P(t(\mathbf{X}) = \mathbf{x}')} \int_{R^p} p(\mathbf{x} \mid Y = 1)P(Y = 1)I_{t(\mathbf{x})=\mathbf{x}'}\, d\mathbf{x} \\
&= \frac{1}{P(t(\mathbf{X}) = \mathbf{x}')} \int_{R^p} \eta(\mathbf{x})I_{t(\mathbf{x})=\mathbf{x}'} p(\mathbf{x})\, d\mathbf{x} \\
&= \frac{E[\eta(\mathbf{X})I_{t(\mathbf{X})=\mathbf{x}'}]}{P(t(\mathbf{X}) = \mathbf{x}')} = E[\eta(\mathbf{X}) \mid t(\mathbf{X}) = \mathbf{x}']\,,
\end{aligned} \tag{2.38}$$

which proves the claim, where we used the fact that, by definition, $E[Z \mid F] = E[ZI_F]/P(F)$, for a random variable Z and an event F. Combining (2.29) and (2.38) gives

$$
\begin{aligned}
\varepsilon^*(\mathbf{X}', Y) &= E[\min\{\eta(\mathbf{X}'), 1 - \eta(\mathbf{X}')\}] \\
&= E[\min\{E[\eta(\mathbf{X}) \mid t(\mathbf{X})], 1 - E[\eta(\mathbf{X}) \mid t(\mathbf{X})]\}] \\
&\geq E[E[\min\{\eta(\mathbf{X}), 1 - \eta(\mathbf{X})\} \mid t(\mathbf{X})]] \\
&= E[\min\{\eta(\mathbf{X}), 1 - \eta(\mathbf{X})\}] = \varepsilon^*(\mathbf{X}, Y),
\end{aligned}
\tag{2.39}
$$

where the inequality follows from Jensen's Inequality (A.66), and the law of total expectation (A.82) was used to obtain the next-to-last equality. Finally, if t is invertible, then apply the result to t and t^{-1} to obtain $\varepsilon^*(\mathbf{X}', Y) \geq \varepsilon^*(\mathbf{X}, Y)$ and $\varepsilon^*(\mathbf{X}, Y) \geq \varepsilon^*(\mathbf{X}', Y)$. ◇

This is a fundamental result in supervised learning, which is used at several points in the book. It states that a transformation on the feature vector never adds discriminatory information (in fact, it often destroys the information). Therefore, the Bayes error cannot possibly decrease after a transformation to the feature vector (and, in fact, typically increases). If the transformation is invertible, then it does not remove any information; hence, the Bayes error stays the same (it is possible for the Bayes error to stay the same even if the transformation is not invertible; examples of that are given in Chapter 9. Unfortunately, many useful and interesting feature vector transformations, such as dimensionality reduction transformations (a topic we discuss in detail in Chapter 9), are not invertible and therefore generally increase the Bayes error. On the other hand, simple transformations such as scalings, translations, and rotations, are invertible and thus do not affect the Bayes error.

Example 2.2. Continuing Example 2.1, we compute the Bayes error, i.e., the error of the classifier (2.23). We do the computation using two distinct methods.

First Method: Using (2.29), we obtain

$$
\begin{aligned}
\varepsilon^* &= E[\min\{\eta(S, H), 1 - \eta(S, H)\}] \\
&= \iint_{\{s, h \geq 0 \mid \eta(s,h) \leq \frac{1}{2}\}} \eta(s, h) p(s, h)\, ds\, dh + \iint_{\{s, h \geq 0 \mid \eta(s,h) > \frac{1}{2}\}} (1 - \eta(s, h)) p(s, h)\, ds\, dh \\
&= \iint_{\{s, h \geq 0 \mid 0 \leq s+h \leq 5 - \ln 2\}} e^{-5}\, ds\, dh + \iint_{\{s, h \geq 0 \mid 5 - \ln 2 < s+h \leq 5\}} (e^{-(s+h)} - e^{-5})\, ds\, dh \\
&= e^{-5} \left[(6 - \ln 2)^2 - \frac{35}{2} \right] \approx 0.0718,
\end{aligned}
\tag{2.40}
$$

where we used the expression for $\eta(s, h)$ in (2.21) and independence to write $p(s, h) = p(s)p(h) = e^{-(s+h)}$, for $s, h > 0$. Hence, the prediction will fail for about 7.2% of the students, who will pass while being predicted to fail or fail while being predicted to pass.

Second Method: The previous method requires double integration. A simpler computation can be obtained if one realizes that the posterior-probability function $\eta(s, h)$ in (2.21), and hence the optimal classifier, depend on s, h only through $s + h$. We say that $U = S + H$ is a *sufficient statistic* for optimal classification based on (S, H). It is shown in Exercise 9.4 that the Bayes error using a sufficient statistic, in this case the univariate feature $U = S + H$, is the same as using the original feature vector (S, H), even though the transformation $s, h \mapsto s + h$ is *not* invertible, and Theorem 2.3 cannot be used to show equality. We can thus write the Bayes error in terms of univariate integrals as follows:

$$
\begin{aligned}
\varepsilon^* &= E[\min\{\eta(U), 1 - \eta(U)\}] \\
&= \int_{\{u \geq 0 \,|\, \eta(u) \leq \frac{1}{2}\}} \eta(u) p(u)\, du + \int_{\{u \geq 0 \,|\, \eta(u) > \frac{1}{2}\}} (1 - \eta(u)) p(u)\, du \\
&= \int_{\{0 \leq u \leq 5 - \ln 2\}} e^{-5} u\, du + \int_{\{5 - \ln 2 < u \leq 5\}} (e^{-u} - e^{-5}) u\, du \\
&= e^{-5} \left[(6 - \ln 2)^2 - \frac{35}{2} \right] \approx 0.0718,
\end{aligned}
\tag{2.41}
$$

where we used the fact that $U = S + H$ is distributed as a Gamma random variable with parameters $\lambda = 1$ and $t = 2$, so that $p(u) = u e^{-u}$, for $u \geq 0$ (see Section A1 for the density of a Gamma r.v.). This is a consequence of the general fact that the sum of n independent exponential random variables with parameter λ is a Gamma random variable with parameters λ and $t = n$. $\diamond$

2.5   Gaussian Model

We consider now the important special case where the class-conditional densities are multivariate Gaussian

$$
p(\mathbf{x} \mid Y = i) = \frac{1}{\sqrt{(2\pi)^d \det(\Sigma_i)}} \exp\left[\frac{1}{2}(\mathbf{x} - \boldsymbol{\mu})^T \Sigma_i^{-1} (\mathbf{x} - \boldsymbol{\mu}_i) \right], \quad i = 0, 1.
\tag{2.42}
$$

The vector means $\boldsymbol{\mu}_0$ and $\boldsymbol{\mu}_1$ are the class centers and the covariance matrices Σ_0 and Σ_1 specify the ellipsoidal shape of the class-conditional densities (see Section A1.7 for a review of the multivariate Gaussian and its properties).

It can be readily verified that the optimal discriminant in (2.27) assumes here the following form:

$$
D^*(\mathbf{x}) = \frac{1}{2}(\mathbf{x} - \boldsymbol{\mu}_0)^T \Sigma_0^{-1} (\mathbf{x} - \boldsymbol{\mu}_0) - \frac{1}{2}(\mathbf{x} - \boldsymbol{\mu}_1)^T \Sigma_1^{-1} (\mathbf{x} - \boldsymbol{\mu}_1) + \frac{1}{2} \ln \frac{\det(\Sigma_0)}{\det(\Sigma_1)}.
\tag{2.43}
$$

We study separately the cases where the covariance matrices are equal and different. In classical statistics, these are known as the *homoskedastic* and *heteroskedastic* models, respectively. In classification, these cases produce linear and quadratic optimal decision boundaries, respectively.

2.5.1 Homoskedastic Case

In the homoskedastic case, $\Sigma_0 = \Sigma_1 = \Sigma$. This case has very nice properties, as we will see next. First, the optimal discriminant in (2.43) can be simplified in this case to

$$D_L^*(\mathbf{x}) = \frac{1}{2}\left(||\mathbf{x} - \boldsymbol{\mu}_0||_\Sigma^2 - ||\mathbf{x} - \boldsymbol{\mu}_1||_\Sigma^2\right), \tag{2.44}$$

where

$$||\mathbf{x}_0 - \mathbf{x}_1||_\Sigma = \sqrt{(\mathbf{x}_0 - \mathbf{x}_1)^T \Sigma^{-1} (\mathbf{x}_0 - \mathbf{x}_1)} \tag{2.45}$$

is the *Mahalanobis distance* between $\mathbf{x}_0$ and $\mathbf{x}_1$. (It can be shown that this is an actual distance metric, provided that Σ is strictly positive definite; see Exercise 2.10.) Notice that if $\Sigma = I_d$, the Mahalanobis distance reduces to the usual Euclidean distance.

It follows from (2.26) and (2.44) that the Bayes classifier can be written as

$$\psi_L^*(\mathbf{x}) = \begin{cases} 1, & ||\mathbf{x} - \boldsymbol{\mu}_1||_\Sigma^2 < ||\mathbf{x} - \boldsymbol{\mu}_0||_\Sigma^2 + 2\ln\frac{P(Y=1)}{P(Y=0)}, \\ 0, & \text{otherwise.} \end{cases} \tag{2.46}$$

If $P(Y = 0) = P(Y = 1)$ (equally-likely classes), then $\psi^*(\mathbf{x}) = 1$ if $||\mathbf{x} - \boldsymbol{\mu}_1||_\Sigma < ||\mathbf{x} - \boldsymbol{\mu}_0||_\Sigma$, otherwise $\psi^*(\mathbf{x}) = 0$. In other words, the optimal prediction is the class with center closest to $\mathbf{x}$, i.e., ψ^* is a Nearest-Mean Classifier, with respect to the Mahalanobis distance. In the case $\Sigma = I_d$, this is a nearest-mean classifier in the usual Euclidean distance sense — clearly, this is still the case if $\Sigma = \sigma^2 I_d$, for an arbitrary variance parameter $\sigma^2 > 0$.

A straightforward algebraic manipulation shows that, despite appearances to the contrary, the discriminant in (2.44) is a linear function of $\mathbf{x}$:

$$D_L^*(\mathbf{x}) = (\boldsymbol{\mu}_1 - \boldsymbol{\mu}_0)^T \Sigma^{-1}\left(\mathbf{x} - \frac{\boldsymbol{\mu}_0 + \boldsymbol{\mu}_1}{2}\right). \tag{2.47}$$

It follows from (2.26) that the Bayes classifier in (2.46) can be written alternatively as

$$\psi_L^*(\mathbf{x}) = \begin{cases} 1, & \mathbf{a}^T\mathbf{x} + b > 0, \\ 0, & \text{otherwise,} \end{cases} \tag{2.48}$$

where

$$\begin{aligned} \mathbf{a} &= \Sigma^{-1}(\boldsymbol{\mu}_1 - \boldsymbol{\mu}_0), \\ b &= (\boldsymbol{\mu}_0 - \boldsymbol{\mu}_1)^T \Sigma^{-1}\left(\frac{\boldsymbol{\mu}_0 + \boldsymbol{\mu}_1}{2}\right) + \ln\frac{P(Y=1)}{P(Y=0)}. \end{aligned} \tag{2.49}$$

Hence, the optimal decision boundary is a hyperplane in R^d, determined by the equation $\mathbf{a}^T\mathbf{x} + b = 0$. If $P(Y = 0) = P(Y = 1)$ (equally-likely classes), then the midpoint $\mathbf{x}_m = (\boldsymbol{\mu}_0 + \boldsymbol{\mu}_1)/2$ between the

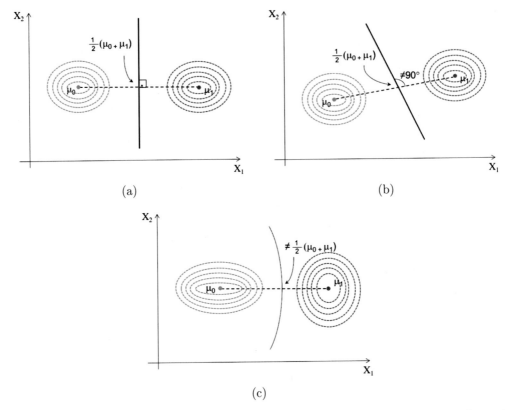

Figure 2.6: Gaussian class-conditional densities and optimal decision boundary in a two-dimensional feature space with $P(Y = 0) = P(Y = 1)$. (a) Homoskedastic case with perpendicular decision boundary ($\boldsymbol{\mu}_1 - \boldsymbol{\mu}_0$ is an eigenvector of Σ). (b) General homoskedastic case. (c) Heteroskedastic case.

class centers satisfies $\mathbf{a}^T \mathbf{x}_m + b = 0$, which shows that the hyperplane passes through $\mathbf{x}_m$. Finally, the hyperplane will be perpendicular to the axis joining the class centers $\boldsymbol{\mu}_0$ and $\boldsymbol{\mu}_1$ provided that the normal vector $\mathbf{a} = \Sigma^{-1}\boldsymbol{\mu}_1 - \boldsymbol{\mu}_0$ is colinear with the vector $\boldsymbol{\mu}_1 - \boldsymbol{\mu}_0$, which can happen in two cases: 1)$\Sigma = \sigma^2 I_d$ (the Euclidean nearest-mean classifier case); 2) $\boldsymbol{\mu}_1 - \boldsymbol{\mu}_0$ is an eigenvector of Σ^{-1}, and thus of Σ. The latter case means that one of the principal axes of the ellipsoidal density contours is aligned with the axis between the class centers. See Figures 2.6(a)-(b) for two-dimensional examples in the case $P(Y = 0) = P(Y = 1)$ (in the general case $P(Y = 1) \neq P(Y = 0)$), the decision boundary is merely pushed *away* from the center of the most likely class).

Another very useful property of the homoskedastic case is that the optimal classification error can be written in closed form, as we show next. First, it follows from (2.47) and the properties of Gaussian random vectors in Section A1.7 that $D_L^*(\mathbf{X}) \mid Y = 0 \sim \mathcal{N}\left(-\frac{1}{2}\delta^2, \delta^2\right)$ and $D_L^*(\mathbf{X}) \mid Y =$

$1 \sim \mathcal{N}\left(\frac{1}{2}\delta^2, \delta^2\right)$, where

$$\delta = \sqrt{(\boldsymbol{\mu}_1 - \boldsymbol{\mu}_0)^T \Sigma^{-1} (\boldsymbol{\mu}_1 - \boldsymbol{\mu}_0)} \tag{2.50}$$

is the Mahalanobis distance between the class centers. It follows that

$$\varepsilon^0[\psi_L^*] = P(D_L^*(\mathbf{X}) > k^* \mid Y = 0) = \Phi\left(\frac{-k^* - \frac{1}{2}\delta^2}{\delta}\right),$$

$$\varepsilon^1[\psi_L^*] = P(D_L^*(\mathbf{X}) \leq k^* \mid Y = 1) = \Phi\left(\frac{k^* - \frac{1}{2}\delta^2}{\delta}\right), \tag{2.51}$$

where $k^* = \ln P(Y = 0)/P(Y = 1)$ and $\Phi(\cdot)$ is the cumulative distribution function of a standard $\mathcal{N}(0, 1)$ Gaussian random variable, and thus the Bayes error is obtained by the formula:

$$\varepsilon_L^* = c\,\Phi\left(\frac{k^* - \frac{1}{2}\delta^2}{\delta}\right) + (1 - c)\Phi\left(\frac{-k^* - \frac{1}{2}\delta^2}{\delta}\right), \tag{2.52}$$

where $c = P(Y = 1)$. In the case $P(Y = 0) = P(Y = 1) = 0.5$, then

$$\varepsilon^0[\psi_L^*] = \varepsilon^1[\psi_L^*] = \varepsilon_L^* = \Phi\left(-\frac{\delta}{2}\right). \tag{2.53}$$

Notice that, regardless of the value of $P(Y = 0)$ and $P(Y = 1)$, the error rates $\varepsilon^0[\psi_L^*]$, $\varepsilon^1[\psi_L^*]$, and ε_L^* are all decreasing functions of the Mahalanobis distance δ between the class centers. In fact, $\varepsilon^0[\psi_L^*], \varepsilon^1[\psi_L^*], \varepsilon_L^* \to 0$ as $\delta \to \infty$.

2.5.2 Heteroskedastic Case

This is the general case, $\Sigma_0 \neq \Sigma_1$. In this case, discrimination has no simple interpretation in terms of nearest-mean classification, and there are no known analytical formulas for the Bayes error rates, which must be computed by numerical integration. However, the shape of the optimal decision boundaries can still be characterized exactly: they are in a family of surfaces known in Geometry as *hyperquadrics*.

In this case, the optimal discriminant in (2.43) is fully quadratic. It follows from (2.26) and (2.43) that the Bayes classifier can be written as

$$\psi_Q^*(\mathbf{x}) = \begin{cases} 1, & \mathbf{x}^T A \mathbf{x} + \mathbf{b}^T \mathbf{x} + c > 0, \\ 0, & \text{otherwise,} \end{cases} \tag{2.54}$$

where

$$A = \frac{1}{2}\left(\Sigma_0^{-1} - \Sigma_1^{-1}\right),$$

$$\mathbf{b} = \Sigma_1^{-1}\boldsymbol{\mu}_1 - \Sigma_0^{-1}\boldsymbol{\mu}_0, \tag{2.55}$$

$$c = \frac{1}{2}(\boldsymbol{\mu}_0^T \Sigma_0^{-1} \boldsymbol{\mu}_0 - \boldsymbol{\mu}_1^T \Sigma_1^{-1} \boldsymbol{\mu}_1) + \frac{1}{2}\ln\frac{\det(\Sigma_0)}{\det(\Sigma_1)} + \ln\frac{P(Y=1)}{P(Y=0)}.$$

In particular, the optimal decision boundary is a hyperquadric surface in R^d, determined by the equation $\mathbf{x}^T A \mathbf{x} + \mathbf{b}^T \mathbf{x} + c = 0$. Depending on the coefficients, the resulting decision boundaries can be hyperspheres, hyperellipsoids, hyperparaboloids, hyperhyperboloids, and even a single hyperplane or pairs of hyperplanes. See Figure 2.6(c) for a two-dimensional example, where the optimal decision boundary is a parabola. Notice that, even in the case $P(Y=0) = P(Y=1)$, the decision boundary does not generally pass through the midpoint between the class centers. See Exercise 2.9 for several additional examples of the heteroskedastic case.

2.6   Additional Topics

2.6.1   Minimax Classification

Absence of knowledge about the prior probabilities $P(Y=1)$ and $P(Y=0)$ implies that the optimal threshold $k = \ln P(Y=0)/P(Y=1)$ in (2.26) is not known, and the Bayes classifier cannot be determined. In this section, we show that an optimal procedure, based on a *minimax* criterion, can still be defined in this case. This will have an impact later when we discuss parametric plug-in classification rules in Chapter 4, in the case where the prior probabilities are unknown and cannot be estimated reliably (or at all).

Given a classifier ψ, the error rates $\varepsilon^0[\psi]$ and $\varepsilon^1[\psi]$ in (2.8) do not depend on $P(Y=1)$ and $P(Y=0)$ and so they are available (even though the overall error rate $\varepsilon[\psi]$ is not). The optimal discriminant D^* in (2.26) is also available, but the optimal threshold $k^* = \ln P(Y=0)/P(Y=1)$ is not. Define classifiers

$$\psi_k^*(\mathbf{x}) = \begin{cases} 1, & D^*(\mathbf{x}) > k, \\ 0, & \text{otherwise,} \end{cases} \tag{2.56}$$

for $k \in R$, with error rates $\varepsilon^0(k) = \varepsilon^0[\psi_k^*]$ and $\varepsilon^1(k) = \varepsilon^1[\psi_k^*]$. Increasing k *decreases* $\varepsilon^0(k)$ but *increases* $\varepsilon^1(k)$, while decreasing k has the opposite effect. A better discriminant possesses uniformly smaller error rates $\varepsilon^0(k)$ and $\varepsilon^1(k)$, as k varies. Accordingly, a plot of $1 - \varepsilon^1(k)$ against $\varepsilon^0(k)$, called a *receiver operating characteristic curve* (ROC) in the signal detection area, is often used to assess discriminatory power. Alternatively, one can use the area under the ROC curve (AUC) to evaluate discriminants; larger AUC corresponds to better discriminants.

Each classifier ψ_k^* is a Bayes classifier for a particular choice of $P(Y = 1)$ and $P(Y = 0)$, with $k = \ln P(Y = 0)/P(Y = 1)$. Recall from (2.34) that the Bayes error $\varepsilon^*(k)$ corresponding to the Bayes classifier ψ_k^* is a linear combination of $\varepsilon^0(k)$ and $\varepsilon^1(k)$; it follows that the maximum value that the Bayes error can take is $\max\{\varepsilon^0(k), \varepsilon^1(k)\}$. The minimax criterion seeks the value k_{mm} that minimizes this maximum possible value for the Bayes error:

$$k_{\mathrm{mm}} = \arg\min_k \max\{\varepsilon^0(k), \varepsilon^1(k)\}, \tag{2.57}$$

which leads to the *minimax classifier*

$$\psi_{\mathrm{mm}}(\mathbf{x}) = \psi_{k_{\mathrm{mm}}}^*(\mathbf{x}) = \begin{cases} 1, & D^*(\mathbf{x}) > k_{\mathrm{mm}}, \\ 0, & \text{otherwise.} \end{cases} \tag{2.58}$$

The following theorem characterizes the minimax threshold k_{mm} and thus the minimax classifier ψ_{mm}, under a mild distributional assumption, namely, the continuity of the error rates.

Theorem 2.4. *Assume that the error rates $\varepsilon^0(k)$ and $\varepsilon^1(k)$ are continuous functions of k.*
(a) The minimax threshold k_{mm} is the unique real value that satisfies

$$\varepsilon^0(k_{\mathrm{mm}}) = \varepsilon^1(k_{\mathrm{mm}}). \tag{2.59}$$

(b) The error of the minimax classifier is equal to the maximum Bayes error rate:

$$\varepsilon^*(k_{\mathrm{mm}}) = \max_k \varepsilon^*(k). \tag{2.60}$$

Proof. (a): For notation convenience, let $c_0 = P(Y = 0)$ and $c_1 = P(Y = 1)$. Suppose that there exists a value k_{mm} such that $\varepsilon^0(k_{\mathrm{mm}}) = \varepsilon^1(k_{\mathrm{mm}})$. In this case, $\varepsilon^*(k_{\mathrm{mm}}) = c_0\varepsilon^0(k_{\mathrm{mm}}) + c_1\varepsilon^1(k_{\mathrm{mm}}) = \varepsilon^0(k_{\mathrm{mm}}) = \varepsilon^1(k_{\mathrm{mm}})$, regardless of the values of c_0, c_1. Now, suppose that there exists $k' > k_{\mathrm{mm}}$ such that $\max\{\varepsilon^0(k'), \varepsilon^1(k')\} < \max\{\varepsilon^0(k_{\mathrm{mm}}), \varepsilon^1(k_{\mathrm{mm}})\} = \varepsilon^*(k_{\mathrm{mm}})$. Since $\varepsilon^0(k)$ and $\varepsilon^1(k)$ are continuous, they are strictly increasing and decreasing functions of k, respectively, so that there are $\delta_0, \delta_1 > 0$ such that $\varepsilon^0(k') = \varepsilon^0(k_{\mathrm{mm}}) + \delta_0 = \varepsilon^*(k_{\mathrm{mm}}) + \delta_0$ and $\varepsilon^1(k') = \varepsilon^1(k_{\mathrm{mm}}) - \delta_1 = \varepsilon^*(k_{\mathrm{mm}}) - \delta_1$. Therefore, $\max\{\varepsilon^0(k'), \varepsilon^1(k')\} = \varepsilon^*(k_{\mathrm{mm}}) + \delta_0 > \varepsilon^*(k_{\mathrm{mm}})$, a contradiction. Similarly, if $k' < k_{\mathrm{mm}}$, there are $\delta_0, \delta_1 > 0$ such that $\varepsilon^0(k') = \varepsilon^*(k_{\mathrm{mm}}) - \delta_0$ and $\varepsilon^1(k') = \varepsilon^*(k_{\mathrm{mm}}) + \delta_1$, so that $\max\{\varepsilon^0(k'), \varepsilon^1(k')\} = \varepsilon^*(k_{\mathrm{mm}}) + \delta_1 > \varepsilon^*(k_{\mathrm{mm}})$, again a contradiction. We conclude that, if there is a k_{mm} such that $\varepsilon^0(k_{\mathrm{mm}}) = \varepsilon^1(k_{\mathrm{mm}})$, then it maximizes $\max\{\varepsilon^0(k), \varepsilon^1(k)\}$. But this point must exist because $\varepsilon^0(k)$ and $\varepsilon^1(k)$ are assumed to be continuous functions of k, and, from (2.8), we have $\varepsilon^0(k) \to 1$ and $\varepsilon^1(k) \to 0$ as $k \to \infty$ and $\varepsilon^0(k) \to 0$ and $\varepsilon^1(k) \to 1$ as $k \to -\infty$. Furthermore, since $\varepsilon^0(k)$ and $\varepsilon^1(k)$ are strictly increasing and decreasing functions of k, respectively, this point is unique.
Part (b): Define

$$\varepsilon(c, c_1) = (1 - c_1)\varepsilon^0\left(\ln\frac{1-c}{c}\right) + c_1 \varepsilon^1\left(\ln\frac{1-c}{c}\right). \tag{2.61}$$

For any given value $k = \ln(1-c_1)/c_1$, we have $\varepsilon^*(k) = \varepsilon(c_1, c_1)$. In addition, $k_{\mathrm{mm}} = \ln(1-c_{\mathrm{mm}})/c_{\mathrm{mm}}$ is the only value such that $\varepsilon^*(k_{\mathrm{mm}}) = \varepsilon(c_{\mathrm{mm}}, c_{\mathrm{mm}}) = \varepsilon(c_{\mathrm{mm}}, c_1)$, for all c_1 (since $\varepsilon^0(k_{\mathrm{mm}}) = \varepsilon^1(k_{\mathrm{mm}})$). Hence, by definition of the Bayes error, $\varepsilon^*(k_{\mathrm{mm}}) = \varepsilon(c_{\mathrm{mm}}, c_1) \geq \varepsilon(c_1, c_1) = \varepsilon^*(k)$, for all k. On the other hand, $\varepsilon^*(k_{\mathrm{mm}}) = \varepsilon(c_{\mathrm{mm}}, c_{\mathrm{mm}}) \leq \max_{c_1} \varepsilon(c_1, c_1)$. This shows that $\varepsilon^*(k_{\mathrm{mm}}) = \max_{c_1} \varepsilon(c_1, c_1) = \max_k \varepsilon^*(k)$. $\diamond$

Let us apply the preceding theory to the Gaussian homoskedastic model discussed in Section 2.5. Using the notation in the current section, we write the error rates in (2.51) as:

$$\varepsilon^0(k) = \Phi\left(\frac{-k - \frac{1}{2}\delta^2}{\delta}\right) \quad \text{and} \quad \varepsilon_1(k) = \Phi\left(\frac{k - \frac{1}{2}\delta^2}{\delta}\right), \tag{2.62}$$

where $\delta = \sqrt{(\boldsymbol{\mu}_1 - \boldsymbol{\mu}_0)^T \Sigma^{-1}(\boldsymbol{\mu}_1 - \boldsymbol{\mu}_0)}$ is the Mahalanobis distance between the class centers, which is assumed known, and $\Phi(\,\cdot\,)$ is the cumulative distribution of the standard normal r.v. Clearly, $\varepsilon^0(k)$ and $\varepsilon^1(k)$ are continuous functions of k, so that, according to Theorem 2.4, the minimax threshold k_{mm} is the unique value that satisfies

$$\Phi\left(\frac{k_{\mathrm{mm}} - \frac{1}{2}\delta^2}{\delta}\right) = \Phi\left(\frac{-k_{\mathrm{mm}} - \frac{1}{2}\delta^2}{\delta}\right). \tag{2.63}$$

Since Φ is monotone, the only solution to this equation is

$$k_{L,\mathrm{mm}} = 0, \tag{2.64}$$

where we use the subscript L to conform with the notation used in Section 2.5 with the homoskedastic case. It follows from (2.47) and (2.58) that the minimax classifier in the Gaussian homoskedastic case is given by

$$\psi_{L,\mathrm{mm}}(\mathbf{x}) = \begin{cases} 1, & (\boldsymbol{\mu}_1 - \boldsymbol{\mu}_0)^T \Sigma^{-1}\left(\mathbf{x} - \frac{\boldsymbol{\mu}_0 + \boldsymbol{\mu}_1}{2}\right) \geq 0, \\ 0, & \text{otherwise.} \end{cases} \tag{2.65}$$

Notice that the minimax classifier corresponds to the Bayes classifier in the case $P(Y = 0) = P(Y = 1)$. The minimax threshold $k_{\mathrm{mm}} = 0$ is conservative in the sense that, in the absence of knowledge about $P(Y = 0)$ and $P(Y = 1)$, it assumes that they are both equal to $1/2$. However, under the heteroskedastic Gaussian model, the minimax threshold can differ substantially from 0.

2.6.2   F-errors

An *F-error* is a generalization of the Bayes error, which can provide a measure of distance between the classes, and thus can be used to measure the quality of the pair $(\mathbf{X}, Y)$ for discrimination. Assuming Y fixed, this means the quality of the feature vector $\mathbf{X}$, in which case F-errors can be used for feature selection and feature extraction, to be discussed in Chapter 9. The idea is that in some situations, some of these F-errors may be easier to estimate from data than the Bayes error. F-errors can also be useful in theoretical arguments about the Bayes error and the difficulty of classification. Finally, they also have historical significance.

Given any concave function $F : [0, 1] \to [0, \infty)$, the corresponding F-error is defined as:

$$d_F(\mathbf{X}, Y) = E[F(\eta(\mathbf{X}))]. \qquad (2.66)$$

Notice that the Bayes error is an F-error:

$$\varepsilon^* = E[\min\{\eta(\mathbf{X}), 1 - \eta(\mathbf{X})\}] = E[F(\eta(\mathbf{X}))] \qquad (2.67)$$

where $F(u) = \min\{u, 1 - u\}$ is nonnegative and concave on the interval $[0, 1]$.

The class of F-errors include many classical measures of discrimination that were introduced independently over the years. The most well-known is probably the *nearest-neighbor distance*:

$$\varepsilon_{\mathrm{NN}} = E[2\eta(\mathbf{X})(1 - \eta(\mathbf{X}))], \qquad (2.68)$$

where $F(u) = 2u(1 - u)$ is nonnegative and concave on $[0, 1]$. The name comes from the fact that this coincides with the asymptotic error of the nearest-neighbor classification rule, which is part of a famous result known as the Cover-Hart Theorem, to be discussed in Chapter 5.

Define the entropy of a binary source as the function

$$\mathcal{H}(u) = -u \ln_2 u - (1 - u) \ln_2(1 - u), \quad 0 \le u \le 1. \qquad (2.69)$$

The *conditional entropy* of Y given $\mathbf{X}$ is a classical metric in information theory, defined by:

$$H(Y \mid \mathbf{X}) = E[\mathcal{H}(\eta(\mathbf{X}))] \qquad (2.70)$$

This is an F-error, since the function $\mathcal{H}(u)$ is nonnegative and concave on $[0, 1]$.

The *Chernoff error* is an F-error resulting from the choice $F_\alpha(u) = u^\alpha (1 - u)^{1-\alpha}$, for $0 < \alpha < 1$. The special case $F_{1/2}(u) = \sqrt{u(1 - u)}$ leads to the so-called *Matsushita error*:

$$\rho = E\left[\sqrt{\eta(\mathbf{X})(1 - \eta(\mathbf{X}))}\right]. \qquad (2.71)$$

See Figure 2.7 for an illustration. Notice in the figure that

$$\sqrt{u(1 - u)} \ge 2u(1 - u) \ge \min\{u, 1 - u\}, \quad 0 \le u \le 1. \qquad (2.72)$$

It follows that $\rho \ge \varepsilon_{\mathrm{NN}} \ge \varepsilon^*$, regardless of the distribution of the data. Similarly, we can show that $H(Y \mid \mathbf{X}) \ge \varepsilon_{\mathrm{NN}} \ge \varepsilon^*$. However, it is not true in general that $H(Y \mid \mathbf{X}) \ge \rho$ or $\rho \ge H(Y \mid \mathbf{X})$ (notice the behavior of the corresponding curves near 0 and 1).

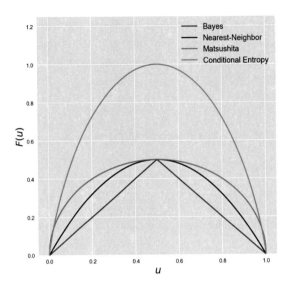

Figure 2.7: F functions corresponding to the Bayes, nearest-neighbor, Matsushita, and conditional entropy errors (plot generated by c02_Ferror.py).

The following theorem collects some useful properties of F-errors. Parts (a)–(c) are immediate, while part (d) can be proved in the same way as Theorem 2.3.

Theorem 2.5.

(a) The F-error d_F is nonnegative, and if $F(u) = 0$ only at $u = 0, 1$, then $d_F = 0$ if and only if $\varepsilon^* = 0$.

(b) If $F(u)$ reaches a maximum at $u = \frac{1}{2}$ then d_F is maximum if and only if $\varepsilon^* = \frac{1}{2}$.

(c) If $F(u) \geq \min(u, 1 - u)$ for all $u \in [0, 1]$, then $d_F \geq \varepsilon^*$.

(d) Let $\mathbf{X}' = t(\mathbf{X})$, where $t : R^d \to R^k$ is a feature-set transformation. Then $d_F(\mathbf{X}', Y) \geq d_F(\mathbf{X}, Y)$, with equality if t is invertible.

Part (a) of the Theorem implies that, for example, $\varepsilon_{\text{NN}} = 0$ if and only if $\varepsilon^* = 0$. Therefore, a small value for ε_{NN} is desirable. In fact, the Cover-Hart Theorem (see Chapter 5), states that $\varepsilon^* \leq \varepsilon_{\text{NN}} \leq 2\varepsilon^*$. Conversely, a large vale of ε_{NN} is undesirable, according to part (b) of the Theorem. This illustrates the relationship of F-errors with the Bayes error, and their potential usefulness in feature selection/extraction problems.

Concavity of F is required to apply Jensen's inequality in the proof of part (d), as was the case in the proof of Theorem 2.3. That is the reason for requiring concavity of F in the definition of an F-error: invariance under an invertible transformation of the feature vector is a very desirable property for any discriminating metric in classification.

2.6.3 Bayes Decision Theory

Bayes decision theory provides a general procedure for decision making that includes optimal classification as a special case.

Suppose that upon observing the feature vector $\mathbf{X} = \mathbf{x}$ one takes an *action* $\alpha(\mathbf{x})$ in a finite set of a possible actions $\alpha(\mathbf{x}) \in \{\alpha_0, \alpha_1, \ldots, \alpha_{a-1}\}$. Assume further that there are c *states of nature*, which are coded into $Y \in \{0, 1, \ldots, c-1\}$. Each action incurs a *loss*

$$\lambda_{ij} = \text{cost of taking action } \alpha_i \text{ when true state of nature is } j. \tag{2.73}$$

Action α_i may be simply deciding that the true state of nature is i, but we may have $a > c$, in which case one of the extra actions might be, for example, *rejecting* to make a decision.

The expected loss upon observing $\mathbf{X} = \mathbf{x}$ is

$$R[\alpha(\mathbf{x}) = \alpha_i] = \sum_{j=0}^{c-1} \lambda_{ij} P(Y = j \mid \mathbf{X} = \mathbf{x}). \tag{2.74}$$

This is called the *conditional risk* given $\mathbf{X} = \mathbf{x}$. The *risk* is given by

$$R = E[R(\alpha(\mathbf{X}))] = \int_{\mathbf{x} \in R^d} R(\alpha(\mathbf{x})) p(\mathbf{x}) \, d\mathbf{x}. \tag{2.75}$$

To minimize R, it is enough to select $\alpha(\mathbf{x}) = \alpha_i$ such that $R[\alpha(\mathbf{x}) = \alpha_i]$ is minimum, *at each value* $\mathbf{x} \in R^d$. This optimal strategy is called the *Bayes decision rule*, with corresponding optimal *Bayes risk* R^*.

In the special case that $a = c = 2$, that is, there are two classes and two actions, we have

$$\begin{aligned} R[\alpha(\mathbf{x}) = \alpha_0] &= \lambda_{00} P(Y = 0 \mid \mathbf{X} = \mathbf{x}) + \lambda_{01} P(Y = 1 \mid \mathbf{X} = \mathbf{x}) \\ R[\alpha(\mathbf{x}) = \alpha_1] &= \lambda_{10} P(Y = 0 \mid \mathbf{X} = \mathbf{x}) + \lambda_{11} P(Y = 1 \mid \mathbf{X} = \mathbf{x}) \end{aligned} \tag{2.76}$$

We decide for action α_0 if $R[\alpha(\mathbf{x}) = \alpha_0] < R[\alpha(\mathbf{x}) = \alpha_1]$, that is, if

$$(\lambda_{10} - \lambda_{00}) P(Y = 0 \mid \mathbf{X} = \mathbf{x}) > (\lambda_{01} - \lambda_{11}) P(Y = 1 \mid \mathbf{X} = \mathbf{x}), \tag{2.77}$$

which is equivalent to

$$\frac{p(\mathbf{x} \mid Y = 0)}{p(\mathbf{x} \mid Y = 1)} > \frac{\lambda_{01} - \lambda_{11}}{\lambda_{10} - \lambda_{00}} \frac{P(Y = 1)}{P(Y = 0)}, \tag{2.78}$$

by an application of Bayes Theorem (assuming that $\lambda_{10} > \lambda_{00}$ and $\lambda_{01} > \lambda_{11}$).

The loss

$$\begin{aligned} \lambda_{00} &= \lambda_{11} = 0 \\ \lambda_{10} &= \lambda_{01} = 1 \end{aligned} \tag{2.79}$$

is called the *0-1 loss*. This is equivalent to the misclassification loss, already mentioned in Chapter 1. In this case, (2.77) is equivalent to the expression for the Bayes classifier in (2.26). Therefore, if action α_i is simply deciding that the state of nature is i, for $i = 0, 1$, then the 0-1 loss case leads to the optimal binary classification case considered before: The (conditional) risk reduces to the (conditional) classification error, and the Bayes decision rule and Bayes risk reduce to the Bayes classifier and Bayes error, respectively.

The general case in (2.77) may be useful in classification in cases where the class-specific error rates are not symmetric; for example, in a diagnostic test, the cost λ_{10} of a false positive involves further testing or unnecessary treatment, but a false negative could be fatal in some cases, so that a much larger cost λ_{01} ought to be assigned to it.

*2.6.4   Rigorous Formulation of the Classification Problem

Our goal is to construct the feature-label distribution $P_{\mathbf{X},Y}$ for the pair $(\mathbf{X}, Y)$, where $\mathbf{X} \in R^d$ and $Y \in R$ takes values in $\{0, 1, \ldots, c-1\}$, for $c \geq 2$ (the arguments below also go through for a countably infinite number of classes). Concepts of probability theory mentioned in this section are reviewed briefly in Section A1 of the Appendix.

Given a Borel set B in R^{d+1}, define its sections $B^y = \{\mathbf{x} \in R^d \mid (\mathbf{x}, y) \in B\}$, for $y = 0, 1, \ldots, c-1$. It can be shown [Billingsley, 1995, Thm 18.1] that each B^y is a Borel set in R^d. An important property used below is that the section of the union is the union of the sections: $(\bigcup B_i)^y = \bigcup B_i^y$.

Consider probability measures μ_y, for $y = 1, \ldots, c-1$ on $(R^d, \mathcal{B}^d)$, and a discrete probability measure on $(R, \mathcal{B})$ putting masses $p_y > 0$ on $y = 0, 1, \ldots, c-1$. Define a set function $P_{\mathbf{X},Y}$ on $\mathcal{B}^{d+1}$ by

$$P_{\mathbf{X},Y}(B) = \sum_{y=0}^{c-1} p_y \mu_y(B^y). \qquad (2.80)$$

Clearly, $P_{\mathbf{X},Y}$ is nonnegative, $P_{\mathbf{X},Y}(R^{d+1}) = \sum_{y=0}^{c-1} p_y \mu_y(R^d) = 1$, and if $B_1, B_2, \ldots$ are pairwise disjoint Borel sets in R^{d+1}, then $P_{\mathbf{X},Y}(\bigcup_{i=1}^{\infty} B_i) = \sum_{y=0}^{c-1} p_y \mu_y(\bigcup_{i=1}^{\infty} B_i^y) = \sum_{y=0}^{c-1} p_y \sum_{i=1}^{\infty} \mu_y(B_i^y) = \sum_{i=1}^{\infty} \sum_{y=0}^{c-1} p_y \mu_y(B_i^y) = \sum_{i=1}^{\infty} P_{\mathbf{X},Y}(B_i)$. Hence, $P_{\mathbf{X},Y}$ is a probability measure on $(R^{d+1}, \mathcal{B}^{d+1})$.

The feature-label pair $(\mathbf{X}, Y)$ is a random vector associated with the distribution $P_{\mathbf{X},Y}$. Note that

$$P(Y = y) = P_{\mathbf{X},Y}(R^d \times \{y\}) = p_y \mu_y(R^d) = p_y \qquad (2.81)$$

are the class prior probabilities, for $y = 0, 1, \ldots, c-1$.

This formulation is entirely general: the measures μ_y can be discrete, singular continuous, absolutely continuous, or a mixture of these. In this chapter, we have considered in detail the case where each measure μ_y is absolutely continuous with respect to Lebesgue measure. In this case, there is a density, i.e., a nonnegative function on R^d integrating to one, which we denote by $p(\mathbf{x} \mid Y = y)$, such that

$$\mu_y(E) = \int_E p(\mathbf{x} \mid Y = y)\lambda(dx), \tag{2.82}$$

for a Borel set $E \subseteq R^d$, in which case, (2.80) gives

$$P(\mathbf{X} \in E, Y = y) = P_{\mathbf{X},Y}(E \times \{y\}) = \int_E P(Y = y)p(\mathbf{x} \mid Y = y)\lambda(dx), \tag{2.83}$$

for each $y = 0, 1, \ldots, c - 1$, which is equation (2.3). Adding the previous equation over y gives

$$P(\mathbf{X} \in E) = \int_E \sum_{y=0}^{c-1} P(Y = y)p(\mathbf{x} \mid Y = y)\lambda(dx), \tag{2.84}$$

which shows that $\mathbf{X}$ itself is absolutely continuous with respect to Lebesgue measure, with density

$$p(\mathbf{x}) = \sum_{y=0}^{c-1} P(Y = y)p(\mathbf{x} \mid Y = y). \tag{2.85}$$

An alternative approach begins instead with a single probability measure μ on $(R^d, \mathcal{B}^d)$, and nonnegative μ-integrable functions η_y on R^d, for $y = 0, 1, \ldots, c - 1$, such that $\sum \eta_y(\mathbf{x}) = 1$, for each $\mathbf{x} \in R^d$. Define $P_{\mathbf{X},Y}$ on $\mathcal{B}^{d+1}$ by

$$P_{\mathbf{X},Y}(B) = \sum_{y=0}^{c-1} \int_{B^y} \eta_y(\mathbf{x})\mu(dx). \tag{2.86}$$

Clearly, $P_{\mathbf{X},Y}$ is nonnegative, and

$$P_{\mathbf{X},Y}(R^{d+1}) = \sum_{y=0}^{c-1} \int_{R^d} \eta_y(\mathbf{x})\mu(dx) = \int_{R^d} \left(\sum_{y=0}^{c-1} \eta_i(\mathbf{x}) \right) \mu(dx) = 1. \tag{2.87}$$

If $B_1, B_2, \ldots$ are pairwise disjoint Borel sets in R^{d+1},

$$P_{\mathbf{X},Y}\left(\bigcup_{i=1}^{\infty} B_i \right) = \sum_{y=0}^{c-1} \int_{\bigcup_{i=1}^{\infty} B_i^y} \eta_y(\mathbf{x})\mu(dx) = \sum_{i=1}^{\infty} \sum_{y=0}^{c-1} \int_{B_i^y} \eta_y(\mathbf{x})\mu(dx) = \sum_{i=1}^{\infty} P_{\mathbf{X},Y}(B_i). \tag{2.88}$$

Hence, $P_{\mathbf{X},Y}$ is a probability measure on $(R^{d+1}, \mathcal{B}^{d+1})$. Notice that

$$P(\mathbf{X} \in E) = P_{\mathbf{X},Y}(E \times R) = \sum_{y=0}^{c-1} \int_E \eta_y(\mathbf{x})\mu(dx) = \int_E \sum_{y=0}^{c-1} \eta_y(\mathbf{x})\mu(dx) = \int_E \mu(dx). \tag{2.89}$$

Hence, μ is the distribution of $\mathbf{X}$. Once again, this formulation is general, where μ, and thus the feature vector $\mathbf{X}$, can be discrete, singular continuous, absolutely continuous, or a mixture of these. Notice that (2.83) here becomes

$$P(\mathbf{X} \in E, Y = y) = P_{\mathbf{X},Y}(E \times \{y\}) = \int_E \eta_y(\mathbf{x})\mu(d\mathbf{x}) = E[\eta_y(\mathbf{X})I_E], \qquad (2.90)$$

for each $y = 0, 1, \ldots, c-1$. The random variable $\eta_y(\mathbf{X})$ therefore has the properties of a conditional probability $P(Y = y \mid \mathbf{X})$ [Rosenthal, 2006, Defn. 13.1.4], and so we write $\eta_y(\mathbf{x}) = P(Y = y \mid \mathbf{X} = \mathbf{x})$. If $\mathbf{X}$ is absolute continuous with density $p(\mathbf{x})$, then (2.90) becomes

$$P(\mathbf{X} \in E, Y = y) = \int_E \eta_y(\mathbf{x})p(\mathbf{x})\lambda(d\mathbf{x}). \qquad (2.91)$$

If in addition μ_y is absolutely continuous, for $i = 0, 1, \ldots, c-1$, then (2.83) and (2.91) yield (2.4):

$$\eta_y(\mathbf{x}) = P(Y = y \mid \mathbf{X} = \mathbf{x}) = \frac{P(Y = y)p(\mathbf{x} \mid Y = y)}{p(\mathbf{x})}. \qquad (2.92)$$

The formulation in (2.80) specifies the distribution of the label Y directly, whereas the one in (2.86) specifies the distribution of the feature vector $\mathbf{X}$ directly. The formulation in (2.80) is perhaps more flexible, since it allows the specification of c probability measures μ_y for each class separately.

2.7   Bibliographical Notes

Chapter 2 of Duda et al. [2001] contains an extensive treatment of optimal classification. In particular, it contains several instructive graphical examples of 2-dim and 3-dim optimal decision boundaries for the Gaussian case.

Our proof of Theorem 2.1 is based on the proof of Theorem 2.1 in Devroye et al. [1996]. Theorem 2.3 is Theorem 3.3 in Devroye et al. [1996], though the proof presented here is different. Examples 2.1 and 2.2 are adapted from the example in Section 2.3 of Devroye et al. [1996].

Precision and *recall* are alternatives to the specificity and sensitivity error rates of a classifier [Davis and Goadrich, 2006]. Recall is simply the sensitivity $P(\psi(\mathbf{X}) = 1 \mid Y = 1)$, but the precision is defined as $P(Y = 1 \mid \psi(\mathbf{x}) = 1)$ and replaces specificity as a measure of true negative rate; the precision asks instead that there is a small number of negatives among the cases classified as positives. This may be appropriate in cases where most of the cases are negatives, as in rare diseases or object detection in imaging. However, precision is affected by the class prior probabilities, whereas the specificity is not [Xie and Braga-Neto, 2019].

Minimax estimation has played an important role in statistical signal processing; e.g. see Poor and Looze [1981]. Theorem 2.4 appears in a different form in Esfahani and Dougherty [2014] and Braga-Neto and Dougherty [2015].

See Devroye et al. [1996] for an extended coverage of F-errors and other alternative class distance metrics, including several additional results.

There are many excellent references on probability theory, and we mention but a few next. At the undergraduate level, Ross [1994] offers a thorough treatment of non-measure-theoretical probability. At the graduate level, Billingsley [1995], Chung [1974], Loève [1977], and Cramér [1999] are classical texts that provide mathematically rigorous expositions of measure-theoretical probability theory, while Rosenthal [2006] presents a modern, concise introduction.

2.8 Exercises

2.1. Suppose that $\mathbf{X}$ is a discrete feature vector, with distribution concentrated over a countable set $D = \{\mathbf{x}^1, \mathbf{x}^2, \ldots\}$ in R^d. Derive the discrete versions of (2.3), (2.4), (2.8), (2.9), (2.11), (2.30), (2.34), and (2.36).

Hint: Note that if $\mathbf{X}$ has a discrete distribution, then integration becomes summation, $P(\mathbf{X} = \mathbf{x}_k)$, for $\mathbf{x}_k \in D$, play the role of $p(\mathbf{x})$, and $P(\mathbf{X} = \mathbf{x}_k \mid Y = y)$, for $\mathbf{x}_k \in D$, play the role of $p(\mathbf{x} \mid Y = y)$, for $y = 0, 1$.

2.2. Redo Examples 2.1 and 2.2 if:

(a) only H is observable.

(b) no observations are available.

Hint: The sum of t independent and identically distributed exponential random variables with common parameter λ is a Gamma random variable X with parameters $\lambda > 0$ and $t = 1, 2, \ldots$ (also known as an Erlang distribution, in this case), with upper tail given by

$$P(X > x) = \left(\sum_{j=0}^{t-1} \frac{(\lambda x)^j}{j!} \right) e^{-\lambda x}, \tag{2.93}$$

for $x \geq 0$ (being trivially equal to 1 if $x < 0$).

2.3. This problem seeks to characterize the case $\varepsilon^* = 0$.

(a) Prove the "Zero-One Law" for perfect discrimination:

$$\varepsilon^* = 0 \iff \eta(\mathbf{X}) = 0 \text{ or } 1 \text{ with probability 1.} \tag{2.94}$$

Hint: Use an argument similar to the one employed to show (2.33).

(b) Show that

$$\varepsilon^* = 0 \ \Leftrightarrow \ \text{there is a function } f \text{ s.t. } Y = f(\mathbf{X}) \text{ with probability } 1 \,. \qquad (2.95)$$

(c) If class-conditional densities exist, and if trivialities are avoided by assuming that $P(Y = 0)P(Y = 1) > 0$, show that

$$\varepsilon^* = 0 \ \Leftrightarrow \ P(p(\mathbf{X} \mid Y = 0)p(\mathbf{X} \mid Y = 1) > 0) = 0 \,, \qquad (2.96)$$

i.e., the class-conditional densities do not "overlap" with probability 1.

2.4. This problem concerns the extension to the multiple-class case of some of the concepts derived in this chapter. Let $Y \in \{0, 1, \ldots, c - 1\}$, where c is the number of classes, and let

$$\eta_i(\mathbf{x}) \ = \ P(Y = i \mid \mathbf{X} = \mathbf{x}) \,, \quad i = 0, 1, \ldots, c - 1 \,,$$

for each $\mathbf{x} \in R^d$. We need to remember that these probabilities are not independent, but satisfy $\eta_0(\mathbf{x}) + \eta_1(\mathbf{x}) + \cdots + \eta_{c-1}(\mathbf{x}) = 1$, for each $\mathbf{x} \in R^d$, so that one of the functions is redundant. In the two-class case, this is made explicit by using a single $\eta(\mathbf{x})$, but using the redundant set above proves advantageous in the multiple-class case, as seen below.
Hint: you should answer the following items in sequence, using the previous answers in the solution of the following ones.

(a) Given a classifier $\psi : R^d \to \{0, 1, \ldots, c - 1\}$, show that its conditional error $P(\psi(\mathbf{X}) \neq Y \mid \mathbf{X} = \mathbf{x})$ is given by

$$P(\psi(\mathbf{X}) \neq Y \mid \mathbf{X} = \mathbf{x}) \ = \ 1 - \sum_{i=0}^{c-1} I_{\psi(\mathbf{x})=i} \, \eta_i(\mathbf{x}) \ = \ 1 - \eta_{\psi(\mathbf{x})}(\mathbf{x}) \,.$$

(b) Assuming that $\mathbf{X}$ has a density, show that the classification error of ψ is given by

$$\varepsilon \ = \ 1 - \sum_{i=0}^{c-1} \int_{\{\mathbf{x} \mid \psi(\mathbf{x})=i\}} \eta_i(\mathbf{x}) p(\mathbf{x}) \, d\mathbf{x} \,.$$

(c) Prove that the Bayes classifier is given by

$$\psi^*(\mathbf{x}) = \arg \max_{i=0,1,\ldots,c-1} \eta_i(\mathbf{x}) \,, \quad \mathbf{x} \in R^d \,.$$

Hint: Start by considering the difference between conditional expected errors $P(\psi(\mathbf{X}) \neq Y \mid \mathbf{X} = \mathbf{x}) - P(\psi^*(\mathbf{X}) \neq Y \mid \mathbf{X} = \mathbf{x})$.

(d) Show that the Bayes error is given by

$$\varepsilon^* = 1 - E\left[\max_{i=0,1,\ldots,c-1} \eta_i(\mathbf{X})\right].$$

(e) Show that the maximum Bayes error possible is $1 - 1/c$.

2.5. In a univariate classification problem, we have the following model

$$Y = T[\cos(\pi X) + N], \quad 0 \le X \le 1, \tag{2.97}$$

where X is uniformly distributed on the interval $[0, 1]$, $N \sim \mathcal{N}(0, \sigma^2)$ is a Gaussian noise term, and $T[\cdot]$ is the standard 0-1 step function. Find the Bayes classifier and the Bayes error. Hint: Use $\int_0^{0.5} \Phi(\cos \pi u)\, du \approx 0.36$, where Φ is the cumulative distribution function of the standard Gaussian distribution.

2.6. Assume the model

$$Y = T\left(\sum_{i=1}^{d} a_i X_i + N\right),$$

where Y is the label, $X \sim \mathcal{N}(0, I_d)$ is the feature vector, $N \sim \mathcal{N}(0, \sigma^2)$ is a noise term, and $T(x) = I_{x>0}$ is the zero-one step function. Assume that X and N are independent, and that $\|a\| = 1$.

(a) Find the Bayes classifier, and show that it is linear.

(b) Find the Bayes error.

Hint: you can use the fact that the sum of independent Gaussian r.v.'s $Z_i \sim \mathcal{N}(\mu_i, \sigma_i^2)$ is again Gaussian, with parameters $(\sum_i \mu_i, \sum_i \sigma_i^2)$, and you can use the formula

$$\int_0^\infty \int_u^\infty e^{-\frac{v^2}{2\sigma^2}}\, dv\, e^{-\frac{u^2}{2}}\, du = \sigma \arctan(\sigma).$$

2.7. Consider the following univariate Gaussian class-conditional densities:

$$p(x \mid Y = 0) = \frac{1}{\sqrt{2\pi}} \exp\left(-\frac{(x-3)^2}{2}\right) \quad \text{and} \quad p(x \mid Y = 1) = \frac{1}{3\sqrt{2\pi}} \exp\left(-\frac{(x-4)^2}{18}\right).$$

Assume that the classes are equally likely, i.e., $P(Y = 0) = P(Y = 1) = \frac{1}{2}$.

(a) Draw the densities and determine the Bayes classifier graphically.

(b) Determine the Bayes classifier.

(c) Determine the specificity and sensitivity of the Bayes classifier.
Hint: use the standard Gaussian CDF $\Phi(x)$.

(d) Determine the overall Bayes error.

2.8. Consider the general heteroskedastic Gaussian model, where

$$p(\mathbf{x} \mid Y = i) \sim N_d(\boldsymbol{\mu}_i, \Sigma_i), \quad i = 0, 1.$$

Given a linear classifier

$$\psi(\mathbf{x}) = \begin{cases} 1, & g(\mathbf{x}) = \mathbf{a}^T\mathbf{x} + b \geq 0 \\ 0, & \text{otherwise.} \end{cases}$$

obtain the classification error of ψ in terms of Φ (the c.d.f. of a standard normal random variable), and the parameters $\mathbf{a}, b, \boldsymbol{\mu}_0, \boldsymbol{\mu}_1, \Sigma_0, \Sigma_1, c_0 = P(Y = 0)$ and $c_1 = P(Y = 1)$.

2.9. Obtain the optimal decision boundary in the Gaussian model with $P(Y = 0) = P(Y = 1)$ and

(a) $\mu_0 = (0,0)^T$, $\mu_1 = (2,0)^T$, $\Sigma_0 = \begin{bmatrix} 2 & 0 \\ 0 & 1 \end{bmatrix}$, $\Sigma_1 = \begin{bmatrix} 2 & 0 \\ 0 & 4 \end{bmatrix}$.

(b) $\mu_0 = (0,0)^T$, $\mu_1 = (2,0)^T$, $\Sigma_0 = \begin{bmatrix} 2 & 0 \\ 0 & 1 \end{bmatrix}$, $\Sigma_1 = \begin{bmatrix} 4 & 0 \\ 0 & 1 \end{bmatrix}$.

(c) $\mu_0 = (0,0)^T$, $\mu_1 = (0,0)^T$, $\Sigma_0 = \begin{bmatrix} 1 & 0 \\ 0 & 1 \end{bmatrix}$, $\Sigma_1 = \begin{bmatrix} 2 & 0 \\ 0 & 2 \end{bmatrix}$.

(d) $\mu_0 = (0,0)^T$, $\mu_1 = (0,0)^T$, $\Sigma_0 = \begin{bmatrix} 2 & 0 \\ 0 & 1 \end{bmatrix}$, $\Sigma_1 = \begin{bmatrix} 1 & 0 \\ 0 & 2 \end{bmatrix}$.

In each case draw the optimal decision boundary, along with the class means and class-conditional density contours, indicating the 0- and 1-decision regions.

2.10. Show that the Mahalanobis distance in (2.45) satisfies all properties of a distance, if Σ is strictly positive definitive:

    i. $||\mathbf{x}_0 - \mathbf{x}_1||_\Sigma > 0$ and $||\mathbf{x}_0 - \mathbf{x}_1||_\Sigma = 0$ if and only if $\mathbf{x}_0 = \mathbf{x}_1$.

    ii. $||\mathbf{x}_0 - \mathbf{x}_1||_\Sigma = ||\mathbf{x}_1 - \mathbf{x}_0||_\Sigma$.

    iii. $||\mathbf{x}_0 - \mathbf{x}_1||_\Sigma \geq ||\mathbf{x}_0 - \mathbf{x}_2||_\Sigma + ||\mathbf{x}_1 - \mathbf{x}_2||_\Sigma$ (triangle inequality).

2.11. The *exponential family* of densities in R^d is of the form

$$p(\mathbf{x} \mid \boldsymbol{\theta}) = \alpha(\boldsymbol{\theta})\beta(\mathbf{x}) \exp\left(\sum_{i=1}^{k} \xi_i(\boldsymbol{\theta})\phi_i(\mathbf{x})\right), \qquad (2.98)$$

where $\boldsymbol{\theta} \in R^m$ is a parameter vector, and $\alpha, \xi_1, \ldots, \xi_k : R^m \to R$ and $\beta, \phi_1, \ldots, \phi_k : R^d \to R$ with $\alpha, \beta \geq 0$.

(a) Assume that the class-conditional densities are $p(\mathbf{x} \mid \boldsymbol{\theta}_0)$ and $p(\mathbf{x} \mid \boldsymbol{\theta}_1)$. Show that the Bayes classifier is of the form

$$
\psi^*(\mathbf{x}) = \begin{cases} 1, & \sum_{i=1}^k a_i(\boldsymbol{\theta}_0, \boldsymbol{\theta}_1)\phi_i(\mathbf{x}) + b(\boldsymbol{\theta}_0, \boldsymbol{\theta}_1) > 0, \\ 0, & \text{otherwise.} \end{cases} \tag{2.99}
$$

This is called a *generalized linear classifier*. The decision boundary is linear in the transformed feature vector $\mathbf{X}' = (\phi_1(\mathbf{X}), \ldots, \phi_k(\mathbf{X})) \in R^k$, but it is generally nonlinear in the original feature space.

(b) Show that the exponential r.v. with parameter $\lambda > 0$, the gamma r.v. with parameters $\lambda, t > 0$, and and the beta r.v. with parameters $a, b > 0$ (see Section A1.4) belong to the exponential family. Obtain the Bayes classifier in each case.

(c) Show that the multivariate Gaussian r.v. with parameters $\boldsymbol{\mu}, \Sigma > 0$ belong to the exponential family. Show that the Bayes classifier obtained from (2.99) coincides with the one in (2.54) and (2.55).

2.12. Suppose that repeated measurements $\mathbf{X}^{(j)} \in R^d$ are made on the same individual, for $j = 1, \ldots, m$. For example, these might be measurements repeated every hour, or every day (this is relatively common in medical studies as well as industrial settings). Assume the following additive model for the multiple measurements:

$$
\mathbf{X}^{(i)} = \mathbf{Z} + \boldsymbol{\varepsilon}^{(i)}, \quad i = 1, \ldots, m, \tag{2.100}
$$

where $\mathbf{Z} \mid Y_i = j \sim \mathcal{N}(\boldsymbol{\mu}_j, \Sigma_j)$, for $j = 0, 1$, and $\boldsymbol{\varepsilon}^{(j)} \sim \mathcal{N}(0, \Sigma_{\mathrm{err}})$, $i = 1, \ldots, m$ are the "signal" and the "noise", respectively. Independence among noise vectors and between noise vectors and signal vectors is assumed. We would like to obtain an optimal classifier in this case. The simplest approach is to stack all measurements up in a single feature vector $\mathbf{X} = (\mathbf{X}^{(1)}, \ldots, \mathbf{X}^{(m)}) \in R^{dm}$.

(a) Show that the optimal discriminant $D^*(\mathbf{x})$ in (2.27), where $\mathbf{x} = (\mathbf{x}^{(1)}, \ldots, \mathbf{x}^{(m)}) \in R^{dm}$ is a point in the stacked feature space, is given by

$$
D^*(\mathbf{x}) = \frac{1}{2}(\bar{\mathbf{x}} - \boldsymbol{\mu}_0)^T (\Sigma_0 + \Sigma_{\mathrm{err}}/m)^{-1}(\bar{\mathbf{x}} - \boldsymbol{\mu}_0) - \frac{1}{2}(\bar{\mathbf{x}} - \boldsymbol{\mu}_1)^T (\Sigma_1 + \Sigma_{\mathrm{err}}/m)^{-1}(\bar{\mathbf{x}} - \boldsymbol{\mu}_1)
$$
$$
+ \frac{1}{2} \ln \frac{\det(\Sigma_0 + \Sigma_{\mathrm{err}}/m)}{\det(\Sigma_1 + \Sigma_{\mathrm{err}}/m)} + \frac{m-1}{2} \mathrm{Trace}(\bar{\Sigma}(\Sigma_0^{-1} - \Sigma_1^{-1})),
$$
$$
\tag{2.101}
$$

where

$$
\bar{\mathbf{x}} = \frac{1}{m} \sum_{j=1}^m \mathbf{x}^{(j)}, \tag{2.102}
$$

and

$$\bar{\Sigma} = \sum_{j=1}^{m} (\mathbf{x}^{(j)} - \bar{\mathbf{x}})(\mathbf{x}^{(j)} - \bar{\mathbf{x}})^T. \tag{2.103}$$

Compare (2.43) and (2.101). What happens in the case $m = 1$?

Hint: Notice that

$$\mathbf{X} \mid Y = k \sim \mathcal{N}(\boldsymbol{\mu}_k \otimes \mathbf{1}_m, \Sigma_k \otimes \mathbf{1}_m \mathbf{1}_m^T + \Sigma_{\text{err}} \otimes I_m), \quad k = 0, 1, \tag{2.104}$$

where $\mathbf{1}_m$ is an $m \times 1$ unit vector and "$\otimes$" denotes the *Kronecker product* of matrices.

(b) Write an expression for the Bayes classifier similar to (2.54) and (2.55).

(c) Specialize items (a) and (b) to the homoskedastic case $\Sigma_0 = \Sigma_1$. Write an expression for the Bayes classifier similar to (2.48) and (2.49). Obtain an expression for the Bayes error similar to (2.52). What happens to the Bayes error as m increases?

2.13. In a univariate pattern recognition problem, the feature X is uniform over the interval $[0, 2]$ if $Y = 0$, and X is uniform over the interval $[1, 3]$ if $Y = 1$. Assuming that the labels are equally likely, compute:

(a) A Bayes classifier ψ^*.

(b) The Bayes error ε^*.

(c) The asymptotic nearest-neighbor error ϵ_{NN}.

2.14. Consider the following measure of distance between classes:

$$\tau = E\left[8\eta(X)^2(1 - \eta(X))^2\right]$$

and consider the proposed relationship $\epsilon^* \leq \tau \leq \epsilon_{NN}$.

(a) Show that $\tau \leq \epsilon_{NN}$ holds regardless of the distribution of (X, Y).
    Hint: You are given that $1 - 5x + 8x^2 - 4x^3 \geq 0$, for $0 \leq x \leq 1$.

(b) Indicate a distribution of (X, Y) for which $\epsilon^* \leq \tau$ fails.
    Hint: Consider what happens when $\eta(X) = 0.1$.

2.15. This problem concerns classification with a rejection option. Assume that there are c classes and $c + 1$ "actions" $\alpha_0, \alpha_1, \ldots, \alpha_c$. For $i = 0, \ldots, c - 1$, action α_i is simply to classify into class i, whereas action α_c is to reject, i.e., abstain from committing to any of the classes, for lack of enough evidence. This can be modeled as a Bayes decision theory problem, where the cost λ_{ij} of taking action α_i when true state of nature is j is given by:

$$\lambda_{ij} = \begin{cases} 0, & i = j, \text{ for } i, j = 0, \ldots, c - 1 \\ \lambda_r, & i = c \\ \lambda_m, & \text{otherwise,} \end{cases}$$

where λ_r is the cost associated with a rejection, and λ_m is the cost of misclassifying a sample. Determine the optimal decision function $\alpha^* : R^d \rightarrow \{\alpha_0, \alpha_1, \ldots, \alpha_c\}$ in terms of the posterior probabilities $\eta_i(\mathbf{x})$ — see the previous problem — and the cost parameters. As should be expected, the occurrence of rejections will depend on the relative cost λ_r/λ_m. Explain what happens when this ratio is zero, 0.5, and greater or equal than 1.

2.9 Python Assignments

2.16. Suppose in Example 2.1 that the model is

$$Y = \begin{cases} 1 \text{ (pass)}, & \text{if } S + H + N > \kappa , \\ 0 \text{ (fail)}, & \text{otherwise}, \end{cases} \tag{2.105}$$

for a given real-valued threshold $\kappa > 0$.

(a) Show that the Bayes classifier is

$$\psi^*(s, h) = \begin{cases} I_{s+h>\kappa-\ln 2}, & \kappa > \ln 2 \\ 1, & 0 < \kappa < \ln 2. \end{cases} \tag{2.106}$$

In particular, if $\kappa < \ln 2$, the optimal prediction is that all students will pass the class.

(b) Show that the Bayes error is

$$\varepsilon^* = \begin{cases} e^{-k}\left[(\kappa+1-\ln 2)^2 - \frac{\kappa(\kappa+2)}{2}\right], & \kappa > \ln 2, \\ 1 - e^{-k}\left[1 + \frac{\kappa(\kappa+2)}{2}\right], & 0 < \kappa < \ln 2. \end{cases} \tag{2.107}$$

Plot this as a function of κ and interpret the resulting graph.

(c) Differentiate (2.107) to find at what value of κ the maximum Bayes error occurs.

(d) Show that

$$c = P(Y = 1) = e^{-k}\left[1 + \frac{\kappa(\kappa + 2)}{2}\right] \tag{2.108}$$

and then show that the bound (2.35) holds, by plotting it in the same graph as the Bayes error in item (b).

2.17. This problem concerns the Gaussian model for synthetic data generation in Section A8.1.

(a) Derive a general expression for the Bayes error for the homoskedastic case with $\boldsymbol{\mu}_0 = (0, \ldots, 0)$, $\boldsymbol{\mu}_1 = (1, \ldots, 1)$, and $P(Y = 0) = P(Y = 1)$. Your answer should be in terms of k, $\sigma_1^2, \ldots, \sigma_k^2$, $l_1, \ldots, l_k$, and $\rho_1, \ldots, \rho_k$.

Hint: Use the fact that

$$
\begin{bmatrix}
1 & \rho & \cdots & \rho \\
\rho & 1 & \cdots & \rho \\
\vdots & \vdots & \ddots & \vdots \\
\rho & \rho & \cdots & 1
\end{bmatrix}_{l\times l}^{-1}
= \frac{1}{(1-\rho)(1+(l-1)\rho)}
\begin{bmatrix}
1+(l-2)\rho & -\rho & \cdots - & \rho \\
-\rho & 1+(l-2)\rho & \cdots - & \rho \\
\vdots & \vdots & \ddots & \vdots \\
-\rho & -\rho & \cdots & 1+(l-2)\rho
\end{bmatrix}.
$$

(2.109)

(b) Specialize the previous formula for equal-sized blocks $l_1 = \cdots = l_k = l$ with equal correlations $\rho_1 = \cdots = \rho_k = \rho$, and constant variance $\sigma_1^2 = \cdots, \sigma_k^2 = \sigma^2$. Write the resulting formula in terms of d, l, σ, and ρ.

 i. Using the python function `norm.cdf` in the `scipy.stats` module, plot the Bayes error as a function of $\sigma \in [0.01, 3]$ for $d = 20$, $l = 4$, and four different correlation values $\rho = 0, 0.25, 0.5, 0.75$ (plot one curve for each value). Confirm that the Bayes error increases monotonically with σ from 0 to 0.5 for each value of ρ, and that the Bayes error for larger ρ is uniformly larger than that for smaller ρ. The latter fact shows that correlation between the features is detrimental to classification.

 ii. Plot the Bayes error as a function of $d = 2, 4, 6, 8, \ldots, 40$, with fixed block size $l = 4$ and variance $\sigma^2 = 1$, and $\rho = 0, 0.25, 0.5, 0.75$ (plot one curve for each value). Confirm that the Bayes error decreases monotonically to 0 with increasing dimensionality, with faster convergence for smaller correlation values.

 iii. Plot the Bayes error as a function of the correlation $\rho \in [0, 1]$ for constant variance $\sigma^2 = 2$ and fixed $d = 20$ with varying block size $l = 1, 2, 4, 10$ (plot one curve for each value). Confirm that the Bayes error increases monotonically with increasing correlation. Notice that the rate of increase is particularly large near $\rho = 0$, which shows that the Bayes error is very sensitive to correlation in the near-independent region.

(c) Use numerical integration to redo the plots in item (b) in a heteroskedastic case, where the features in class 0 are always uncorrelated.

 i. For parts i and ii, discretize the range for σ in steps of 0.02, and use $\rho = 0$ for class 0 and $\rho = 0.25, 0.5, 0.75$ for class 1 (plot one curve for each value).

 ii. For part iii, use $\rho = 0$ for class 0 and $\rho \in [0, 1]$ for class 1. Discretize the range for ρ in steps of 0.02.

Compare to the results obtained in the homoskedastic case.

Hint: Generate a large sample from the synthetic model, apply the optimal Gaussian discriminant to it, and form an empirical error estimate.

2.18. The univariate Student's t random variable with $\nu > 0$ degrees of freedom provides a model for "heavy-tailed" unimodal distributions, with a density given by:

$$f_\mu(x) = K(\nu) \left(1 + \frac{x^2}{\nu} \right)^{-\frac{\nu+1}{2}}, \qquad (2.110)$$

where $K(\nu) > 0$ is a normalization constant to make the density integrate to 1. The smaller ν is, the heavier the tails are, and the fewer moments exist. The case $\nu = 1$ corresponds to the Cauchy r.v., which has no moments. Conversely, larger ν leads to thinner tails and the existence of more moments. It can be shown that the univariate Gaussian is the limiting case as $\nu \to \infty$.

Let the class-conditional densities in a classification problem be modeled by shifted and scaled univariate t distributions:

$$p(x \mid Y = i) = f_\nu \left(\frac{x - a_i}{b} \right), \qquad i = 0, 1, \qquad (2.111)$$

where a_0, a_1, and $b > 0$ play the role of μ_0, μ_1, and σ, respectively, in the Gaussian case. Assume $P(Y = 0) = P(Y = 1)$.

(a) Determine the Bayes classifier.

(b) Determine the Bayes error as a function of the parameters a_0, a_1, and b, and ν. You may assume $a_0 < a_1$, without loss of generality, and express your answer in terms of the CDF $F_\nu(t)$ of a standard Student's t random variable with ν degrees of freedom.

(c) Using the python function t.cdf in the scipy.stats module, plot the Bayes error as a function of $(a_1 - a_0)/b$, for $\nu = 1, 2, 4, 100$ (the case $\nu = 100$ corresponds essentially to the Gaussian case). How does the value ν affect the Bayes error? Where do the maximum and minimum values of the Bayes error occur?

Chapter 3

Sample-Based Classification

> "I often say that when you can measure what you are speaking about, and express it in numbers, you know something about it; but when you cannot express it in numbers, your knowledge is of a meagre and unsatisfactory kind; it may be the beginning of knowledge, but you have scarcely, in your thoughts, advanced to the stage of science."
> – Lord Kelvin, *Popular Lectures and Addresses*, 1889.

Optimal classification requires full knowledge of the feature-label distribution. In practice, that is a relatively rare scenario, and a combination of distributional knowledge and sample data must be employed to obtain a classifier. In this chapter, we introduce the basic concepts related to sample-based classification, including designed classifiers and error rates, and consistency. The chapter includes a section showing that distribution-free classification rules have important limitations. The material in this chapter provides the foundation for the next several chapters on sample-based classification.

3.1 Classification Rules

The *training data* $S_n = \{(\mathbf{X}_1, Y_1), \ldots, (\mathbf{X}_n, Y_n)\}$ of n consists of sample feature vectors and their associated labels, which are typically produced by performing a vector measurement $\mathbf{X}_i$ on each of n specimens in an experiment, and then having an "expert" produce a label Y_i for each specimen. We assume that S_n is an *independent and identically distributed (i.i.d.)* sample from the feature-label distribution $P_{\mathbf{X},Y}$; i.e., the set of sample points is independent and each sample point has distribution $P_{\mathbf{X},Y}$ (but see Section 3.5.2 for a different scenario). For each specimen corresponding to feature vector $\mathbf{X}_i$, the "expert" produces a label $Y_i = 0$ with probability $P(Y = 0 \mid \mathbf{X}_i)$ and a label $Y_i = 1$ with probability $P(Y = 1 \mid \mathbf{X}_i)$. Hence, the labels in the training data are not the "true" labels, as in general there is no such a thing; the labels are simply assumed to be assigned with the correct probabilities. In addition, notice that the numbers $N_0 = \sum_{i=1}^n I_{Y_i=0}$ and $N_1 = \sum_{i=1}^n I_{Y_i=1}$ of sample points from class 0 and class 1 are binomial random variables with parameters $(n, 1-p)$ and (n, p), respectively, where $p = P(Y = 1)$. Obviously, N_0 and N_1 are not independent, since $N_0 + N_1 = n$.[1]

Given the training data S_n as input, a classification rule is an operator that outputs a trained classifier ψ_n. The subscript "n" reminds us that ψ_n is a function of the data S_n (it plays a similar role to the hat notation used for estimators in classical statistics). It is important to understand the difference between a classifier and a classification rule; the latter does not output class labels, but rather classifiers.

Formally, let $\mathcal{C}$ denote the set of all classifiers, i.e., all (Borel-measurable) functions from R^d into $\{0, 1\}$. Then a classification rule is defined as a mapping $\Psi_n : [R^d \times \{0, 1\}]^n \to \mathcal{C}$. In other words, Ψ_n maps sample data $S_n \in [R^d \times \{0, 1\}]^n$ into a classifier $\psi_n = \Psi_n(S_n) \in \mathcal{C}$.

Example 3.1. (*Nearest-Centroid Classification Rule.*) Consider the following simple classifier:

$$\psi_n(\mathbf{x}) = \begin{cases} 1, & \|\mathbf{x} - \hat{\boldsymbol{\mu}}_1\| < \|\mathbf{x} - \hat{\boldsymbol{\mu}}_0\|, \\ 0, & \text{otherwise,} \end{cases} \tag{3.1}$$

where

$$\hat{\boldsymbol{\mu}}_0 = \frac{1}{N_0} \sum_{i=1}^n \mathbf{X}_i I_{Y_i=0} \quad \text{and} \quad \hat{\boldsymbol{\mu}}_1 = \frac{1}{N_1} \sum_{i=1}^n \mathbf{X}_i I_{Y_i=1} \tag{3.2}$$

are the sample means for each class. In other words, the classifier assigns to the test point $\mathbf{x}$ the label of the nearest (sample) class mean. It is easy to see that this classification rule produces hyperplane decision boundaries. By replacing the sample mean with other types of centroids (e.g., the sample median), a family of like-minded classification rules can be obtained. Notice the similarity between (2.46) and (3.1) — more on this in Chapter 4. ◇

Example 3.2. (*Nearest-Neighbor Classification Rule.*) Another simple classifier is given by

$$\psi_n(\mathbf{x}) = Y_{(1)}(\mathbf{x}), \tag{3.3}$$

where $(\mathbf{X}_{(1)}(\mathbf{x}), Y_{(1)}(\mathbf{x}))$ is the nearest training point:

$$\mathbf{X}_{(1)}(\mathbf{x}) = \arg \min_{\mathbf{X}_1,\dots,\mathbf{X}_n} \|\mathbf{X} - \mathbf{x}\|. \tag{3.4}$$

---

[1]All these concepts can be immediately extended to any number of classes $c > 2$.

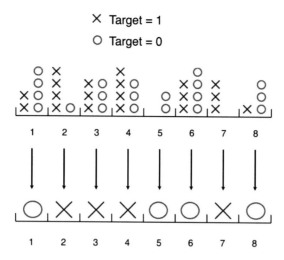

Figure 3.1: Discrete histogram rule: top shows distribution of the sample data in the bins; bottom shows the designed classifier.

(If there is a tie, take the point with the smallest index i.) In other words, the classifier assigns to the test point $\mathbf{x}$ the label of the nearest neighbor in the training data. The decision boundaries produced by this classification rule are very complex. A family of similar classification rules is obtained by replacing the Euclidean norm with other metrics (e.g., the correlation). In addition, a straightforward generalization of this classification rule is obtained by assigning to the test point the majority label in the set of k nearest training points ($k = 1$ yielding the previous case), with odd k to avoid ties. This is called the k-nearest neighbor classification rule, which is studied in detail in Chapter 5. ◇

Example 3.3. (*Discrete Histogram Classification Rule.*) Assume that the distribution of $\mathbf{X}$ is concentrated over a finite number of points $\{\mathbf{x}^1, \ldots, \mathbf{x}^b\}$ in R^d. This corresponds to the case where the measurement $\mathbf{X}$ can yield only a finite number of different values. Let U_j and V_j be the number of training points with $(\mathbf{X}_i = \mathbf{x}^j, Y_i = 0)$ and $(\mathbf{X}_i = \mathbf{x}^j, Y_i = 1)$, respectively, for $j = 1, \ldots, b$. The discrete histogram rule is given by

$$\psi_n(\mathbf{x}^j) = \begin{cases} 1, & U_j < V_j \\ 0, & \text{otherwise,} \end{cases} \tag{3.5}$$

for $j = 1, \ldots, b$. In other words, the discrete histogram rule assigns to $\mathbf{x}^j$ the majority label among the training points that coincide with $\mathbf{x}^j$. In the case of a tie, the label is set to zero. See Figure 3.1 for an illustration. ◇

One of the problems with small training sample size in the occurrence of ties between U_j and V_j

(including missing categories, i.e., $U_j = V_j = 0$). In addition to setting the classifier arbitrarily to zero (or one), once could use the majority label in the overall data, or assign zero if the value of $U_j = V_j$ is even, and one if it is odd. If random factors are allowed in the definition of a classification rule, then one can also attribute the label randomly, with 50%-50% probabilities, or using the observed values N_0/n and N_1/n as probabilities. In this case, one has a random classifier: repeating the procedure generates different classifiers for the same training data. Unless otherwise stated, all classification rules considered in the sequel are nonrandom. (See Section 3.5.1 for additional examples of random classification rules.)

3.2   Classification Error Rates

Given a classification rule Ψ_n, the error of a designed classifier $\psi_n = \Psi_n(S_n)$ trained on data S_n is given by

$$\varepsilon_n = P(\Psi_n(S_n)(X) \neq Y \mid S_n) = P(\psi_n(X) \neq Y \mid S_n). \tag{3.6}$$

Here, (X, Y) can be considered to be a test point, which is *independent* of S_n. Notice that ε_n is similar to the classifier error defined in (2.7). However, there is a fundamental difference between the two error rates, as ε_n is a function of the random sample data S_n and therefore a *random variable*. On the other hand, assuming that Ψ_n is nonrandom, the classification error ε_n is an ordinary real number once the data S_n is specified and fixed. The error ε_n is sometimes called the *conditional error*, since it is conditioned on the data.

Another important error rate in sample-based classification is the *expected error*:

$$\mu_n = E[\varepsilon_n] = P(\psi_n(X) \neq Y). \tag{3.7}$$

This error rate is nonrandom. It is sometimes called the *unconditional error*.

Comparing ε_n and μ_n, we observe that the conditional error ε_n is usually the one of most practical interest, since it is the error of the classifier designed on the actual sample data at hand. Nevertheless, μ_n can be of interest because it is *data-independent*: it is a function only of the classification rule. Therefore, μ_n can be used to define global properties of classification rules, such as consistency (see the next section). In addition, since it is nonrandom, μ_n can be bounded, tabulated and plotted. This can be convenient in both analytical and empirical (simulation) studies. Finally, the most common criterion for comparing the performance of classification rules is to pick the one with smallest expected error μ_n, for a fixed given sample size n.

Similarly to what was done in (2.8), we can define class-specific error rates:

$$\varepsilon_n^0 = P(\psi_n(\mathbf{X}) = 1 \mid Y = 0, S_n),$$
$$\varepsilon_n^1 = P(\psi_n(\mathbf{X}) = 0 \mid Y = 1, S_n),$$

(3.8)

with classification error as in (2.9):

$$\begin{aligned}
\varepsilon[\psi_n] &= P(\psi_n(\mathbf{X}) \neq Y \mid S_n) \\
&= P(\psi_n(\mathbf{X}) = 1 \mid Y = 0, S_n)P(Y = 0) + P(\psi_n(\mathbf{X}) = 0 \mid Y = 1, S_n)P(Y = 1) \\
&= P(Y = 0)\,\varepsilon_n^0[\psi] + P(Y = 1)\,\varepsilon_n^1[\psi].
\end{aligned}$$

(3.9)

*3.3 Consistency

Consistency has to do with the natural requirement that, as the sample size increases, the classification error should approach the optimal error. Accordingly, the classification rule[2] is said to be *consistent* if, as $n \to \infty$,

$$\varepsilon_n \to \varepsilon^*, \quad \text{in probability,}$$

(3.10)

that is, given any $\tau > 0$, $P(|\varepsilon_n - \varepsilon^*| > \tau) \to 0$. (See Section A1.8 for a review of modes of convergence for random variables.) In other words, for a large sample size n, ε_n will be near ε^* with a large probability. The classification rule Ψ_n is said to be *strongly consistent* if, as $n \to \infty$,

$$\varepsilon_n \to \varepsilon^*, \quad \text{with probability 1,}$$

(3.11)

that is, $P(\varepsilon_n \to \varepsilon^*) = 1$. Since convergence with probability 1 implies convergence in probability, strong consistency implies ordinary ("weak") consistency. Strong consistency is a much more demanding criterion than ordinary consistency. It roughly requires ε_n to converge to ε^* for almost all possible sequences of training data $\{S_n; n = 1, 2, \ldots\}$. In a very real sense, however, ordinary consistency is enough for practical purposes. Furthermore, all commonly used consistent classification rules turn out, interestingly, to be strongly consistent as well.

The previous definitions hold for a given fixed feature-label distribution $P_{\mathbf{X},Y}$. So a classification rule can be consistent under a feature-label distribution but not under another. A *universally* (strong) consistent classification rule is consistent under any distribution; hence, universal consistency is a property of the classification rule alone.

It should be kept in mind that consistency is a large-sample property, and therefore, is not generally indicative of classification performance under small sample sizes. Universally consistent rules tend to produce complex classifiers and could thus produce a "Scissors Effect," as discussed in Section 1.6.

---

[2] Throughout this section, what we call a classification rule Ψ_n is actually a sequence $\{\Psi_n; n = 1, 2, \ldots\}$.

Example 3.4. (Consistency of the Nearest-Centroid Classification Rule.) The classifier in (3.1) can be written as:

$$\psi_n(\mathbf{x}) = \begin{cases} 1, & \mathbf{a}_n^T \mathbf{x} + b_n > 0, \\ 0, & \text{otherwise,} \end{cases} \tag{3.12}$$

where $\mathbf{a}_n = \hat{\boldsymbol{\mu}}_1 - \hat{\boldsymbol{\mu}}_0$ and $b_n = (\hat{\boldsymbol{\mu}}_1 - \hat{\boldsymbol{\mu}}_0)(\hat{\boldsymbol{\mu}}_1 + \hat{\boldsymbol{\mu}}_0)/2$ (use the fact that $||\mathbf{x} - \hat{\boldsymbol{\mu}}||^2 = (\mathbf{x} - \hat{\boldsymbol{\mu}})^T(\mathbf{x} - \hat{\boldsymbol{\mu}})$). Now, assume that the feature-label distribution of the problem is specified by multivariate spherical Gaussian densities $p(\mathbf{x} \mid Y = 0) \sim \mathcal{N}_d(\boldsymbol{\mu}_0, I_d)$ and $p(\mathbf{x} \mid Y = 1) \sim \mathcal{N}_d(\boldsymbol{\mu}_1, I_d)$, with $\boldsymbol{\mu}_0 \neq \boldsymbol{\mu}_1$ and $P(Y = 0) = P(Y = 1)$. The classification error is given by

$$\begin{aligned} \varepsilon_n &= P(\psi_n(\mathbf{X}) = 1 \mid Y = 0)P(Y = 0) + P(\psi_n(\mathbf{X}) = 0 \mid Y = 1)P(Y = 1) \\ &= \frac{1}{2}\left(P(\mathbf{a}_n^T\mathbf{X} + b_n > 0 \mid Y = 0) + P(\mathbf{a}_n^T\mathbf{X} + b_n \leq 0 \mid Y = 1) \right) \\ &= \frac{1}{2}\left(\Phi\left(\frac{\mathbf{a}_n^T\boldsymbol{\mu}_0 + b_n}{||\mathbf{a}_n||}\right) + \Phi\left(-\frac{\mathbf{a}_n^T\boldsymbol{\mu}_1 + b_n}{||\mathbf{a}_n||}\right) \right), \end{aligned} \tag{3.13}$$

where $\Phi(\cdot)$ is the CDF of a standard Gaussian and we used the fact that $\mathbf{a}_n^T\mathbf{X} + b_n \mid Y = i \sim \mathcal{N}(\mathbf{a}_n^T\boldsymbol{\mu}_i + b_n, ||\mathbf{a}_n||^2)$, for $i = 0, 1$ (see Section A1.7 for the properties of the multivariate Gaussian distribution). We also know from (2.53) that the Bayes error for this problem is

$$\varepsilon^* = \Phi\left(-\frac{||\boldsymbol{\mu}_1 - \boldsymbol{\mu}_0||}{2}\right). \tag{3.14}$$

Now, by the vector version of the Law of Large Numbers (see Thm. A.12), we know that, with probability 1, $\hat{\boldsymbol{\mu}}_0 \to \boldsymbol{\mu}_0$ and $\hat{\boldsymbol{\mu}}_1 \to \boldsymbol{\mu}_1$, so that $\mathbf{a}_n \to \mathbf{a} = \boldsymbol{\mu}_1 - \boldsymbol{\mu}_0$ and $b_n \to b = (\boldsymbol{\mu}_1 - \boldsymbol{\mu}_0)(\boldsymbol{\mu}_1 + \boldsymbol{\mu}_0)/2$, as $n \to \infty$. Furthermore, ε_n in (3.13) is a continuous function of $\mathbf{a}_n$ and b_n. Hence, by the Continuous Mapping Theorem (see Thm. A.6),

$$\varepsilon_n(\mathbf{a}_n, b_n) \to \varepsilon_n(\mathbf{a}, b) = \Phi\left(-\frac{||\boldsymbol{\mu}_1 - \boldsymbol{\mu}_0||}{2}\right) = \varepsilon^* \quad \text{with probability 1,} \tag{3.15}$$

as can be easily verified. Hence, the nearest-centroid classification rule is strongly consistent under spherical Gaussian densities with the same variance and equally-likely classes. ⋄

The nearest-centroid classification rule is not universally consistent; if the covariance matrices are not spherical, or the class-conditional densities are not Gaussian, then the classification error does not converge in general to the Bayes error as sample size increases. But, if the classes are known to be at least approximately Gaussian with spherical shapes, then the nearest-centroid rule is "approximately consistent," and in fact can perform quite well even under small sample sizes.

Example 3.5. (Consistency of the Discrete Histogram Rule.) With $c_0 = P(Y = 0)$, $c_1 = P(Y = 1)$, $p_j = P(\mathbf{X} = \mathbf{x}^j \mid Y = 0)$, and $q_j = P(\mathbf{X} = \mathbf{x}^j \mid Y = 1)$, for $j = 1, \ldots, b$, we have that

$$\eta(\mathbf{x}^j) = P(Y = 1 \mid \mathbf{X} = \mathbf{x}^i) = \frac{c_1 q_j}{c_0 p_j + c_1 q_j}, \tag{3.16}$$

for $j = 1, \ldots, b$. Therefore, the Bayes classifier is

$$\psi^*(\mathbf{x}^j) = I_{\eta(\mathbf{x}^j)>1/2} = I_{c_1 q_j > c_0 p_j}, \tag{3.17}$$

for $j = 1, \ldots, b$, with Bayes error

$$\varepsilon^* = E[\min\{\eta(\mathbf{X}), 1 - \eta(\mathbf{X})\}] = \sum_{j=1}^{b} \min\{c_0 p_j, c_1 q_j\}. \tag{3.18}$$

Now, the error of the classifier in (3.5) can be written as:

$$\varepsilon_n = P(\psi_n(\mathbf{X}) \neq Y) = \sum_{j=1}^{b} P(\mathbf{X} = \mathbf{x}^j, \psi_n(\mathbf{x}^j) \neq Y)$$

$$= \sum_{j=1}^{b} \left[P(\mathbf{X} = \mathbf{x}^j, Y = 0) I_{\psi_n(\mathbf{x}^j)=1} + P(\mathbf{X} = \mathbf{x}^j, Y = 1) I_{\psi_n(\mathbf{x}^j)=0} \right] \tag{3.19}$$

$$= \sum_{j=1}^{b} \left[c_0 p_j I_{V_j > U_j} + c_1 q_j I_{U_j \geq V_j} \right].$$

Clearly U_j is a binomial random variable with parameters $(n, c_0 p_j)$. To see this, note that U_j is the number of times that one of the n training points independently goes into the "bin" ($\mathbf{X} = \mathbf{x}^j, Y = 0$) with probability $c_0 p_i$. Thus $U_j = \sum_{i=1}^{n} Z_{ji}$, where the Z_{ji} are i.i.d. Bernoulli random variables with parameter $c_0 p_j$, and it follows from the Law of Large Numbers (see Thm. A.12) that $U_j/n \xrightarrow{a.s.} c_0 p_j$ as $n \to \infty$. Similarly, V_j is a binomial random variable with parameters $(n, c_1 q_j)$ and $V_j/n \xrightarrow{a.s.} c_1 q_j$ as $n \to \infty$. By the Continuous Mapping Theorem (see Thm. A.6), it follows that $I_{V_j/n>U_j/n} \xrightarrow{a.s.} I_{c_1 q_j > c_0 p_j}$, provided that $c_1 q_j \neq c_0 p_j$, as the function $I_{u-v>0}$ is continuous everywhere except at $u - v = 0$. But notice that we can rewrite (3.18) and (3.19), respectively, as

$$\varepsilon^* = \sum_{\substack{j=1 \\ c_0 p_j = c_1 q_j}}^{b} c_0 p_j + \sum_{\substack{j=1 \\ c_0 p_j \neq c_1 q_j}}^{b} \left[c_0 p_j I_{c_1 q_j > c_0 p_j} + c_1 q_j \left(1 - I_{c_1 q_j > c_0 p_j}\right) \right] \tag{3.20}$$

and

$$\varepsilon_n = \sum_{\substack{j=1 \\ c_0 p_j = c_1 q_j}}^{b} c_0 p_j + \sum_{\substack{j=1 \\ c_0 p_j \neq c_1 q_j}}^{b} \left[c_0 p_j I_{V_j/n>U_j/n} + c_1 q_j \left(1 - I_{V_j/n>U_j/n}\right) \right], \tag{3.21}$$

from which it follows that $\varepsilon_n \xrightarrow{a.s.} \varepsilon^*$ and the discrete histogram rule is universally strongly consistent (over the class of all discrete feature-label distributions). ◇

The following result, which is a simple application of the Thm A.10, shows that consistency can be fully characterized by the behavior of the expected classification error as sample size increases.

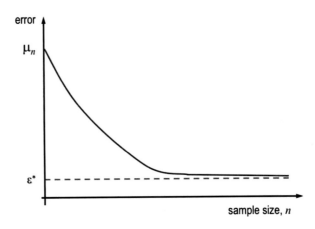

Figure 3.2: Representation of the expected classification error vs. sample size for a consistent classification rule.

Theorem 3.1. *The classification rule Ψ_n is consistent if and only if, as $n \to \infty$,*

$$E[\varepsilon_n] \to \varepsilon^*. \tag{3.22}$$

Proof. Note that $\{\varepsilon_n; \ n = 1, 2, \ldots\}$ is a uniformly bounded random sequence, as $0 \leq \varepsilon_n \leq 1$ for all n. Since $\varepsilon_n - \varepsilon^* > 0$, it follows from Thm A.10 that $\varepsilon_n \to \varepsilon^*$ in probability is equivalent to

$$E[\varepsilon_n - \varepsilon^*] = E[|\varepsilon_n - \varepsilon^*|] \to 0, \tag{3.23}$$

i.e., $E[\varepsilon_n] \to \varepsilon^*$. $\diamond$

Theorem 3.1 demonstrates the remarkable fact that consistency is characterized entirely by the first moment of the random variable ε_n as n increases. This is not sufficient for strong consistency, which in general depends on the behavior of the entire distribution of ε_n. Notice that $\{\mu_n; n = 1, 2, \ldots\}$ is a sequence of real numbers (not random variables) and the convergence in (3.22) is ordinary convergence, and thus can be plotted to obtain a graphical representation of consistency. See Figure 3.2 for an illustration, where the expected classification error is represented as a continuous function of n for ease of interpretation.

Example 3.6. (Consistency of the Nearest-Neighbor Classification Rule.) In Chapter 5, it is shown that the expected error of the nearest-neighbor classification rule of Example 3.2 satisfies $\lim_{n \to \infty} E[\varepsilon_n] \leq 2\varepsilon^*$. Assume that the feature-label distribution is such that $\varepsilon^* = 0$. Then, by Theorem 3.1, the nearest-neighbor classification rule is consistent. $\diamond$

The condition $\varepsilon^* = 0$ is quite restrictive, as it implies that perfect discrimination is achievable, which requires nonoverlapping classes (see Exercise 2.3). In fact, the k-nearest-neighbor classification rule is not universally consistent, for any fixed $k = 1, 2, \ldots$ However, we will see in Chapter 5 that the k-nearest neighbor classification rule is universally consistent, provided that k is allowed to increase with n at a specified rate.

Consistency is a *large sample* property, and could be irrelevant in *small-sample* cases, as non-consistent classification rules are typically better than consistent classification rules when the training data size is small. The reason is that consistent classification rules, especially universal ones, tend to be complex, while non-consistent ones are often simpler. We saw this counterintuitive phenomenon represented in the "scissors plot" of Figure Fig-basic(b). As was mentioned previously, the blue curve in the plot represents the expected error of a consistent classification rule, while the green one does not. However, the non-consistent classification rule is still better at small-sample sizes ($n < N_0$ in the plot), in which case the performance of the complex consistent rule degrades due to *overfitting*. The precise value of N_0 is very difficult to pinpoint, as it depends on the complexity of the classification rules, the dimensionality of the feature vector, dimensionality, and the feature-label distribution. We will have more to say about this topic in later chapters.

3.4 No-Free-Lunch Theorems

Universal consistency is a remarkable property in that it appears to imply that no knowledge at all about the feature-label distribution is needed to obtain optimal performance, i.e., a purely data-driven approach obtains performance arbitrarily close to the optimal performance if one has a large enough sample.

The next two theorems by Devroye and collaborators show that this is deceptive. They are sometimes called "No-Free-Lunch" theorems, as they imply that some knowledge about the feature-label distribution must be obtained to guarantee acceptable performance (or at least, to avoid terrible performance), after all. The proofs are based on finding simple feature-label distributions (in fact, discrete ones with zero Bayes error) that are "bad" enough.

The first theorem states that all classification rules can be arbitrarily bad at finite sample sizes. In the case of universally consistent classification rules, this means that one can never know if their finite-sample performance will be satisfactory, no matter how large n is (unless one knows something about the feature-label distribution). For a proof, see [Devroye et al., 1996, Thm 7.1].

Theorem 3.2. *For every $\tau > 0$, integer n, and classification rule Ψ_n, there exists a feature-label distribution $P_{\mathbf{X},Y}$ (with $\varepsilon^* = 0$) such that*

$$E[\varepsilon_n] \geq \frac{1}{2} - \tau. \tag{3.24}$$

The feature-label distribution in the previous theorem may have to be different for different n. The next remarkable theorem applies to a fixed feature-label distribution and implies that, though one may get $E[\varepsilon_n] \to \varepsilon^*$ in a distribution-free manner, one must know something about the feature-label distribution in order to guarantee a rate of convergence. For a proof, see [Devroye et al., 1996, Thm 7.2].

Theorem 3.3. *For every classification rule Ψ_n, there exists a monotonically decreasing sequence $a_1 \geq a_2 \geq \ldots$ converging to zero such that there is a feature-label distribution $P_{\mathbf{X},Y}$ (with $\varepsilon^* = 0$) for which*

$$E[\varepsilon_n] \geq a_n, \tag{3.25}$$

for all $n = 1, 2, \ldots$

For the discrete histogram rule with $\varepsilon^* = 0$, one can show that there exists a constant $r > 0$ such that $E[\varepsilon_n] < e^{-rn}$, for $n = 1, 2, \ldots$ (see Exercise 3.2). This however does not contradict Theorem 3.3, because the constant $r > 0$ is distribution-dependent. In fact, it can be made as close to zero as wanted by choosing the distribution.

For another example of no-free-lunch result, see Exercise 3.4.

3.5   Additional Topics

3.5.1   Ensemble Classification

Ensemble classification rules combine the decision of multiple classification rules by majority voting. This is an application of the "wisdom of the crowds" principle, which can reduce overfitting and increase accuracy over the component classification rules (which are called in some contexts *weak learners*).

Formally, given a set of classification rules $\{\Psi_n^1, \ldots, \Psi_n^m\}$, an *ensemble classification rule* $\Psi_{n,m}^E$ produces a classifier

$$\psi_{n,m}^E(\mathbf{x}) = \begin{cases} 1, & \frac{1}{m} \sum_{j=1}^m \Psi_n^j(S_n)(\mathbf{x}) > \frac{1}{2} \\ 0, & \text{otherwise.} \end{cases} \tag{3.26}$$

In other words, the ensemble classifier assigns label 1 to the test point $\mathbf{x}$ if a majority of the component classification rules produce label 1 on $\mathbf{x}$; otherwise, it assigns label 0.

In practice, ensemble classifiers are almost always produced by *resampling*. (This is a general procedure, which will be important again when we discuss error estimation in Chapter 7.) Consider an operator $\tau : (S_n, \xi) \mapsto S_n^*$, which applies a "perturbation" to the training data S_n and produces a modified data set S_n^*. The variable ξ represents random factors, rendering S_n^* random, given the data S_n. A base classification rule Ψ_n is selected, and the components classification rules are defined by $\Psi_n^j(S_n) = \Psi_n(\tau(S_n)) = \Psi_n(S_n^*)$. Notice that this produces *random classification rules* due to the randomness of τ. It follows that the ensemble classification rule $\psi_{n,m}^E$ in (3.26) is likewise random. This means that different classifiers result from repeated application of $\psi_{n,m}^E$ to the same training data S_n. We will not consider random classification rules in detail in this book.

The perturbation τ may consist of taking random subsets of the data, adding small random noise to the training points, flipping the labels randomly, and more. Here we consider in the detail an example of perturbation called *bootstrap* sampling. Given a fixed data set $S_n = \{(\mathbf{X}_1 = \mathbf{x}_1, Y_1 = y1), \ldots, (\mathbf{X}_n = \mathbf{x}_n, Y_n = y_n)\}$, the *empirical feature-label distribution* is a discrete distribution, with probability mass function given by $\hat{P}(\mathbf{X} = \mathbf{x}_i, Y = y_i) = \frac{1}{n}$, for $i = 1, \ldots, n$. A *bootstrap sample* is a sample S_n^* from the empirical distribution; it consists of n equally-likely draws with replacement from the original sample S_n. Some sample points will appear multiple times, whereas others will not appear at all. The probability that any given sample point will not appear in S_n^* is $(1 - 1/n)^n \approx e^{-1}$. It follows that a bootstrap sample of size n contains on average $(1 - e^{-1})n \approx 0.632n$ of the original sample points. The ensemble classification rule in (3.26) is called in this case a *bootstrap aggregate* and the procedure is called "bagging."

3.5.2 Mixture Sampling vs. Separate Sampling

The assumption being made thus far is that the training data S_n is an i.i.d. sample from the feature-label distribution $P_{\mathbf{X},Y}$; i.e., the set of sample points is independent and each sample point has distribution $P_{\mathbf{X},Y}$. In this case, each $\mathbf{X}_i$ is distributed as $p(\mathbf{x} \mid Y = 0)$ with probability $P(Y = 0)$ or $p(\mathbf{x} \mid Y = 1)$ with probability $P(Y = 1)$. It is common to say then that each $\mathbf{X}_i$ is sampled from a *mixture* of the *populations* $p(\mathbf{x} \mid Y = 0)$ and $p(\mathbf{x} \mid Y = 1)$, with mixing proportions $P(Y = 0)$ and $P(Y = 1)$, respectively.

This sampling design is pervasive in the literature — most papers and textbooks assume it, often tacitly. However, suppose sampling is not from the mixture of populations, but rather from each population separately, such that a *nonrandom* number n_0 of sample points are drawn from $p(\mathbf{x} \mid Y = 0)$, while a *nonrandom* number n_1 of points are drawn from $p(\mathbf{x} \mid Y = 1)$, where $n_0 + n_1 = n$. This *separate sampling* case is quite distinct from unconstrained random sampling, where the numbers N_0 and N_1 of sample points from each class are binomial random variables (see Section 3.1). In addition, in separate sampling, the labels $Y_1, \ldots, Y_n$ are no longer independent: knowing that, say, $Y_1 = 0$, is informative about the status of Y_2, since the number of points from class 0 is fixed. A key fact in the separate sampling case is that the class prior probabilities $p_0 = P(Y = 0)$ and $p_1 = P(Y = 1)$ *are not estimable from the data.* In the random sampling case, $\hat{p}_0 = \frac{N_0}{n}$ and $\hat{p}_1 = \frac{N_1}{n}$ are unbiased estimators of p_0 and p_1, respectively; they are also consistent estimators, i.e., $\hat{p}_0 \to p_0$ and $\hat{p}_1 \to p_1$ with probability 1 as $n \to \infty$, by virtue of the Law of Large Numbers (Theorem A.12). Therefore, provided that the sample size is large enough, $\hat{c}_0$ and $\hat{c}_1$ provide decent estimates of the prior probabilities. However, $\hat{c}_0$ and $\hat{c}_1$ are clearly nonsensical estimators in the separate sampling case. As a matter of fact, there are no sensible estimate of p_0 and p_1 under separate sampling. There is simply no information about p_0 and p_1 in the training data.

Separate sampling is a very common scenario in observational case-control studies in biomedicine, which typically proceed by collecting data from the populations separately, where the separate sample sizes n_0 and n_1, with $n_0 + n_1 = n$, are pre-determined experimental design parameters. The reason is that one of the populations is often small (e.g., a rare disease). Sampling from the mixture of healthy and diseased populations would produce a very small number of diseased subjects (or none). Therefore, *retrospective* studies employ a fixed number of members of each population, by assuming that the outcomes (labels) are known in advance.

Separate sampling can be seen as an example of restricted random sampling, where in this case, the restriction corresponds to conditioning on $N_0 = n_0$, or equivalently, $N_1 = n_1$. Failure to account for this restriction in sampling will manifest itself in two ways. First, it will affect the design of classifiers that require, directly or indirectly, estimation of p_0 and p_1, in which case alternative procedures, such as minimax classification (see Section 2.6.1) need to be employed; an example of this will be seen in Chapter 4. Secondly, it will affect population-wide average performance metrics, such as the expected classification error rates. Under a separate sampling restriction, the expected error rate is given by

$$\mu_{n_0, n_1} = E[\varepsilon_n \mid N = n_0] = P(\psi_n(X) \neq Y \mid N = n_0). \tag{3.27}$$

This is in general different, and sometimes greatly so, from the unconstrained expected error rate μ_n. Failure to account for the sampling mechanism used to acquire the data can have practical negative consequences in the analysis of performance of classification algorithms, and even in the accuracy of the classifiers themselves.

3.6 Bibliographical Notes

Tibshirani et al. [2002] proposed the "nearest-shrunken centroids" algorithm, a popular version of the nearest-centroid classification rule for high-dimensional gene-expression data, where the class means are estimated by "shrunken centroids," i.e., regularized sample means estimators driven towards zero (this resembles the LASSO, to be discussed in Chapter 11, also proposed by Tibshirani et al. [2002]).

Glick [1973] gave a general (distribution-dependent) bound on the rate of convergence of $E[\varepsilon_n - \varepsilon^*]$ to zero for the discrete histogram rule. This does not contradict Theorem 3.3 since the bound is distribution-dependent.

The equivalence of consistent and strongly consistent rules for "well-behaved" classification rules is discussed in Devroye et al. [1996].

Wolpert [2001] proved well-known "no-free lunch" theorems for classification, though using a different setting (random distributions and "out-of-sample" classification error).

The bootstrap was proposed by Efron [1979], while bagging was introduced by Breiman [1996]. *boosting* is a different ensemble approach where the different classifiers are trained sequentially, instead of in parallel, and decisions made by early classifiers affect decisions made by later classifiers by means of weights assigned to the training points [Freund, 1990].

For an extensive treatment of the mixture vs. separate sampling issue, see McLachlan [1992]. This topic is also discussed at length in Braga-Neto and Dougherty [2015]. For more on separate sampling in biomedical studies, see Zolman [1993].

3.7 Exercises

3.1. Show that the expected classification error of the discrete histogram rule is given by:

$$E[\varepsilon_n] = c_1 + \sum_{j=1}^{b} (c_0 p_j - c_1 q_j) \sum_{\substack{k,l=0 \\ k<l \\ k+l \leq n}}^{n} \frac{n!}{k!l!(n-k-l)!} \left(c_0 p_j \right)^k \left(c_1 q_j \right)^l \left(1 - c_0 p_j - c_1 q_j \right)^{n-k-l} \quad (3.28)$$

Hint: first show that

$$E[\varepsilon_n] = \sum_{j=1}^{b} \left[c_0 p_j \, P(V_j > U_j) + c_1 q_j \, P(U_j \geq V_j) \right] . \quad (3.29)$$

3.2. For the discrete histogram rule with $\varepsilon^* = 0$ and ties broken in the direction of class 0, show that

$$E[\varepsilon_n] < e^{-rn}, \tag{3.30}$$

for $n = 1, 2, \ldots$, where $r > 0$ is given by

$$r = \ln\left(\frac{1}{1 - cs}\right), \tag{3.31}$$

with $c = P(Y = 1)$ and $s = \min\{s_j = P(X = \mathbf{x}^j \mid Y = 1) \text{ s.t. } s_j \neq 0\}$. If ties are broken in the direction of class 1, then the same result holds, with $Y = 1$ replaced by $Y = 0$ everywhere. Hint: use (3.29) and the fact that the distributions $\{p_j\}$ and $\{q_j\}$ do not overlap if $\varepsilon^* = 0$. (Indeed, the only source of error in this case comes from tie-breaking over empty cells.)

3.3. A classification rule is called *smart* if the sequence of expected classification errors $\{\mu_n; n = 1, 2, \ldots\}$ is nonincreasing for any feature-label distribution $P_{\mathbf{X},Y}$. This expresses the natural requirement that, as the sample size increases, the expected classification error should never increase, regardless of the feature-label distribution.

   (a) Consider a simple univariate classification rule such that $\psi_n(x) = I_{x>0}$ if $\sum_{i=1}^n I_{X_i>0,Y_i=1} > \sum_{i=1}^n I_{X_i>0,Y_i=0}$, otherwise $\psi_n(x) = I_{x\leq0}$ (this classifier assigns to x the majority label among the training points that have the same sign as x). Show that this classification rule is smart.

   (b) Show that the nearest-neighbor classification rule is *not* smart.

      Hint: consider the univariate feature-label distribution such that (X, Y) is equal to $(0, 1)$ with probability $p < 1/5$ and is equal to $(Z, 0)$ with probability $1 - p$, where Z is a uniform random variable on the interval $[-1000, 1000]$. Now compute $E[\varepsilon_1]$ and $E[\varepsilon_2]$. (This example is due to Devroye et al. [1996].)

3.4. (No *super classification rule*.) Show that for every classification rule Ψ_n, there is another classification rule Ψ'_n, with classification error ε'_n, and a feature-label distribution $P_{\mathbf{X},Y}$ (with $\varepsilon^* = 0$) such that

$$E[\varepsilon'_n] < E[\varepsilon_n], \quad \text{for all } n. \tag{3.32}$$

Hint: Find a feature-label distribution $P_{\mathbf{X},Y}$ such that $\mathbf{X}$ is concentrated over a finite number of points over R^d and Y is a deterministic function of $\mathbf{X}$.

3.5. In the standard sampling case, $P(Y_i = 0) = p_0 = P(Y = 0)$ and $P(Y_i = 1) = p_1 = P(Y = 1)$, for $i = 1, \ldots, n$. Show that in the separate sampling case (see Section 3.5.2) we have instead

$$P(Y_i = 0 \mid N_0 = n_0) = \frac{n_0}{n} \text{ and } P(Y_i = 1 \mid N_0 = n_0) = \frac{n_1}{n}, \tag{3.33}$$

for $i = 1, \ldots, n$.

Hint: Under the restriction $N_0 = n_0$, only the order of the labels $Y_1, \ldots, Y_n$ may be random. Thus, $f(Y_1, \ldots, Y_n \mid N_0 = n_0)$ is a discrete uniform distribution over all $\binom{n}{n_0}$ possible orderings.

3.8 Python Assignments

3.6. Using the synthetic data model in Section A8.1 for the homoskedastic case with $\boldsymbol{\mu}_0 = (0, \ldots, 0)$, $\boldsymbol{\mu}_1 = (1, \ldots, 1)$, $P(Y = 0) = P(Y = 1)$, and $k = d$ (independent features), generate a large number (e.g., $M = 1000$) of training data sets for each sample size $n = 20$ to $n = 100$, in steps of 10, with $d = 2, 5, 8$, and $\sigma = 1$. Obtain an approximation of the expected classification error $E[\varepsilon_n]$ of the nearest centroid classifier in each case by averaging ε_n, computed using the exact formula (3.13), over the M synthetic training data sets. Plot $E[\varepsilon_n]$ as a function of the sample size, for $d = 2, 5, 8$ (join the individual points with lines to obtain a smooth curve). Explain what you see.

3.7. (Learning with an unreliable teacher.) Consider that the labels in the training data $S_n = \{(\mathbf{X}_1, Y_1), \ldots, (\mathbf{X}_n, Y_n)\}$ can be flipped with probability $t < 1/2$. That is, the observed data is actually $\bar{S}_n = \{(\mathbf{X}_1, Z_1), \ldots, (\mathbf{X}_n, Z_n)\}$, where $Z_i = 1 - Y_i$ with probability p, otherwise $Z_i = Y_i$, independently for each $i = 1, \ldots, n$.

(a) Repeat Problem 3.6 with $t = 0.1$ to $t = 0.5$ in steps of 0.1. Plot in the same graph $E[\varepsilon_n]$ as a function of n for each value of t plus the original result ($t = 0$), for $d = 2, 5, 8$. Explain what you see.

(b) Here we try to recover from the unreliability of the labels by using a rejection procedure: compute the distance $d(\mathbf{X}_i, \hat{\boldsymbol{\mu}}_j)$ from each training point to its ostensible class centroid; if $d(\mathbf{X}_i, \hat{\boldsymbol{\mu}}_j) > 2\sigma$, then flip the corresponding label Z_i, otherwise accept the label. Update the centroid after each flip (this means that the order in which you go through the data may change the result). Repeat item (a) with the correction procedure and compare the results.

Chapter 4

Parametric Classification

In this chapter and the next, we discuss simple classification rules that are based on estimating the feature-label distribution from the data. If ignorance about the distribution is confined to a few numerical parameters, then these algorithms are called parametric classification rules. After presenting the general definition of a parametric classification rule, we discuss the important Gaussian discriminant case, including Linear and Quadratic Discriminant Analysis and their variations, and then logistic classification is examined. Additional topics include an extension of Gaussian discriminant analysis, called Regularized Discriminant Analysis, and Bayesian parametric classification.

4.1 Parametric Plug-in Rules

In the parametric approach, we assume that knowledge about the feature-label distribution is coded into a family of probability density functions $\{p(\mathbf{x} \mid \boldsymbol{\theta}) \mid \boldsymbol{\theta} \in \Theta \subseteq R^m\}$, such that class-conditional densities are $p(\mathbf{x} \mid \boldsymbol{\theta}_0^*)$ and $p(\mathbf{x} \mid \boldsymbol{\theta}_1^*)$, for "true" parameter values $\boldsymbol{\theta}_0^*, \boldsymbol{\theta}_1^* \in R^m$. Let $\boldsymbol{\theta}_{0,n}$ and $\boldsymbol{\theta}_{1,n}$ be estimators of $\boldsymbol{\theta}_0^*$ and $\boldsymbol{\theta}_1^*$ based on the sample data $S_n = \{(\mathbf{X}_1, Y_1), \ldots, (\mathbf{X}_n, Y_n)\}$. The *parametric plug-in* sample-based discriminant is obtained by plugging $\boldsymbol{\theta}_{0,n}$ and $\boldsymbol{\theta}_{1,n}$ in the expression for the optimal discriminant in (2.27):

$$D_n(\mathbf{x}) = \ln \frac{p(\mathbf{x} \mid \boldsymbol{\theta}_{1,n})}{p(\mathbf{x} \mid \boldsymbol{\theta}_{0,n})}. \tag{4.1}$$

© Springer Nature Switzerland AG 2020
U. Braga-Neto, *Fundamentals of Pattern Recognition and Machine Learning*,
https://doi.org/10.1007/978-3-030-27656-0_4

A parametric plug-in classifier results from plugging this in the Bayes classifier formula in (2.26):

$$\psi_n(\mathbf{x}) = \begin{cases} 1, & D_n(\mathbf{x}) > k_n, \\ 0, & \text{otherwise.} \end{cases} \tag{4.2}$$

Example 4.1. Consider the exponential family of densities (see Exercise 2.11):

$$p(\mathbf{x} \mid \boldsymbol{\theta}) = \alpha(\boldsymbol{\theta})\beta(\mathbf{x}) \exp\left(\sum_{i=1}^{k} \xi_i(\boldsymbol{\theta})\phi_i(\mathbf{x})\right). \tag{4.3}$$

It is easy to see that the sample-based discriminant is given by

$$D_n(\mathbf{x}) = \sum_{i=1}^{k} [\xi_i(\boldsymbol{\theta}_{1,n}) - \xi_i(\boldsymbol{\theta}_{0,n})]\phi_i(\mathbf{x}) + \ln\frac{\alpha(\boldsymbol{\theta}_{1,n})}{\alpha(\boldsymbol{\theta}_{0,n})}. \tag{4.4}$$

In particular, the discriminant does not depend on $\beta(\mathbf{x})$. This highlights the fact that only the discriminatory information in the class-conditional densities is relevant for classification. ◇

There are a few choices for obtaining the threshold k_n, in rough order of preference:

(1) If the true prevalences $c_0 = P(Y = 0)$ and $c_1 = P(Y = 1)$ are known (e.g., from public health records in the case of disease classification), one should use the optimal threshold value $k^* = \ln c_0/c_1$, as given by (2.28).

(2) If c_0 and c_1 are not known, but the training sample size is moderate to large and sampling is random (i.i.d.), then, following (2.28), one can use the estimate

$$k_n = \ln\frac{N_0/n}{N_1/n} = \ln\frac{N_0}{N_1}, \tag{4.5}$$

where $N_0 = \sum_{i=1}^{n} I_{Y_i=0}$ and $N_1 = \sum_{i=1}^{n} I_{Y_i=1}$ are the class-specific sample sizes.

(3) If c_0 and c_1 are not known, and the sample size is small or the sampling is not random (e.g., see Section 3.5.2), then one could use the minimax approach (see Section 2.6.1) to obtain k_n.

(4) Vary k_n and search for the best value using error estimation methods (see Chapter 7).

(5) In some applications, the sample size may be small due to the cost of labeling the data, but there may be an abundance of unlabeled data (e.g., this is the case in many imaging applications). Assuming that the unlabeled data is an i.i.d. sample from the same distribution as the training data, a method for estimating c_0 and c_1, and thus k_n, is discussed in Exercise 4.1.

The method in item (4) above produces an estimate a Receiver Operating Characteristic (ROC) curve (see Section 2.6.1). For many classification rules, it is not as straightforward to build a ROC; this is an advantage of the parametric classifier in (4.2)

Another method to obtain parametric plug-in classifiers is to assume that the posterior-probability function $\eta(\mathbf{x} \mid \boldsymbol{\theta}^*)$ is a member of a parametric family $\{\eta(\mathbf{x} \mid \boldsymbol{\theta}) \mid \boldsymbol{\theta} \in \Theta \subseteq R^m\}$. A classifier results from plugging an estimator $\boldsymbol{\theta}_n$ of $\boldsymbol{\theta}^*$ in the Bayes classifier formula in (2.15):

$$\psi_n(\mathbf{x}) = \begin{cases} 1, & \eta(\mathbf{x} \mid \boldsymbol{\theta}_n) > \frac{1}{2}, \\ 0, & \text{otherwise.} \end{cases} \tag{4.6}$$

Notice that this approach avoids dealing with multiple parameters $\boldsymbol{\theta}_0^*$ and $\boldsymbol{\theta}_1^*$ as well as the choice of the discriminant threshold k_n. The disadvantage is that in practice it is often easier to model the class-conditional densities of a problem than the posterior-probability function directly. We will see in the next sections examples of both parametric approaches.

4.2 Gaussian Discriminant Analysis

The most important class of parametric classification rules correspond to the choice of multivariate Gaussian densities parametrized by the mean vector $\boldsymbol{\mu}$ and covariance matrix Σ:

$$p(\mathbf{x} \mid \boldsymbol{\mu}, \Sigma) = \frac{1}{\sqrt{(2\pi)^d \det(\Sigma)}} \exp\left[-\frac{1}{2}(\mathbf{x} - \boldsymbol{\mu})^T \Sigma^{-1}(\mathbf{x} - \boldsymbol{\mu})\right]. \tag{4.7}$$

This case is referred to as Gaussian Discriminant Analysis. It is the sample-based plug-in version of the optimal Gaussian case discussed in Section 2.5. The parametric discriminant in (4.2) corresponds to plugging in estimators $(\hat{\boldsymbol{\mu}}_0, \hat{\Sigma}_0)$ and $(\hat{\boldsymbol{\mu}}_1, \hat{\Sigma}_1)$[1] for the true parameters $(\boldsymbol{\mu}_0, \Sigma_0)$ and $(\boldsymbol{\mu}_1, \Sigma_1)$ in the expression for the optimal discriminant in (2.43):

$$D_n(\mathbf{x}) = \frac{1}{2}(\mathbf{x} - \hat{\boldsymbol{\mu}}_0)^T \hat{\Sigma}_0^{-1}(\mathbf{x} - \hat{\boldsymbol{\mu}}_0) - \frac{1}{2}(\mathbf{x} - \hat{\boldsymbol{\mu}}_1)^T \hat{\Sigma}_1^{-1}(\mathbf{x} - \hat{\boldsymbol{\mu}}_1) + \frac{1}{2}\ln\frac{\det(\hat{\Sigma}_0)}{\det(\hat{\Sigma}_1)}. \tag{4.8}$$

Different assumptions about the parameters $(\boldsymbol{\mu}_0, \Sigma_0)$ and $(\boldsymbol{\mu}_1, \Sigma_1)$ lead to different classification rules, which we examine next.

---

[1] We employ in this case the classical statistical notation $\hat{\boldsymbol{\mu}}$, $\hat{\Sigma}$, instead of $\boldsymbol{\mu}_n$, Σ_n.

4.2.1   Linear Discriminant Analysis

This is the sample-based version of the homoskedastic Gaussian case in Section 2.5. The maximum-likelihood estimators of the mean vectors are given by the sample means:

$$\hat{\boldsymbol{\mu}}_0 \;=\; \frac{1}{N_0} \sum_{i=1}^{n} \mathbf{X}_i I_{Y_i=0} \quad \text{and} \quad \hat{\boldsymbol{\mu}}_1 \;=\; \frac{1}{N_1} \sum_{i=1}^{n} \mathbf{X}_i I_{Y_i=1}\,, \tag{4.9}$$

where $N_0 = \sum_{i=1}^{n} I_{Y_i=0}$ and $N_1 = \sum_{i=1}^{n} I_{Y_i=1}$ are the class-specific sample sizes. Under the homoskedastic assumption that $\Sigma_0 = \Sigma_1 = \Sigma$, the maximum likelihood estimator of Σ is

$$\hat{\Sigma}^{ML} \;=\; \frac{N_0}{n}\,\hat{\Sigma}_0^{ML} + \frac{N_1}{n}\,\hat{\Sigma}_1^{ML}\,, \tag{4.10}$$

where

$$\hat{\Sigma}_0^{ML} \;=\; \frac{1}{N_0} \sum_{i=1}^{n} (\mathbf{X}_i - \hat{\boldsymbol{\mu}}_0)(\mathbf{X}_i - \hat{\boldsymbol{\mu}}_0)^T I_{Y_i=0}\,, \tag{4.11}$$

$$\hat{\Sigma}_1^{ML} \;=\; \frac{1}{N_1} \sum_{i=1}^{n} (\mathbf{X}_i - \hat{\boldsymbol{\mu}}_1)(\mathbf{X}_i - \hat{\boldsymbol{\mu}}_1)^T I_{Y_i=1}\,. \tag{4.12}$$

In order to obtain unbiased estimators, it is usual to consider the *sample covariance estimators* $\hat{\Sigma}_0 = (N_0/N_0 - 1)\hat{\Sigma}_0^{ML}$, $\hat{\Sigma}_1 = (N_1/N_1 - 1)\hat{\Sigma}_1^{ML}$, and $\hat{\Sigma} = n/(n-2)\hat{\Sigma}^{ML}$, so that one can write

$$\hat{\Sigma} \;=\; \frac{(N_0 - 1)\hat{\Sigma}_0 + (N_1 - 1)\hat{\Sigma}_1}{n - 2}\,. \tag{4.13}$$

This estimator is known as the *pooled sample covariance matrix*. It reduces to $\frac{1}{2}(\hat{\Sigma}_0 + \hat{\Sigma}_1)$ if $N_0 = N_1$, i.e., if the sample is *balanced*.

The *LDA discriminant* is obtained by substituting the pooled sample covariance matrix $\hat{\Sigma}$ for $\hat{\Sigma}_0$ and $\hat{\Sigma}_1$ in (4.8), which leads to

$$D_{L,n}(\mathbf{x}) \;=\; (\hat{\boldsymbol{\mu}}_1 - \hat{\boldsymbol{\mu}}_0)^T \hat{\Sigma}^{-1}\left(\mathbf{x} - \frac{\hat{\boldsymbol{\mu}}_0 + \hat{\boldsymbol{\mu}}_1}{2}\right). \tag{4.14}$$

The LDA classifier is then given by (4.2), with $D_n = D_{L,n}$. This discriminant $D_{L,n}$ is also known as *Anderson's W statistic*.

Similarly as in the homoskedastic Gaussian case in Section 2.5, the LDA classifier produces a hyperplane decision boundary, determined by the equation $\mathbf{a}_n^T \mathbf{x} + b_n = k_n$, where

$$\begin{aligned} \mathbf{a}_n &= \hat{\Sigma}^{-1}(\hat{\boldsymbol{\mu}}_1 - \hat{\boldsymbol{\mu}}_0)\,, \\ b_n &= (\hat{\boldsymbol{\mu}}_0 - \hat{\boldsymbol{\mu}}_1)^T \hat{\Sigma}^{-1}\left(\frac{\hat{\boldsymbol{\mu}}_0 + \hat{\boldsymbol{\mu}}_1}{2}\right). \end{aligned} \tag{4.15}$$

It was shown in Section 2.6.1 that the minimax threshold for the homoskedastic Gaussian case is $k_{\mathrm{mm}} = 0$. Accordingly, if the true prevalences $P(Y = 0)$ and $P(Y = 1)$ are not known, and the sample size is small or the sampling is nonrandom, a common choice of threshold for LDA is $k_n = 0$. (See the discussion on the choice of threshold in Section 4.1.) In that case, the decision hyperplane passes through the midpoint $\hat{\mathbf{x}}_m = (\hat{\boldsymbol{\mu}}_0 + \hat{\boldsymbol{\mu}}_1)/2$ between the sample means.

Estimation of the full pooled sample covariance matrix $\hat{\Sigma}$ involves $d + d(d-1)/2$ parameters (since the covariance matrix is symmetric). In small-sample cases, where the number of training points n is small compared to the dimensionality d, this may cause problems. If d approaches n, large eigenvalues of Σ tend to be overestimated, while small eigenvalues of Σ tend to be underestimated. The latter fact means that $\hat{\Sigma}$ becomes nearly singular, which renders its computation numerically intractable. The following variants of LDA, ordered in terms of increasing restrictions on the estimator of the pooled covariance matrix Σ, may perform better than the standard LDA in small-sample cases.

(1) *Diagonal LDA*: The estimator of Σ is constrained to be a diagonal matrix $\hat{\Sigma}_D$. The diagonal elements are the univariate pooled sample variances along each dimension, i.e.:

$$(\hat{\Sigma}_D)_{jj} = (\hat{\Sigma})_{jj}, \tag{4.16}$$

with $(\hat{\Sigma}_D)_{jk} = 0$ for $j \neq k$.

(2) *Nearest-Mean Classifier (NMC)*: The estimator of Σ is constrained to be a diagonal matrix $\hat{\Sigma}_M$ with equal diagonal elements. The common value is the sample variance across all dimensions:

$$(\hat{\Sigma}_M)_{jj} = \hat{\sigma}^2 = \sum_{k=1}^{d} (\hat{\Sigma})_{kk}, \tag{4.17}$$

with $(\hat{\Sigma}_M)_{jk} = 0$ for $j \neq k$. From the expression of $\mathbf{a}_n$ in (4.15), it is clear that the decision boundary is perpendicular to the line joining the sample means. Furthermore, the term $1/\hat{\sigma}^2$ appears in both $\mathbf{a}_n$ and b_n. With the choice $k_n = 0$, this term drops out and the resulting classifier does not depend on $\hat{\sigma}$ (which need not be estimated); moreover, the classifier assigns to a test point $\mathbf{x}$ the label of the sample mean closest to $\mathbf{x}$. Only the sample means need to be estimated in this case.

(3) *Covariance Plug-In*. If Σ is assumed to be known or can be guessed, it can be used in place of $\hat{\Sigma}$ in (4.14). As in the case of the MNC, only the sample means need to be estimated.

Other such *shrinkage* estimators of Σ exist to balance the degrees of freedom of the model with small sample sizes; see the Bibliographical Notes.

Example 4.2. The training data in a classification problem is:

$$S_n = \{((1,2)^T,0),((2,2)^T,0),((2,4)^T,0),((3,4)^T,0),((4,5)^T,1),((6,4)^T,1),((6,6)^T,1),((8,5)^T,1)\}.$$

Assuming $k_n = 0$, we obtain the LDA, DLDA and NMC decision boundaries below. First, we compute the sample estimates:

$$\hat{\boldsymbol{\mu}}_0 = \frac{1}{4}\left(\begin{bmatrix}1\\2\end{bmatrix}+\begin{bmatrix}2\\2\end{bmatrix}+\begin{bmatrix}2\\4\end{bmatrix}+\begin{bmatrix}3\\4\end{bmatrix}\right) = \begin{bmatrix}2\\3\end{bmatrix},$$

$$\hat{\boldsymbol{\mu}}_1 = \frac{1}{4}\left(\begin{bmatrix}4\\5\end{bmatrix}+\begin{bmatrix}6\\4\end{bmatrix}+\begin{bmatrix}6\\6\end{bmatrix}+\begin{bmatrix}8\\5\end{bmatrix}\right) = \begin{bmatrix}6\\5\end{bmatrix},$$

$$\hat{\Sigma}_0 = \frac{1}{3}\left(\begin{bmatrix}-1\\-1\end{bmatrix}[-1\ -1]+\begin{bmatrix}0\\-1\end{bmatrix}[0\ -1]+\begin{bmatrix}0\\1\end{bmatrix}[0\ 1]+\begin{bmatrix}1\\1\end{bmatrix}[1\ 1]\right) = \frac{2}{3}\begin{bmatrix}1&1\\1&2\end{bmatrix}, \tag{4.18}$$

$$\hat{\Sigma}_1 = \frac{1}{3}\left(\begin{bmatrix}-2\\0\end{bmatrix}[-2\ 0]+\begin{bmatrix}0\\-1\end{bmatrix}[0\ -1]+\begin{bmatrix}0\\1\end{bmatrix}[0\ 1]+\begin{bmatrix}2\\0\end{bmatrix}[2\ 0]\right) = \frac{2}{3}\begin{bmatrix}4&0\\0&1\end{bmatrix},$$

$$\hat{\Sigma} = \frac{1}{2}\left(\hat{\Sigma}_0+\hat{\Sigma}_1\right) = \frac{1}{3}\begin{bmatrix}5&1\\1&3\end{bmatrix} \Rightarrow \hat{\Sigma}^{-1} = \frac{3}{14}\begin{bmatrix}3&-1\\-1&5\end{bmatrix}.$$

The 2-dimensional LDA decision boundary is given by $a_{n,1}x_1 + a_{n,2}x_2 + b_n = 0$, where

$$\begin{bmatrix}a_{n,1}\\a_{n,2}\end{bmatrix} = \hat{\Sigma}^{-1}(\hat{\boldsymbol{\mu}}_1-\hat{\boldsymbol{\mu}}_0) = \frac{3}{14}\begin{bmatrix}3&-1\\-1&5\end{bmatrix}\begin{bmatrix}4\\2\end{bmatrix} = \frac{3}{7}\begin{bmatrix}5\\3\end{bmatrix},$$

$$b_n = (\hat{\boldsymbol{\mu}}_0-\hat{\boldsymbol{\mu}}_1)^T\hat{\Sigma}^{-1}\left(\frac{\hat{\boldsymbol{\mu}}_0+\hat{\boldsymbol{\mu}}_1}{2}\right) = \frac{3}{14}[-4\ -2]\begin{bmatrix}3&-1\\-1&5\end{bmatrix}\begin{bmatrix}4\\4\end{bmatrix} = -\frac{96}{7}. \tag{4.19}$$

Thus, the LDA decision boundary is given by $5x_1 + 3x_2 = 32$. On the other hand, Σ_D is obtained from Σ by zeroing the off-diagonal elements, so that

$$\hat{\Sigma}_D = \frac{1}{3}\begin{bmatrix}5&0\\0&3\end{bmatrix} \Rightarrow \hat{\Sigma}_D^{-1} = \frac{1}{5}\begin{bmatrix}3&0\\0&5\end{bmatrix}. \tag{4.20}$$

The DLDA decision boundary is given by $c_{n,1}x_1 + c_{n,2}x_2 + d_n = 0$, where

$$\begin{bmatrix}c_{n,1}\\c_{n,2}\end{bmatrix} = \hat{\Sigma}_D^{-1}(\hat{\boldsymbol{\mu}}_1-\hat{\boldsymbol{\mu}}_0) = \frac{1}{5}\begin{bmatrix}3&0\\0&5\end{bmatrix}\begin{bmatrix}4\\2\end{bmatrix} = \frac{2}{5}\begin{bmatrix}6\\5\end{bmatrix},$$

$$d_n = (\hat{\boldsymbol{\mu}}_0-\hat{\boldsymbol{\mu}}_1)^T\hat{\Sigma}^{-1}\left(\frac{\hat{\boldsymbol{\mu}}_0+\hat{\boldsymbol{\mu}}_1}{2}\right) = \frac{1}{5}[-4\ -2]\begin{bmatrix}3&0\\0&5\end{bmatrix}\begin{bmatrix}4\\4\end{bmatrix} = -\frac{88}{5}. \tag{4.21}$$

Thus, the DLDA decision boundary is given by $6x_1 + 5x_2 = 44$. As for the NMC decision boundary, it is given by $e_{n,1}x_1 + e_{n,2}x_2 + f_n = 0$, where

$$\begin{bmatrix}e_{n,1}\\c_{n,2}\end{bmatrix} = \hat{\boldsymbol{\mu}}_1-\hat{\boldsymbol{\mu}}_0 = \begin{bmatrix}4\\2\end{bmatrix}, \tag{4.22}$$

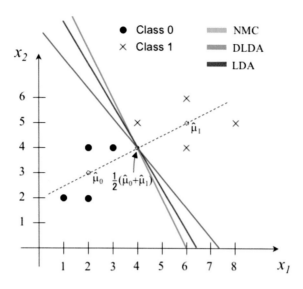

Figure 4.1: Sample data and linear decision boundaries in Example 4.2.

$$d_n = (\hat{\boldsymbol{\mu}}_0 - \hat{\boldsymbol{\mu}}_1)^T \left(\frac{\hat{\boldsymbol{\mu}}_0 + \hat{\boldsymbol{\mu}}_1}{2} \right) = [-4 \ -2] \begin{bmatrix} 4 \\ 4 \end{bmatrix} = -24. \qquad (4.23)$$

Thus, the NMC decision boundary is given by $2x_1 + x_2 = 12$.

Figure 4.1 displays the training data with superimposed LDA, DLDA, and NMC decision boundaries. We can see that, as a result of the choice $k_n = 0$, all three decision boundaries go through the midpoint between the two class means; however, only the NMC decision boundary is perpendicular to the line joining the two sample means. ◇

4.2.2 Quadratic Discriminant Analysis

This is the sample-based version of the heteroskedastic Gaussian case in Section 2.5. The *QDA discriminant* is simply the general discriminant in (4.8) using the sample means $\hat{\boldsymbol{\mu}}_0$ and $\hat{\boldsymbol{\mu}}_1$ and the sample covariance matrices $\hat{\Sigma}_0$ and $\hat{\Sigma}_1$ introduced in the previous subsection.

It is easy to show that the QDA classifier decision boundary produces a hyperplane decision boundary, determined by the equation $x^T A_n x + b_n^T x + c_n = k_n$, where

$$
\begin{aligned}
A_n &= -\frac{1}{2}\left(\hat{\Sigma}_1^{-1} - \hat{\Sigma}_0^{-1}\right) \\
b_n &= \hat{\Sigma}_1^{-1}\hat{\mu}_1 - \hat{\Sigma}_0^{-1}\hat{\mu}_0 \\
c_n &= -\frac{1}{2}\left(\hat{\mu}_1^T\hat{\Sigma}_1^{-1}\hat{\mu}_1 - \hat{\mu}_0^T\hat{\Sigma}_0^{-1}\hat{\mu}_0\right) - \frac{1}{2}\ln\frac{|\hat{\Sigma}_1|}{|\hat{\Sigma}_0|}
\end{aligned}
\tag{4.24}
$$

Clearly, the LDA parameters in (4.15) result from (4.24) in the case $\hat{\Sigma}_0 = \hat{\Sigma}_1 = \hat{\Sigma}$. The optimal QDA decision boundaries are hyperquadric surfaces in R^d, as discussed in Section 2.5.2: hyperspheres, hyperellipsoids, hyperparaboloids, hyperhyperboloids, and single or double hyperplanes. Differently than LDA, the decision boundary does not generally pass through the midpoint between the class means, even if the classes are equally-likely. As in the case of LDA, corresponding diagonal, spherical, and covariance plug-in restrictions, and other shrinkage constraints, can be placed on the covariance matrix estimators $\hat{\Sigma}_0$ and $\hat{\Sigma}_1$ in order to cope with small sample sizes.

Example 4.3. The training data in a classification problem is:

$$S_n = \{((1,0)^T, 0), ((0,1)^T, 0), ((-1,0)^T, 0), ((0,-1)^T, 0), ((2,0)^T, 1), ((0,2)^T, 1), ((-2,0)^T, 1), ((0,-2)^T, 1)\}$$

Assuming $k_n = 0$, we obtain the QDA decision boundary below. As in Example 4.2, we first compute the sample estimates:

$$
\begin{aligned}
\hat{\mu}_0 &= \frac{1}{4}\left(\begin{bmatrix}1\\0\end{bmatrix} + \begin{bmatrix}0\\1\end{bmatrix} + \begin{bmatrix}-1\\0\end{bmatrix} + \begin{bmatrix}0\\-1\end{bmatrix}\right) = \begin{bmatrix}0\\0\end{bmatrix}, \\
\hat{\mu}_1 &= \frac{1}{4}\left(\begin{bmatrix}2\\0\end{bmatrix} + \begin{bmatrix}0\\2\end{bmatrix} + \begin{bmatrix}-2\\0\end{bmatrix} + \begin{bmatrix}0\\-2\end{bmatrix}\right) = \begin{bmatrix}0\\0\end{bmatrix}, \\
\hat{\Sigma}_0 &= \frac{1}{3}\left(\begin{bmatrix}1\\0\end{bmatrix}[1\ 0] + \begin{bmatrix}0\\1\end{bmatrix}[0\ 1] + \begin{bmatrix}-1\\0\end{bmatrix}[-1\ 0] + \begin{bmatrix}0\\-1\end{bmatrix}[0\ -1]\right) = \frac{2}{3}I_2 \Rightarrow \hat{\Sigma}_0^{-1} = \frac{3}{2}I_2, \\
\hat{\Sigma}_1 &= \frac{1}{3}\left(\begin{bmatrix}2\\0\end{bmatrix}[2\ 0] + \begin{bmatrix}0\\2\end{bmatrix}[0\ 2] + \begin{bmatrix}-2\\0\end{bmatrix}[-2\ 0] + \begin{bmatrix}0\\-2\end{bmatrix}[0\ -2]\right) = \frac{8}{3}I_2 \Rightarrow \hat{\Sigma}_1^{-1} = \frac{3}{8}I_2,
\end{aligned}
\tag{4.25}
$$

The 2-dimensional QDA decision boundary is given by $a_{n,11}x_1^2 + 2a_{n,12}x_1x_2 + a_{n,22}x_2^2 + b_{n,1}x_1 + b_{n,2}x_2 + c_n = 0$, where

$$
\begin{aligned}
\begin{bmatrix}a_{n,11} & a_{n,12}\\ a_{n,12} & a_{n,22}\end{bmatrix} &= -\frac{1}{2}\left(\hat{\Sigma}_1^{-1} - \hat{\Sigma}_0^{-1}\right) = \frac{9}{16}\begin{bmatrix}1 & 0\\0 & 1\end{bmatrix}, \\
\begin{bmatrix}b_{n,1}\\ b_{n,2}\end{bmatrix} &= \hat{\Sigma}_1^{-1}\hat{\mu}_1 - \hat{\Sigma}_0^{-1}\hat{\mu}_0 = \begin{bmatrix}0\\0\end{bmatrix}, \\
c_n &= -\frac{1}{2}\left(\hat{\mu}_1^T\hat{\Sigma}_1^{-1}\hat{\mu}_1 - \hat{\mu}_0^T\hat{\Sigma}_0^{-1}\hat{\mu}_0\right) - \frac{1}{2}\ln\frac{|\hat{\Sigma}_1|}{|\hat{\Sigma}_0|} = -2\ln 2.
\end{aligned}
\tag{4.26}
$$

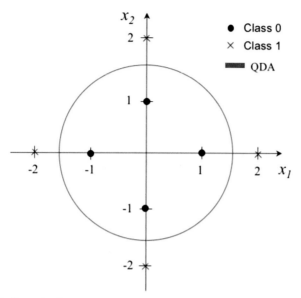

Figure 4.2: Sample data and quadratic decision boundary in Example 4.3.

Thus, the QDA decision boundary is given by

$$\frac{9}{16}x_1^2 + \frac{9}{16}x_2^2 - 2\ln 2 = 0 \quad \Rightarrow \quad x_1^2 + x_2^2 = \frac{32}{9}\ln 2\,, \tag{4.27}$$

which is the equation of a circle with center at the origin and radius equal to $4\sqrt{2\ln 2}/3 \approx 1.57$. Figure 4.2 displays the training data with the superimposed QDA decision boundary. This classifier is completely determined by the different variances between the classes, since the class means coincide. Obviously, neither LDA, DLDA or NMC could achieve any degree of discrimination in this case. In fact, no linear classifier (i.e., a classifier with a line for a decision boundary) could do a good job. This is an example of a *nonlinearly-separable data set*. ⋄

4.3 Logistic Classification

Logistic classification is an example of the second kind of parametric classification rule discussed in Section 4.1. First, define the "logit" transformation

$$\mathrm{logit}(p) \;=\; \ln\left(\frac{p}{1-p}\right), \quad 0 < p < 1\,, \tag{4.28}$$

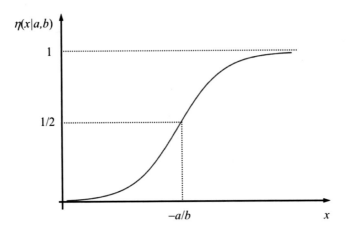

Figure 4.3: Univariate logistic curve with parameters $a > 0$ and b.

which maps the $[0, 1]$ interval onto the real line. In logistic classification, the posterior-probability function is linear in logit space, with parameters $(\mathbf{a}, b)$:

$$\text{logit}(\eta(\mathbf{x} \mid \mathbf{a}, b)) = \ln\left(\frac{\eta(\mathbf{x} \mid \mathbf{a}, b)}{1 - \eta(\mathbf{x} \mid \mathbf{a}, b)}\right) = \mathbf{a}^T \mathbf{x} + b. \tag{4.29}$$

Note that the logit transformation is necessary since a probability, which must be bounded between 0 and 1, could not be modeled by a linear function otherwise.

Inverting (4.29) yields

$$\eta(\mathbf{x} \mid \mathbf{a}, b) = \frac{e^{\mathbf{a}^T \mathbf{x} + b}}{1 + e^{\mathbf{a}^T \mathbf{x} + b}} = \frac{1}{1 + e^{-(\mathbf{a}^T \mathbf{x} + b)}}. \tag{4.30}$$

This function is called the *logistic curve* with parameters $(\mathbf{a}, b)$. The logistic curve is strictly increasing, constant, or strictly decreasing in the direction of x_j according to whether a_j is positive, zero, or negative, respectively. See Figure 4.3 for a univariate example with parameter $a > 0$ (so the curve is strictly increasing).

Estimates $\mathbf{a}_n$ and b_n of the logistic curve coefficients are typically obtained by maximizing the conditional log-likelihood $L(\mathbf{a}, b \mid S_n)$ of observing the data S_n (with the training feature vectors $\mathbf{X}_1, \ldots, \mathbf{X}_n$ assumed fixed) under the parametric assumption $\eta(\mathbf{x} \mid \mathbf{a}, b)$:

$$\begin{aligned} L(\mathbf{a}, b \mid S_n) &= \ln\left(\prod_{i=1}^{n} P(Y = Y_i \mid \mathbf{X} = \mathbf{X}_i)\right) = \sum_{i=1}^{n} \ln\left(\eta(\mathbf{X}_i \mid \mathbf{a}, b)^{Y_i}(1 - \eta(\mathbf{X}_i \mid \mathbf{a}, b))^{1-Y_i}\right) \\ &= \sum_{i=1}^{n} Y_i \ln(1 + e^{-(a^T \mathbf{X}_i + b)}) + (1 - Y_i) \ln(1 + e^{a^T \mathbf{X}_i + b}) \end{aligned} \tag{4.31}$$

This function is strictly concave so that the solution $(\mathbf{a}_n, b_n)$, if it exists, is unique and satisfies the equations:

$$
\begin{aligned}
\frac{\partial L}{\partial a_j}(\mathbf{a}_n, b_n) &= \sum_{i=1}^{n} Y_i \frac{e^{-(\mathbf{a}_n^T \mathbf{X}_i + b_n)} a_{n,j}}{1 + e^{-(\mathbf{a}_n^T \mathbf{X}_i + b_n)}} + (1 - Y_i) \frac{e^{\mathbf{a}_n^T \mathbf{X}_i + b_n} a_{n,j}}{1 + e^{\mathbf{a}_n^T \mathbf{X}_i + b_n}} \\
&= \sum_{i=1}^{n} Y_i \, \eta(\mathbf{X}_i \mid -\mathbf{a}_n, -b_n) \, a_{n,j} + (1 - Y_i) \, \eta(\mathbf{X}_i \mid \mathbf{a}_n, b_n) \, a_{n,j} = 0, \quad j = 1, \dots, d, \\
\frac{\partial L}{\partial b}(\mathbf{a}_n, b_n) &= \sum_{i=1}^{n} Y_i \frac{e^{-(\mathbf{a}_n^T \mathbf{X}_i + b_n)}}{1 + e^{-(\mathbf{a}_n^T \mathbf{X}_i + b_n)}} + (1 - Y_i) \frac{e^{\mathbf{a}_n^T \mathbf{X}_i + b_n}}{1 + e^{\mathbf{a}_n^T \mathbf{X}_i + b_n}} \\
&= \sum_{i=1}^{n} Y_i \, \eta(\mathbf{X}_i \mid -\mathbf{a}_n, -b_n) + (1 - Y_i) \, \eta(\mathbf{X}_i \mid \mathbf{a}_n, b_n) = 0.
\end{aligned}
\tag{4.32}
$$

This is a system of $d + 1$ highly nonlinear equations, which must be solved by iterative numerical methods.

The logistic classifier is then obtained by plugging in the estimates $\mathbf{a}_n$ and b_n into (4.30) and (4.6):

$$
\psi_n(\mathbf{x}) = \begin{cases} 1, & \frac{1}{1 + e^{-(\mathbf{a}_n^T \mathbf{x} + b_n)}} > \frac{1}{2}, \\ 0, & \text{otherwise,} \end{cases} = \begin{cases} 1, & \mathbf{a}_n^T \mathbf{x} + b_n > 0, \\ 0, & \text{otherwise.} \end{cases}
\tag{4.33}
$$

This reveals that, perhaps surprisingly, the logistic classifier is a *linear* classifier, with a hyperplane decision boundary determined by the parameters $\mathbf{a}_n$ and b_n.

4.4 Additional Topics

4.4.1 Regularized Discriminant Analysis

The NMC, LDA, and QDA classification rules discussed in Section 4.2 can be combined to obtain hybrid classification rules with intermediate characteristics.

In particular, one of the problems faced in parametric classification is the estimation of the parameters. When the sample size is small compared to the number of dimensions, estimation becomes poor and so do the designed classifiers. For example, $\hat{\Sigma}$ becomes a poor estimator of Σ under small sample size: small eigenvalues of Σ tend to be underestimated, while large eigenvalues of Σ tend to be overestimated. It is also possible that all eigenvalues of $\hat{\Sigma}$ are too small and the matrix cannot be inverted.

Among the three rules, QDA demands the most data, followed by LDA, and then NMC. LDA can be seen as an attempt to *regularize* or *shrink* QDA by pooling all the available data to estimate a

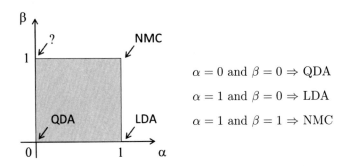

Figure 4.4: Parameter space for Regularized Discriminant Analysis.

single sample covariance matrix. The shrinkage from QDA to LDA can be controlled by introducing a parameter $0 \leq \alpha \leq 1$ and setting

$$\hat{\Sigma}_i^R(\alpha) = \frac{N_i(1-\alpha)\hat{\Sigma}_i + n\alpha\hat{\Sigma}}{N_i(1-\alpha) + n\alpha} , \qquad (4.34)$$

for $i = 0, 1$, where $\hat{\Sigma}$ is the pooled sample covariance matrix, and $\hat{\Sigma}_i$ and N_i are the individual sample covariance matrices and sample sizes. Notice that $\alpha = 0$ leads to QDA, while $\alpha = 1$ leads to LDA. Intermediate values $0 < \alpha < 1$ produce hybrid classification rules between QDA and LDA.

To get more regularization while not overly increasing bias, one can further shrink $\hat{\Sigma}_i^R(\alpha)$ towards its average eigenvalue multiplied by the identity matrix, by introducing a further parameter $0 \leq \beta \leq 1$ (this has the effect of decreasing large eigenvalues and increasing small eigenvalues, thereby offsetting the biasing effect mentioned earlier):

$$\hat{\Sigma}_i^R(\alpha,\beta) = (1-\beta)\hat{\Sigma}_i^R(\alpha) + \beta \frac{\text{trace}(\hat{\Sigma}_i(\alpha))}{d} I_d , \qquad (4.35)$$

for $i = 0, 1$. Note that $\alpha = \beta = 1$ leads to NMC. Hence, this rule ranges from QDA to LDA to NMC, and intermediate cases, depending on the selected values of α and β. This is called *regularized discriminant analysis*. See Figure 4.4 for an illustration. The "unknown" vertex $\alpha = 0$ and $\beta = 1$ corresponds to $\hat{\Sigma}_0^R = m_0 I_d$ and $\hat{\Sigma}_1^R = m_1 I_d$, where $m_i = \text{trace}(\hat{\Sigma}_i)/d \geq 0$, for $i = 0, 1$. It can be shown that this leads to a *spherical* decision boundary for $m_0 \neq m_1$ ($m_0 = m_1$ yields the plain NMC):

$$\left\| X - \frac{m_1\hat{\boldsymbol{\mu}}_0 - m_0\hat{\boldsymbol{\mu}}_1}{m_1 - m_0} \right\|^2 = \frac{m_1 m_0}{m_1 - m_0} \left(\frac{\|\hat{\boldsymbol{\mu}}_1 - \hat{\boldsymbol{\mu}}_0\|^2}{m_1 - m_0} + d \ln \frac{m_1}{m_0} \right). \qquad (4.36)$$

Each value of α and β corresponds to a different classification rule. Therefore, these parameters are not of the same kind as, for example, the means and covariance matrices in Gaussian discriminant analysis. Picking the value of α and β corresponds to picking a classification rule; this process is called *model selection* (see Chapter 8).

*4.4.2 Consistency of Parametric Rules

It might be expected that if the estimators $\boldsymbol{\theta}_{0,n}$ and $\boldsymbol{\theta}_{1,n}$ are consistent, meaning that $\boldsymbol{\theta}_{0,n} \to \boldsymbol{\theta}_0^*$ and $\boldsymbol{\theta}_{1,n} \to \boldsymbol{\theta}_1^*$ in probability as $n \to \infty$, and additionally the prior probabilities $P(Y = 0)$ and $P(Y = 1)$ are known or can also be estimated consistently, then $\varepsilon_n \to \varepsilon^*$ in probability, i.e., the corresponding parametric classification rule is consistent. This is indeed not the case, as additional smoothness conditions are required. This is shown by the following example.

Example 4.4. Consider parametric classification with a family of univariate Gaussian distributions specified by

$$p(x \mid \theta) = \begin{cases} \mathcal{N}(\theta, 1), & \theta \leq 1, \\ \mathcal{N}(\theta + 1, 1), & \theta \gtrless 1, \end{cases} \tag{4.37}$$

for $\theta \in \mathbb{R}$. Assume that the class prior probabilities are known to be $P(Y = 0) = P(Y = 1) = 1/2$ and that the (unknown) true values of the parameters are $\theta_0^* = -1$ and $\theta_1^* = 1$. The Bayes error is found using (2.53):

$$\varepsilon_L^* = \Phi\left(-\frac{|\theta_1^* - \theta_0^*|}{2}\right) = \Phi(-1) \approx 0.1587. \tag{4.38}$$

A simple computation reveals that $D_n(x) = \ln p(x \mid \theta_{1,n})/p(x \mid \theta_{0,n}) = a_n x + b_n$ is a linear discriminant, with parameters

$$a_n = (\theta_{1,n} - \theta_{0,n}) + (I_{\theta_{1,n}>1} - I_{\theta_{0,n}>1})$$
$$b_n = -a_n \left(\frac{\theta_{0,n} + \theta_{1,n}}{2} + \frac{I_{\theta_{0,n}>1} + I_{\theta_{1,n}>1}}{2}\right) \tag{4.39}$$

and the classification error is equal to (see Exercise 4.3):

$$\varepsilon_n = \frac{1}{2}\left[\Phi\left(\frac{a_n\theta_0 + b_n}{|a_n|}\right) + \Phi\left(-\frac{a_n\theta_1 + b_n}{|a_n|}\right)\right]. \tag{4.40}$$

The natural estimators $\theta_{0,n}$ and $\theta_{1,n}$ are the usual sample means. By the Law of Large Numbers (see Thm. A.12), $\theta_{0,n} \to \theta_0^* = -1$ and $\theta_{1,n} \to \theta_1^* = 1$ as $n \to \infty$, with probability 1. The function $I_{\theta>1}$ is continuous at $\theta_0^* = -1$, and this implies that $I_{\theta_{0,n}>1} \to 0$ with probability 1. However, $I_{\theta>1}$ is not continuous at $\theta_1^* = 1$, and $I_{\theta_{1,n}>1}$ does not converge, as $\liminf I_{\theta_{1,n}>1} = 0$ and $\limsup I_{\theta_{1,n}>1} = 1$ with probability one. In fact, since Φ is a continuous increasing function, it follows from (4.40) that, with probability one,

$$\liminf \varepsilon_n = \frac{1}{2}\left[\Phi\left(\liminf \frac{a_n\theta_0 + b_n}{|a_n|}\right) + \Phi\left(-\liminf \frac{a_n\theta_1 + b_n}{|a_n|}\right)\right] = \Phi(-1) = \varepsilon^* \approx 0.1587, \tag{4.41}$$

but

$$\limsup \varepsilon_n = \frac{1}{2}\left[\Phi\left(\limsup \frac{a_n\theta_0 + b_n}{|a_n|}\right) + \Phi\left(-\limsup \frac{a_n\theta_1 + b_n}{|a_n|}\right)\right] = \Phi\left(-\frac{1}{2}\right) \approx 0.3085. \tag{4.42}$$

As there are no subsequences along which there is convergence of $I_{\theta_1,n>1}$, and thus of ε_n, with probability one, it follows from Theorem A.9 that ε_n does not converge to ε^* in probability either.
◇

The trouble in the previous example is that the function $p(x \mid \theta)$ is not continuous at $\theta = \theta_1$ for any value of $x \in R$. Without further ado, we have the following theorem on the consistency of parametric classification rules.

Theorem 4.1. *If the parametric class-conditional density function $p(\mathbf{X} \mid \boldsymbol{\theta})$ is continuous at the true parameter values $\boldsymbol{\theta}_0^*$ and $\boldsymbol{\theta}_1^*$ almost everywhere in the measure of $\mathbf{X}$, and if $\boldsymbol{\theta}_{0,n} \to \boldsymbol{\theta}_0^*$, $\boldsymbol{\theta}_{1,n} \to \boldsymbol{\theta}_1^*$, and $k_n \to k^*$ in probability, then $\varepsilon_n \to \varepsilon^*$ in probability and the parametric classification rule given by (4.2) is consistent.*

Proof. Given $\mathbf{x} \in \mathbf{X}$, if $p(\mathbf{x} \mid \boldsymbol{\theta})$ is continuous at $\boldsymbol{\theta}_0^*$ and $\boldsymbol{\theta}_1^*$ and $\boldsymbol{\theta}_{0,n} \xrightarrow{P} \boldsymbol{\theta}_0^*$ and $\boldsymbol{\theta}_{1,n} \xrightarrow{P} \boldsymbol{\theta}_1^*$, then

$$D_n(\mathbf{x}) \,=\, \ln \frac{p(\mathbf{x} \mid \boldsymbol{\theta}_{1,n})}{p(\mathbf{x} \mid \boldsymbol{\theta}_{0,n})} \xrightarrow{P} \ln \frac{p(\mathbf{x} \mid \boldsymbol{\theta}_1^*)}{p(\mathbf{x} \mid \boldsymbol{\theta}_0^*)} \,=\, D^*(\mathbf{x})\,, \qquad (4.43)$$

by the Continuous Mapping Theorem (see Thm. A.6). Now, the conditional classification error of the parametric classifier (4.2) is given by

$$\begin{aligned}
\varepsilon[\psi_n \mid \mathbf{X} = \mathbf{x}] \,=\,& P(D_n(\mathbf{X}) > k_n \mid \mathbf{X} = \mathbf{x}, Y = 0, S_n)P(Y = 0) \\
&+ P(D_n(\mathbf{X}) \le k_n \mid \mathbf{X} = \mathbf{x}, Y = 1, S_n)P(Y = 1) \\
=\,& E[I_{D_n(\mathbf{x})-k_n>0} \mid \mathbf{X} = \mathbf{x}, Y = 0, S_n]P(Y = 0) + E[I_{D_n(\mathbf{x})-k_n\le 0} \mid \mathbf{X} = \mathbf{x}, Y = 1, S_n]P(Y = 1) \\
\xrightarrow{P}\,& E[I_{D^*(\mathbf{x})-k^*>0} \mid \mathbf{X} = \mathbf{x}, Y = 0]P(Y = 0) + E[I_{D^*(\mathbf{x})-k^*\le 0} \mid \mathbf{X} = \mathbf{x}, Y = 1]P(Y = 1) \\
=\,& P(D^*(\mathbf{x}) - k^* > 0 \mid \mathbf{X} = \mathbf{x}, Y = 0]P(Y = 0) + P(D^*(\mathbf{x}) - k^* \le 0 \mid \mathbf{X} = \mathbf{x}, Y = 1]P(Y = 1) \\
=\,& \varepsilon[\psi^* \mid \mathbf{X} = \mathbf{x}]
\end{aligned}$$

$$(4.44)$$

by the Bounded Convergence Theorem (see Thm. A.11), since $I_{D_n(\mathbf{x})-k_n} \xrightarrow{P} I_{D^*(\mathbf{x})-k}$. But since this holds with probability one over $\mathbf{X}$, another application of the Bounded Convergence Theorem gives $\varepsilon_n = E[\varepsilon[\psi_n \mid \mathbf{X}]] \xrightarrow{P} E[\varepsilon[\psi^* \mid \mathbf{X}]] = \varepsilon^*$. ◇

Note that the previous theorem includes the case where $k^* = \ln P(Y = 0)/P(Y = 1)$ is known (in this case, set $k_n \equiv k^*$). As an application, consider Gaussian Discriminant Analysis. In this case, $p(\mathbf{x} \mid \boldsymbol{\mu}, \Sigma)$ is continuous at any values of $\mathbf{x}$, $\boldsymbol{\mu}$, and Σ. Since the sample means are sample covariance matrix estimators are consistent (e.g., see Casella and Berger [2002]), and using $k_n = \ln N_0/N_1$, which is consistent under random sampling, it follows from Theorem 4.1 that the NMC, LDA, DLDA, and QDA classification rules are consistent if their respective distributional assumptions hold (consistency of the NMC was shown directly in Chapter 3).

More generally, the exponential density $p(\mathbf{x} \mid \boldsymbol{\theta})$ in (4.3) satisfies the conditions in Theorem 4.1 if and only if the functions α and ξ_i, for $i = 1, \ldots, k$, are continuous at $\boldsymbol{\theta}_0^*$ and $\boldsymbol{\theta}_1^*$. Using consistent estimators then leads to consistent classification rules.

If the parametric classification rule is instead specified as in (4.6), then a similar result to Theorem 4.1 holds. The proof of the following theorem is left as an exercise.

Theorem 4.2. *If the parametric posterior-probability function $\eta(\mathbf{x} \mid \boldsymbol{\theta})$ is continuous at the true parameter value $\boldsymbol{\theta}^*$ almost everywhere in the measure of $\mathbf{X}$, and if $\boldsymbol{\theta}_n \to \boldsymbol{\theta}^*$ in probability, then $\varepsilon_n \to \varepsilon^*$ in probability and the parametric classification rule in (4.6) is consistent.*

The logistic function $\eta(\mathbf{x} \mid \mathbf{a}, b)$ in (4.30) is clearly continuous at all values of $\mathbf{a}$ and b, for all values of $\mathbf{x} \in R^d$. Furthermore, the maximum-likelihood estimators $\mathbf{a}_n$ and b_n discussed in Section 4.3 are consistent, in principle (if one ignores the approximation introduced in their numerical computation). Therefore, by Theorem 4.2, the logistic classification rule is consistent under the parametric assumption (4.29).

4.4.3 Bayesian Parametric Rules

A Bayesian approach to estimation of the parameters leads to the Bayesian parametric classification rules. These classification rules have found a lot of application (see the Bibliographical notes). They allow the introduction of prior knowledge into the classification problem, by means of the "prior" distributions of the parameters. Bayesian classification rules are particularly effective in small-sample and high-dimensional cases, even in the case where the priors are uninformative. Bayesian statistics is a classical theory with a long history, to which we do not have space to do justice here (but see the Bibliographical notes). In this Section, we will only provide a brief glimpse of its application to parametric classification.

We still have a a family of probability density functions $\{p(\mathbf{x} \mid \boldsymbol{\theta}) \mid \boldsymbol{\theta} \in \Theta \subseteq R^m\}$. But now the true parameter values $\boldsymbol{\theta}_0$ and $\boldsymbol{\theta}_1$ (note that we remove the star) are *assumed to be random variables*. That means that there is a joint distribution $p(\boldsymbol{\theta}_0, \boldsymbol{\theta}_1)$, which is called the *prior* distribution (or more compactly, just "priors"). It is common to assume, as we will do here, that $\boldsymbol{\theta}_0$ and $\boldsymbol{\theta}_1$ are independent prior to observing the data, in which case $p(\boldsymbol{\theta}_0, \boldsymbol{\theta}_1) = p(\boldsymbol{\theta}_0)p(\boldsymbol{\theta}_1)$ and there are individual priors for $\boldsymbol{\theta}_0$ and $\boldsymbol{\theta}_1$. In what follows, we will assume for simplicity that $c = P(Y = 1)$ is known (or can be estimated accurately), but it is possible to include it as a Bayesian parameter as well.

The idea behind Bayesian inference is to update the prior distribution with the data S_n, using Bayes

Theorem, to obtain the *posterior* distribution:

$$p(\boldsymbol{\theta}_0, \boldsymbol{\theta}_1 \mid S_n) = \frac{p(S_n \mid \boldsymbol{\theta}_0, \boldsymbol{\theta}_1)p(\boldsymbol{\theta}_0, \boldsymbol{\theta}_1)}{\int_{\boldsymbol{\theta}_0, \boldsymbol{\theta}_1} p(S_n \mid \boldsymbol{\theta}_0, \boldsymbol{\theta}_1)p(\boldsymbol{\theta}_0, \boldsymbol{\theta}_1)d\boldsymbol{\theta}_0 d\boldsymbol{\theta}_1}. \tag{4.45}$$

The distribution

$$p(S_n \mid \boldsymbol{\theta}_0, \boldsymbol{\theta}_1) = \Pi_{i=1}^{n}(1 - c)^{1-y_i}p(\mathbf{x}_i \mid \boldsymbol{\theta}_0)^{1-y_i}c^{y_i}p(\mathbf{x}_i \mid \boldsymbol{\theta}_1)^{y_i} \tag{4.46}$$

is the likelihood of observing the data under the model specified by $\boldsymbol{\theta}_0$ and $\boldsymbol{\theta}_1$. By assumption, a closed-form analytical expression exists to compute $p(S_n \mid \boldsymbol{\theta}_0, \boldsymbol{\theta}_1)$; this constitutes the parametric (model-based) assumption in Bayesian statistics (some modern Bayesian approaches attempt to estimate the likelihood from the data, see the Bibliographical notes). We note that the denominator in (4.45) acts as a normalization constant (it is not a function of the parameters), and therefore is often omitted by writing $p(\boldsymbol{\theta}_0, \boldsymbol{\theta}_1 \mid S_n) \propto p(S_n \mid \boldsymbol{\theta}_0, \boldsymbol{\theta}_1)p(\boldsymbol{\theta}_0, \boldsymbol{\theta}_1)$. In addition, it can be shown that if $\boldsymbol{\theta}_0$ and $\boldsymbol{\theta}_1$ are assumed to be independent prior to observing S_n, they remain so afterwards, hence $p(\boldsymbol{\theta}_0, \boldsymbol{\theta}_1 \mid S_n) = p(\boldsymbol{\theta}_0 \mid S_n)p(\boldsymbol{\theta}_1 \mid S_n)$ (see Exercise 7.12).

Bayesian classification is based on the idea of computing *predictive densities* $p_0(\mathbf{x} \mid S_n)$ and $p_1(\mathbf{x} \mid S_n)$ for each label by integrating out the parameters:

$$p_0(\mathbf{x} \mid S_n) = \int_{R^m} p(\mathbf{x} \mid \boldsymbol{\theta}_0)p(\boldsymbol{\theta}_0 \mid S_n)d\boldsymbol{\theta}_0 \quad \text{and} \quad p_1(\mathbf{x} \mid S_n) = \int_{R^m} p(\mathbf{x} \mid \boldsymbol{\theta}_1)p(\boldsymbol{\theta}_1 \mid S_n)d\boldsymbol{\theta}_1. \tag{4.47}$$

A sample-based discriminant can be defined in the usual way:

$$D_n(\mathbf{x}) = \ln \frac{p_1(\mathbf{x} \mid S_n)}{p_0(\mathbf{x} \mid S_n)} \tag{4.48}$$

and the Bayesian classifier can be defined as

$$\psi_n(\mathbf{x}) = \begin{cases} 1, & D_n(\mathbf{x}) > \ln \frac{c_0}{c_1} \\ 0, & \text{otherwise.} \end{cases} \tag{4.49}$$

Example 4.5. The Gaussian case for Bayesian parametric classification has been extensively studied. We consider here the multivariate Gaussian parametric family $p(\mathbf{x} \mid \boldsymbol{\mu}, \Sigma)$ in (4.7), where the class means $\boldsymbol{\mu}_0$ and $\boldsymbol{\mu}_1$ and class covariance matrices Σ_0 and Σ_1 are general and unknown, with "vague" priors $p(\boldsymbol{\mu}_i) \propto 1$ and $p(\Sigma^{-1}) \propto |\Sigma_i|^{\frac{d-1}{2}}$ (note that the prior is defined on $\Lambda = \Sigma^{-1}$, also known as the "precision matrix"). These priors are improper, that is, they do not integrate to 1 as a usual density. They are also "uninformative," meaning roughly that they are not biased towards any particular value of the parameters. Indeed, in the case of the means, the priors are "uniform" over the entire space R^d. However, the posterior densities obtained from these priors are proper.

It can be shown that the predictive densities in this case are given by

$$p_i(\mathbf{x} \mid S_n) = t_{N_i-d}\left(\mathbf{x} \mid \hat{\boldsymbol{\mu}}_i, \frac{N_i^2 - 1}{N_i - d}\hat{\Sigma}_i\right), \quad \text{for } i = 0, 1, \tag{4.50}$$

where $\hat{\boldsymbol{\mu}}_0$, $\hat{\boldsymbol{\mu}}_1$, $\hat{\Sigma}_0$, and $\hat{\Sigma}_1$ are the usual sample means and sample covariance matrices, and $t_\nu(\mathbf{a}, B)$ is a multivariate t density with μ degrees of freedom:

$$t_\nu(\mathbf{a}, B) = \frac{\Gamma(\frac{\nu+d}{2})}{\Gamma(\frac{\nu}{2})(\nu\pi)^{\frac{p}{2}}|B|^{\frac{1}{2}}} \left[1 + \frac{1}{\nu}(\mathbf{x} - \mathbf{a})^T B^{-1}(\mathbf{x} - \mathbf{a}) \right]^{-\frac{\nu+d}{2}}, \tag{4.51}$$

where $\Gamma(t)$ is the Gamma function (see Section A1). It can be seen that the decision boundary for this classifier is polynomial. In fact, it can be shown that it reduces to a quadratic decision boundary, as in the ordinary heteroskedastic Gaussian case, under further assumptions on the parameters. ◇

4.5 Bibliographical Notes

Linear Discriminant Analysis has a long history, having been originally based on an idea by Fisher [1936] (the "Fisher's discriminant", see Chapter 9), developed by Wald [1944], and given the form known today by Anderson [1951]. There is a wealth of results on the properties of the classification error of LDA under Gaussian class-conditional densities. For example, the exact distribution and expectation of the true classification error were determined by John [1961] in the univariate case, and in the multivariate case by assuming that the covariance matrix Σ is known in the formulation of the discriminant. The case when Σ is not known and the sample covariance matrix S is used in discrimination is very difficult, and John [1961] gave only an asymptotic approximation to the distribution of the true error. In publications that appeared in the same year, Bowker [1961] provided a statistical representation of the LDA discriminant, whereas Sitgreaves [1961] gave the exact distribution of the discriminant in terms of an infinite series when the numbers of samples in each class are equal. Several classical papers have studied the distribution and moments of the true classification error of LDA under a parametric Gaussian assumption, using exact and approximate methods [Harter, 1951; Sitgreaves, 1951; Bowker and Sitgreaves, 1961; Teichroew and Sitgreaves, 1961; Okamoto, 1963; Kabe, 1963; Hills, 1966; Raudys, 1972; Anderson, 1973; Sayre, 1980]. McLachlan [1992] and Anderson [1984] provide extensive surveys of these methods, whereas Wyman et al. [1990] provide a compact survey and numerical comparison of several of the asymptotic results in the literature on the true error for Gaussian discrimination. A survey of similar results published in the Russian literature has been given by [Raudys and Young, 2004, Section 3]. McFarland and Richards [2001, 2002] gave a statistical representation of the QDA discriminant in the multivariate heteroskedastic case.

In addition, to nearest-mean classification and diagonal LDA, other approaches based on shrinkage of the covariance matrix include the cases where 1) the matrix has constant diagonal elements $\hat{\sigma}^2$ and constant off-diagonal (constant covariance) elements $\hat{\rho}$, 2) the diagonal elements $\hat{\sigma}_i^2$ are different and the off-diagonal elements are nonzero but are not estimated, being given by the "perfect" covariance

$\hat{\sigma}_i \hat{\sigma}_j$; 3) the diagonal elements $\hat{\sigma}_i^2$ are different but the off-diagonal elements are constant and equal to $\hat{\rho}$. See Schafer and Strimmer [2005] for for an extensive discussion of such shrinkage methods. In extreme high-dimensional cases, shrinkage of the sample means themselves might be useful, as is the case of the nearest-shrunken centroid method proposed by Tibshirani et al. [2002], already mentioned in the Bibliographical Notes section of Chapter 3.

The covariance plug-in classifier is due to John [1961]. It is called John's Linear Discriminant in the book by Braga-Neto and Dougherty [2015].

For details on the numerical solution of the system of equations (4.32), see Section 12.6.4 of Casella and Berger [2002].

The classical reference on bayesian statistics is Jeffreys [1961]. A modern, comprehensive treatment is found in Robert [2007]. According to McLachlan [1992], citing Aitchison and Dunsmore [1975], Bayesian parametric classification was proposed for the first time by Geisser [1964], who studied extensively the univariate and multivariate Gaussian cases, and stated that the idea of a Bayesian predictive density goes back to Jeffreys [1961] and even, using a fiducial argument, to Fisher [1935]. Dalton and Dougherty [2013] showed that this approach, which they call the *Optimal Bayesian Classifier* (OBC), is optimal in the sense of minimizing the posterior expected classification error over the class of all models. See this latter reference for more details on Example 4.5; e.g., the conditions under which the decision boundary of the Bayesian classifier becomes quadratic. Braga-Neto et al. [2018] provide a survey of the application of Bayesian classification algorithms in a number of problems in Bioinformatics.

4.6 Exercises

4.1. As discussed in Section 4.1, there are a few options for setting the threshold k_n in (4.2). One of them is to form estimates $\hat{c}_0$ and $\hat{c}_1$ from the data S_n and set $k_n = \ln \hat{c}_0/\hat{c}_1$. However, if the data S_n are obtained under separate sampling, estimation of c_0 and c_1 is not possible, since the data contain no information about them. Estimation of these probabilities is important also in other contexts, such as determining disease prevalence. But suppose that a large amount of unlabeled data $S_m^u = \{\mathbf{X}_{n+1}, \ldots, \mathbf{X}_{n+m}\}$ obtained by mixture sampling is available (this is common in many applications), and let ψ be a fixed classifier (e.g., it might be a classifier trained on the original data S_n). Then the proportions $R_{0,m}$ and $R_{1,m}$ of points in S_m^u assigned by ψ to classes 0 and 1 are estimators of c_0 and c_1, respectively.

(a) Show that

$$E[R_{0,m}] = c_0(1 - \varepsilon^0) + c_1\varepsilon^1 \,,$$
$$E[R_{1,m}] = c_0\varepsilon^0 + c_1(1 - \varepsilon^1) \,. \tag{4.52}$$

where

$$\varepsilon^0 = P(\psi(\mathbf{X}) = 1 \mid Y = 0) \,,$$
$$\varepsilon^1 = P(\psi(\mathbf{X}) = 0 \mid Y = 1) \,. \tag{4.53}$$

Conclude that the estimators $R_{0,m}$ and $R_{1,m}$ are generally biased, unless the classifier ψ is perfect, or $\varepsilon^1/\varepsilon^0 = c_1/c_0$.

Hint: Notice that $mR_{0,m}$ and $mR_{1,m}$ are binomially distributed.

(b) Show that

$$\mathrm{Var}(R_{0,m}) = \mathrm{Var}(R_{1,m}) = \frac{1}{m}(c_0(1 - \varepsilon^0) + c_1\varepsilon^1)(c_1(1 - \varepsilon^1) + c_0\varepsilon^0) \,. \tag{4.54}$$

(c) Now suppose that ε^0 and ε^1 are known or can be accurately estimated (estimation of ε^0 and ε^1 does not require knowledge of c_0 and c_1). Show that

$$c_{0,m} = \frac{R_{0,m} - \varepsilon^1}{1 - \varepsilon^0 - \varepsilon^1} \quad \text{and} \quad c_{1,m} = \frac{R_{1,m} - \varepsilon^0}{1 - \varepsilon^0 - \varepsilon^1} \tag{4.55}$$

are unbiased estimators of c_0 and c_1, respectively, such that $c_{1,m} = 1 - c_{0,m}$. Furthermore, use the result of item (b) to show that $c_{0,m}$ and $c_{1,m}$ are consistent estimators of c_0 and c_1, respectively, i.e., $c_{0,m} \to c_0$ and $c_{1,m} \to c_1$ in probability as $m \to \infty$. This leads to a good estimator $\ln c_{0,m}/c_{1,m}$ for the threshold k_n in (4.2), provided that m is not too small.

Hint: First show that an unbiased estimator with an asymptotically vanishing variance is consistent.

4.2. A common method to extend binary classification rules to K classes, $K > 2$, is the *one-vs-one approach*, in which $K(K-1)$ classifiers are trained between all pairs of classes, and a majority vote of assigned labels is taken.

(a) Formulate a multiclass version of parametric plug-in classification using the one-vs-one approach.

(b) Show that if the threshold $k_{ij,n}$ between classes i and j is given by $\ln \hat{c}_j / \ln \hat{c}_i$, then the one-vs-one parametric classification rule is equivalent to the simple decision

$$\psi_n(\mathbf{x}) = \arg\max_{k=1,\ldots,K} \hat{c}_k \, p(\mathbf{x} \mid \boldsymbol{\theta}_{k,n}) \,, \quad \mathbf{x} \in R^d \,. \tag{4.56}$$

(For simplicity, you may ignore the possibility of ties.)

(c) Applying the approach in items (a) and (b), formulate a multiclass version of Gaussian discriminant analysis. In the case of multiclass NMC, with all thresholds equal to zero, how does the decision boundary look like?

4.3. Under the general Gaussian model $p(\mathbf{x} \mid Y = 0) \sim \mathcal{N}_d(\boldsymbol{\mu}_0, \Sigma_0)$ and $p(\mathbf{x} \mid Y = 1) \sim \mathcal{N}_d(\boldsymbol{\mu}_1, \Sigma_1)$, the classification error $\varepsilon_n = P(\psi_n(\mathbf{X}) \neq Y) \mid S_n)$ of *any* linear classifier in the form

$$\psi_n(\mathbf{x}) = \begin{cases} 1, & \mathbf{a}_n^T \mathbf{x} + b_n > 0, \\ 0, & \text{otherwise,} \end{cases} \tag{4.57}$$

(examples discussed so far include LDA and its variants, and the logistic classifier) can be readily computed in terms of Φ (the c.d.f. of a standard normal random variable), the classifier parameters $\mathbf{a}_n$ and b_n, and the distributional parameters $c = P(Y = 1)$, $\boldsymbol{\mu}_0$, $\boldsymbol{\mu}_1$, Σ_0, and Σ_1.

(a) Show that

$$\varepsilon_n = (1 - c) \, \Phi \left(\frac{\mathbf{a}_n^T \boldsymbol{\mu}_0 + b_n}{\sqrt{\mathbf{a}_n^T \Sigma_0 \mathbf{a}_n}} \right) + c \, \Phi \left(-\frac{\mathbf{a}_n^T \boldsymbol{\mu}_1 + b_n}{\sqrt{\mathbf{a}_n^T \Sigma_1 \mathbf{a}_n}} \right). \tag{4.58}$$

Hint: the discriminant $\mathbf{a}_n^T \mathbf{x} + b_n$ has a simple Gaussian distribution in each class.

(b) Compute the errors of the NMC, LDA, and DLDA classifiers in Example 4.2 if $c = 1/2$,

$$\boldsymbol{\mu}_0 = \begin{bmatrix} 2 \\ 3 \end{bmatrix}, \boldsymbol{\mu}_1 = \begin{bmatrix} 6 \\ 5 \end{bmatrix}, \Sigma_0 = \begin{bmatrix} 1 & 1 \\ 1 & 2 \end{bmatrix}, \text{ and } \Sigma_1 = \begin{bmatrix} 4 & 0 \\ 0 & 1 \end{bmatrix}.$$ Which classifier does the best?

4.4. Even in the Gaussian case, the classification error of quadratic classifiers in general require numerical integration for its computation. In some special simple cases, however, it is possible to obtain exact solutions. Assume a two-dimensional Gaussian problem with $P(Y = 1) = 1/2$, $\boldsymbol{\mu}_0 = \boldsymbol{\mu}_1 = \mathbf{0}$, $\Sigma_0 = \sigma_0^2 I_2$, and $\Sigma_1 = \sigma_1^2 I_2$. For definiteness, assume that $\sigma_0 < \sigma_1$.

(a) Show that the Bayes classifier is given by

$$\psi^*(\mathbf{x}) = \begin{cases} 1, & \|\mathbf{x}\| > r^*, \\ 0, & \text{otherwise,} \end{cases} \quad \text{where } r^* = \sqrt{2 \left(\frac{1}{\sigma_0^2} - \frac{1}{\sigma_1^2} \right)^{-1} \ln \frac{\sigma_1^2}{\sigma_0^2}}. \tag{4.59}$$

In particular, the optimal decision boundary is a circle of radius r^*.

(b) Show that the corresponding Bayes error is given by

$$\varepsilon^* = \frac{1}{2} - \frac{1}{2} \left(\sigma_1^2/\sigma_0^2 - 1 \right) e^{-(1 - \sigma_0^2/\sigma_1^2)^{-1} \ln \sigma_1^2/\sigma_0^2}. \tag{4.60}$$

In particular, the Bayes error is a function only of the ratio of variances σ_1^2/σ_0^2, and $\varepsilon^* \to 0$ as $\sigma_1^2/\sigma_0^2 \to \infty$.

Hint: Use polar coordinates to solve the required integrals analytically.

(c) Compare the optimal classifier to the QDA classifier in Example 4.3. Compute the error of the QDA classifier and compare to the Bayes error.

4.5. (Detection of signal in noise.) Suppose that a (nonrandom) message $\mathbf{s} = (s_1, \ldots, s_d)$ is to be sent through a noisy channel. At time $t = k$, if the message is present, the receiver reads

$$\mathbf{X}_k = \mathbf{s} + \boldsymbol{\varepsilon}_k, \tag{4.61}$$

otherwise, the receiver reads

$$\mathbf{X}_k = \boldsymbol{\varepsilon}_k, \tag{4.62}$$

where $\boldsymbol{\varepsilon}_k \sim \mathcal{N}(0, \Sigma)$ is a noise term (note that the noise can be correlated). The problem is to detect whether there is a message at time t or there is only noise. Suppose that the first n time points are used to "train" the receiver, e.g., by sending the message at $t = 1, \ldots, n/2$ and nothing over $t = n/2 + 1, \ldots, n$.

(a) Formulate this problem as a Linear Discriminant Analysis problem. Find the LDA detector (write the LDA coefficients in terms of the signal and the noise values).

(b) Find the LDA detection error, in terms of the classifier coefficients and the true values of $\mathbf{s} = (s_1, \ldots, s_d)$ and Σ.

(c) Say as much as you can about the difficulty of the problem in terms of the values of $\mathbf{s} = (s_1, \ldots, s_d), \Sigma$, and the training sample size n.

4.6. Consider the parametric classification rule in (4.6).

(a) Prove that

$$\varepsilon_n - \varepsilon^* \leq 2E\left[|\eta(\mathbf{X} \mid \boldsymbol{\theta}_n) - \eta(\mathbf{X} \mid \boldsymbol{\theta}^*)| \mid S_n\right]. \tag{4.63}$$

Hint: Show that, given S_n,

$$\varepsilon[\psi_n \mid \mathbf{X} = \mathbf{x}] - \varepsilon[\psi^* \mid \mathbf{X} = \mathbf{x}] = 2\left|\eta(\mathbf{x} \mid \boldsymbol{\theta}^*) - \frac{1}{2}\right| I_{\psi_n(\mathbf{x}) \neq \psi^*(\mathbf{x})} \leq 2|\eta(\mathbf{x} \mid \boldsymbol{\theta}_n) - \eta(\mathbf{x} \mid \boldsymbol{\theta}^*)|, \tag{4.64}$$

then take expectation over $\mathbf{X}$.

(b) Conclude that if $\eta(\mathbf{X} \mid \boldsymbol{\theta})$ is continuous at $\boldsymbol{\theta}^*$, in the sense that $\boldsymbol{\theta}_n \xrightarrow{P} \boldsymbol{\theta}^*$ implies $\eta(\mathbf{X} \mid \boldsymbol{\theta}_n) \xrightarrow{L^1} \eta(\mathbf{X} \mid \boldsymbol{\theta}^*)$, then the classification rule is consistent if $\boldsymbol{\theta}_n$ is. Compare with Theorem 4.2.

4.7 Python Assignments

4.7. Using the synthetic data model in Section A8.1 for the 2-D homoskedastic case, with $\boldsymbol{\mu}_0 = (0,0)$, $\boldsymbol{\mu}_1 = (1,1)$, $P(Y = 0) = P(Y = 1)$, $\sigma = 0.7$ and independent features, set `np.random.seed` to 0 for reproducibility, and generate training data sets for each sample size $n = 20$ to $n = 100$, in steps of 10. Obtain the LDA, DLDA and NMC decision boundaries corresponding to these data, by calculating the sample means and sample covariance matrices using `np.mean` and `np.cov`, and applying the corresponding formulas. Also determine the optimal classifier for this problem.

(a) Plot the data (using O's for class 0 and X's for class 1), with the superimposed decision boundaries for the optimal classifier and the designed LDA, DLDA, and NMC classifiers. Describe what you see.

(b) Compute the errors of all classifiers, using two methods:

    i. the formulas (2.53) for the Bayes error and (4.58) for the other errors.

    ii. Computing the proportion of errors committed on a test set of size $M = 100$. Compare the errors among the classifiers as well as between the exact (item i) and approximate (item ii) methods of computation.

(c) Generate a large number (e.g., $N = 1000$) of synthetic training data sets for each sample size $n = 20$ to $n = 100$, in steps of 10. For each data set, obtain the errors of the LDA, DLDA, and NMC classifiers using the exact formula (no need to plot any of the classifiers), and average over the $N = 1000$ data sets to obtain an accurate approximation of $E[\varepsilon_n]$ for each classifier. Plot this estimated expected error as a function of n for each classifier. Repeat for $\sigma = 0.5$ and $\sigma = 1$. What do you observe?

4.8. Apply linear discriminant analysis to the stacking fault energy (SFE) data set (see Section A8.4), already mentioned in Chapter 1. Categorize the SFE values into two classes, low (SFE ≤ 35) and high (SFE ≥ 45), excluding the middle values.

(a) Apply the preprocessing steps in `c01_matex.py` to obtain a data matrix of dimensions 123 (number of sample points) $\times 7$ (number of features), as described in Section 1.8.2. Define low (SFE ≤ 35) and high (SFE ≥ 45) labels for the data. Pick the first 20% of the sample points to be the training data and the remaining 80% to be test data.

(b) Using the function `ttest_ind` from the `scipy.stats` module, apply Welch's two-sample t-test on the training data, and produce a table with the predictors, T statistic, and p-value, ordered with largest absolute T statistics at the top.

(c) Pick the top two predictors and design an LDA classifier. (This is an example of *filter feature selection*, to be discussed in Chapter 9.) Plot the training data with the superimposed LDA decision boundary. Plot the testing data with the superimposed previously-obtained LDA decision boundary. Estimate the classification error rate on the training and test data. What do you observe?

(d) Repeat for the top three, four, and five predictors. Estimate the errors on the training and testing data (there is no need to plot the classifiers). What can you observe?

Chapter 5

Nonparametric Classification

> "The task of science is both to extend the range of
> our experience and to reduce it to order."
> – Niels Bohr, *Atomic Theory and
> the Description of Nature*, 1934.

Nonparametric classification rules differ from parametric classification rules in one key respect: no assumption about the shapes of the distributions are made, but rather these are approximated by a process of *smoothing.* This allows these rules to be distribution-free and produce complex decision boundaries, and thus achieve universal consistency in many cases. The price paid for this flexibility is a larger data requirement than in the parametric case and an increased risk of overfitting. In addition, all nonparametric classification rules introduce free parameters that control the amount of smoothing, and these must be set using model selection criteria. In this chapter, after discussing general aspects of nonparametric classification, we present the histogram, nearest-neighbor, and kernel classification rules. We also discuss the famous Cover-Hart and Stone Theorems on the asymptotic performance of nonparametric classification rules.

5.1 Nonparametric Plug-in Rules

The nonparametric approach is based on smoothing the available data to obtain approximations $p_n(\mathbf{x} \mid Y = 0)$ and $p_n(\mathbf{x} \mid Y = 1)$ of the class-conditional densities, or an approximation $\eta_n(\mathbf{x}) = P_n(Y = 1 \mid \mathbf{X} = \mathbf{x})$ of the posterior-probability function, and then plug it in the definition of the Bayes classifier. A critical aspect of this process is selecting the "right" amount of smoothing given the sample size and complexity of the distribution.

© Springer Nature Switzerland AG 2020
U. Braga-Neto, *Fundamentals of Pattern Recognition and Machine Learning,*
https://doi.org/10.1007/978-3-030-27656-0_5

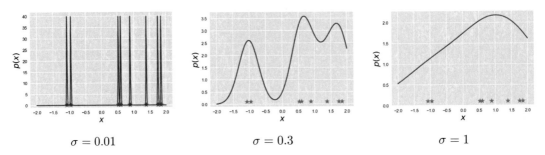

$$\sigma = 0.01 \qquad\qquad \sigma = 0.3 \qquad\qquad \sigma = 1$$

Figure 5.1: Sample data and kernel-based smoothing using different kernel standard deviations (plots generated by c05_kern_dnst.py).

Example 5.1. Figure 5.1 illustrates smoothing using a kernel-based approximation. If too little smoothing is applied (leftmost plot), i.e., the kernel standard deviation (also known as the *bandwidth*) is too small, then a poor approximation is obtained; the approximation skews too close to the data, leading to overfitting. On the other hand, if too much smoothing is applied (rightmost plot), i.e., the kernel standard deviation is too large, then an equally poor approximation results, where the structure of the data cannot be captured, leading to underfitting. In this example, it appears that the center plot would correspond to the appropriate kernel standard deviation and amount of smoothing. The kernel bandwidth in this case is a free parameter that needs to be set by the user; the existence of such parameters is a constant of all nonparametric classification methods. ⋄

We will focus here on the case of smoothing estimates $\eta_n(\mathbf{x})$ of the posterior-probability function $\eta(\mathbf{x})$ (this covers the most common examples of nonparametric classification rules). Given the training data $S_n = \{(\mathbf{X}_1, Y_1), \ldots, (\mathbf{X}_n, Y_n)\}$, we consider general approximations of the following kind:

$$\eta_n(\mathbf{x}) = \sum_{i=1}^{n} W_{n,i}(\mathbf{x}, \mathbf{X}_1, \ldots, \mathbf{X}_n) I_{Y_i=1}, \tag{5.1}$$

where the weights $W_{n,i}(\mathbf{x}, \mathbf{X}_1, \ldots, \mathbf{X}_n)$ satisfy

$$W_{n,i}(\mathbf{x}, \mathbf{X}_1, \ldots, \mathbf{X}_n) \geq 0, \ i = 1, \ldots, n \ \text{ and } \ \sum_{i=1}^{n} W_{n,i}(\mathbf{x}, \mathbf{X}_1, \ldots, \mathbf{X}_n) = 1, \quad \text{for all } \mathbf{x} \in R^d. \tag{5.2}$$

Each $W_{n,i}(\mathbf{x}, \mathbf{X}_1, \ldots, \mathbf{X}_n)$ weights the corresponding training point $(\mathbf{X}_i, Y_i)$ in forming the estimate $\eta_n(\mathbf{x})$. Being a function of $\mathbf{x}$, the weights can change from point to point, while being a function of $\mathbf{X}_1, \ldots, \mathbf{X}_n$, the weights depend on the entire spatial configuration of the training feature vectors. For ease of notation, from this point on, we will denote the weights by $W_{n,i}(\mathbf{x})$, with the dependence on $\mathbf{X}_1, \ldots, \mathbf{X}_n$ being implicit.

As can be easily verified, the plug-in classifier is given by

$$\psi_n(\mathbf{x}) = \begin{cases} 1, & \eta_n(\mathbf{x}) > \frac{1}{2}, \\ 0, & \text{otherwise}, \end{cases} = \begin{cases} 1, & \sum_{i=1}^n W_{n,i}(\mathbf{x})I_{Y_i=1} > \sum_{i=1}^n W_{n,i}(\mathbf{x})I_{Y_i=0}, \\ 0, & \text{otherwise}. \end{cases} \tag{5.3}$$

This can be seen as adding the "influences" of each data point $(\mathbf{X}_i, Y_i)$ on $\mathbf{x}$ and assigning the label of the most "influent" class (with ties broken in an arbitrary deterministic manner, here in the direction of class 0). Typically, $W_{n,i}(\mathbf{x})$ is inversely related to the Euclidean distance $||\mathbf{x} - \mathbf{X}_i||$; hence, the "influence" of training point $(\mathbf{X}_i, Y_i)$ is larger if $\mathbf{X}_i$ is closer to $\mathbf{x}$, and smaller, otherwise. This spatial coherence assumption is a key premise of nonparametric classification and, indeed, of all smoothing methods: points that are spatially close to each other are more likely to come from the same class than points that are far away from each other. In the next few sections we consider specific examples of nonparametric classification rules based on this approach.

5.2 Histogram Classification

Histogram rules are based on *partitions* of the feature space. A partition is a mapping A from R^d to the power set $\mathcal{P}(R^d)$, i.e., the family of all subsets of R^d, such that

$$A(\mathbf{x}_1) = A(\mathbf{x}_2) \text{ or } A(\mathbf{x}_1) \cap A(\mathbf{x}_2) = \emptyset, \quad \text{for all } \mathbf{x}_1, \mathbf{x}_2 \in R^d, \tag{5.4}$$

and

$$R^d = \bigcup_{\mathbf{x} \in R^p} A(\mathbf{x}). \tag{5.5}$$

A partition consists therefore of non-overlapping zones $A(\mathbf{x})$ that tile the entire feature space R^d.

The general histogram classification rule is in the form (5.3), with weights

$$W_{n,i}(\mathbf{x}) = \begin{cases} \frac{1}{N(\mathbf{x})}, & \mathbf{X}_i \in A(\mathbf{x}) \\ 0, & \text{otherwise}, \end{cases} \tag{5.6}$$

where $N(\mathbf{x})$ denotes the total number of training points in $A(\mathbf{x})$. Only the points inside the same zone as the test point contribute (equally) to the posterior probability estimate, and therefore to the classification.

We can see that the histogram classifier assigns to each zone of the partition the majority label among the training points that fall into the zone. Following (5.3), ties are broken in a deterministic manner, always in the direction of class 0. Notice that this is equivalent to quantizing, or discretizing, the feature space using the partition, and then applying the discrete histogram rule of Example 3.3. In the *cubic histogram rule*, the partition is a regular grid, and each zone is a (hyper)cube with a side of length l.

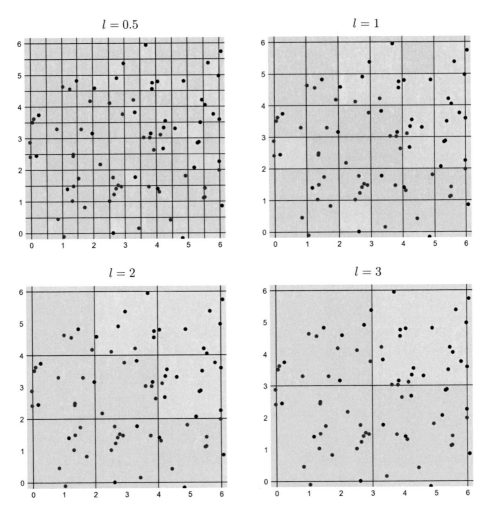

Figure 5.2: Classifiers produced by the cubic histogram classification rule on synthetic Gaussian training data (plots generated by c05_cubic.py).

Example 5.2. Figure 5.2 displays classifiers produced by the cubic histogram classification on data sampled from spherical Gaussian densities centered at $(2, 2)$ and $(4, 4)$ with variance equal to 4. The sample size is 50 points per class. The optimal decision boundary is therefore a line with slope -45° going through the point $(3, 3)$. We can see that the histogram rule cannot provide a good approximation of the optimal classifier, since the orientation of its grid does not match it (this is a similar problem faced by the CART rule, discussed in Chapter 6). Notice that l plays the role of smoothing parameter: if is too small for the sample size, there is not enough points inside each zone to estimate correctly the posterior probability, while if is too large the estimate is too coarse.

At very fine grids, such as the case $l = 0.5$, there are many empty zones. This creates "0-0" ties, which are artificially broken towards class 0 (In Python Assignment 5.8, the effect of using random tie-breaking instead is considered.) ◇

5.3 Nearest-Neighbor Classification

The k-nearest-neighbor (kNN) classification rule is the oldest, most well-known (and most studied) nonparametric classification rule. It consists of assigning the majority label among the k nearest neighbors to the test point in the training data. By simply using odd values of k, the possibility of ties between the labels is removed, avoiding one of the main problems with the histogram rule. In addition, the kNN classification rule is able to adapt to any shape or orientation of the data, unlike the histogram rule. If $k = 1$, the classifier simply assigns to the test point the label of the nearest training point. This simple nearest-neighbor (1NN) rule can be surprisingly good if the sample size is sufficiently large, as shown by the famous Cover-Hart Theorem, to be discussed below. However, the 1NN rule can be very bad in small-sample cases due to overfitting. It also tends to perform badly in case the classes overlap substantially. There is theoretical and empirical evidence that the 3NN and 5NN rules are much better than 1NN, under both large and small sample sizes.

The kNN classification rule is in the form (5.3) with weights

$$W_{n,i}(\mathbf{x}) = \begin{cases} \frac{1}{k}, & \mathbf{X}_i \text{ is among the } k \text{ nearest neighbors of } \mathbf{x}, \\ 0, & \text{otherwise.} \end{cases} \tag{5.7}$$

Hence, only the points that are among the k nearest neighbors of the test point contribute (equally) to the posterior probability estimate, and therefore to the classification. Weights that assign different importance according to distance rank among the k-nearest neighbors can also be used, which leads to *weighted kNN* classification rules. Distances other than Euclidean can also be used. The choice of k is a model selection problem, but as already mentioned, $k = 3$ or $k = 5$ are usually good choices in most practical situations.

Example 5.3. Figure 5.3 displays classifiers produced by the standard kNN classification rule on the training data in Figure 5.2. The plots show that the decision boundaries in all cases are complex, but this is true especially in the 1NN case. In fact, one can see that the 1NN classifier badly overfits the data, due to the small sample size and the overlapping densities. The 3NN and 5NN rules produce much better classifiers. The 7NN rule produces some improvement, but not much, over 5NN. At this point, the kNN classification rule produces a decision boundary that is very close to the optimal one (a line with slope -45° going through the center of the plot, as mentioned in the previous section). Note that k is the smoothing parameter and has the same interpretation as in

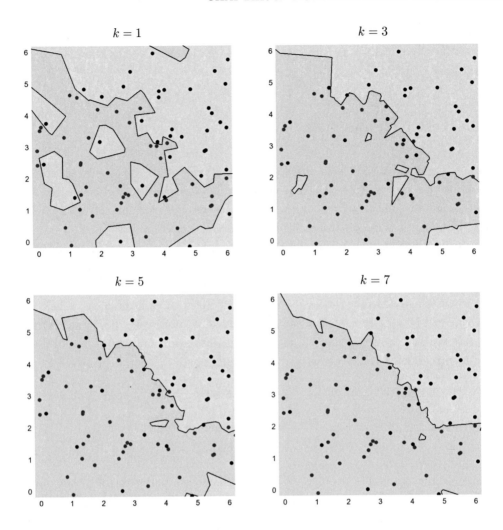

Figure 5.3: Classifiers produced by the kNN classification rule on the training data in Figure 5.2 (plots generated by c05_knn.py).

other nonparametric classification rules: if it is too small (e.g. $k = 1$) compared to the sample size/complexity of the problem, there is overfitting due to lack of smoothing, while if it is too large, the posterior probability estimate may become too coarse. The choice of k is a model selection problem (see Chapter 8). Problem 5.9 shows that the value of k should be increased further in this example. ◇

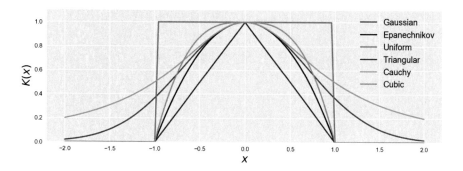

Figure 5.4: Univariate kernels used in kernel classification (plot generated by c05_kern_univ.py).

5.4 Kernel Classification

A *kernel* is a nonnegative function $k : R^d \rightarrow R$. Kernel rules are very general nonparametric classification rules, where the weighting functions $W_{n,i}$ in (5.3) and (5.1) are expressed in terms of kernels. A kernel that is a monotonically decreasing function of $||\mathbf{x}||$ is also known as a *radial basis function* (RBF). Some examples are: are:

- Gaussian kernel: $k(\mathbf{x}) = e^{-||\mathbf{x}||^2}$.

- Cauchy kernel: $k(\mathbf{x}) = \frac{1}{1+||\mathbf{x}||^{d+1}}$.

- Triangle kernel: $k(\mathbf{x}) = (1 - ||\mathbf{x}||)\, I_{\{||\mathbf{x}|| \leq 1\}}$.

- Epanechnikov kernel: $k(\mathbf{x}) = (1 - ||\mathbf{x}||^2)\, I_{\{||\mathbf{x}|| \leq 1\}}$.

- Uniform (spherical) kernel: $k(\mathbf{x}) = I_{\{||\mathbf{x}|| \leq 1\}}$.

- Uniform (cubic) kernel:

$$k(\mathbf{x}) = \begin{cases} 1, & |x_j| \leq \frac{1}{2}, \text{ for all } j = 1, \ldots, d, \\ 0, & \text{otherwise.} \end{cases} \tag{5.8}$$

All the previous kernels, with the exception of the uniform cubic kernel, are RBFs. In addition, all the previous kernels have bounded support, with the exception of the Gaussian and Cauchy kernels. See Figure 5.4 for an illustration. Notice that in the univariate case, the spherical and cubic kernels coincide.

The kernel classification rule is in the form (5.3) with weighting function

$$W_{n,i}(\mathbf{x}) = \frac{k\left(\frac{\mathbf{x}-\mathbf{X}_i}{h}\right)}{\sum_{i=1}^n k\left(\frac{\mathbf{x}-\mathbf{X}_i}{h}\right)}, \tag{5.9}$$

where the smoothing parameter h is the kernel bandwidth. The nonparametric classifier in (5.3) can be written in this case as

$$\psi_n(\mathbf{x}) = \begin{cases} 1, & \sum_{i=1}^n k\left(\frac{\mathbf{x}-\mathbf{X}_i}{h}\right) I_{\{Y_i=1\}} \geq \sum_{i=1}^n k\left(\frac{\mathbf{x}-\mathbf{X}_i}{h}\right) I_{\{Y_i=0\}}, \\ 0, & \text{otherwise.} \end{cases} \tag{5.10}$$

Hence, the class with larger cumulative kernel values at the test point $\mathbf{x}$ assigns its label to $\mathbf{x}$. This has a physics analogy, in which the training points from different classes represent charges of opposite signs, and a resultant electrostatic potential is computed at $\mathbf{x}$, the assigned class depending on the sign of the potential.

Example 5.4. Figure 5.5 displays classifiers produced by the Gaussian-RBF kernel classification rule on the training data in Figures 5.2 and 5.3. We can see the the decision boundaries become progressively smooth as the bandwidth increases. If the bandwidth h is too small, namely, $h = 0.1$ and $h = 0.3$, the rule is too "local" (only the closest points exert influence on the test point $\mathbf{x}$) which leads to overfitting. At $h = 1$, the decision boundary is very close to the optimal one (a line with slope -45° going through the center of the plot). It is instructive to compare the plots in Figures 5.3 and 5.5. We can see that the kernel classifier with $h = 0.1$ closely resembles the 1NN classifier; at such a small bandwidth, the influence of the nearest training point to the test point is much larger than that of any other training point, and the kernel classifier is essentially a 1NN classifier. As h increases, however, the behaviors are different: we can see that the kernel classification rule, with a Gaussian RBF, produces more smooth decision boundaries than the kNN classification rule. If h becomes too large, the kernel rule can become too "global" (far points exert influence on $\mathbf{x}$), which can lead to underfitting. In general, The best value of h must in general be chosen using model selection criteria (see Chapter 8). Problem 5.10 shows that the value of h should be increased further in this example. ◇

Notice that the requirement that $k(\mathbf{x}) \geq 0$, for all $\mathbf{x} \in R^d$, is necessary to guarantee that the weights in (5.9) are nonnegative, and make (5.1) a "sensible" estimate of the posterior-probability function. However, the expression of the kernel nonparametric classifier in (5.10) does *not* require nonnegative of k, i.e., the kernel can take negative values and still define a valid classifier. For example, the *Hermite kernel*:

$$k(\mathbf{x}) = (1 - ||\mathbf{x}||^2) e^{-||\mathbf{x}||^2}, \tag{5.11}$$

and the *sinc kernel*:

$$k(\mathbf{x}) = \frac{\sin(\pi||\mathbf{x}||)}{\pi||\mathbf{x}||}, \tag{5.12}$$

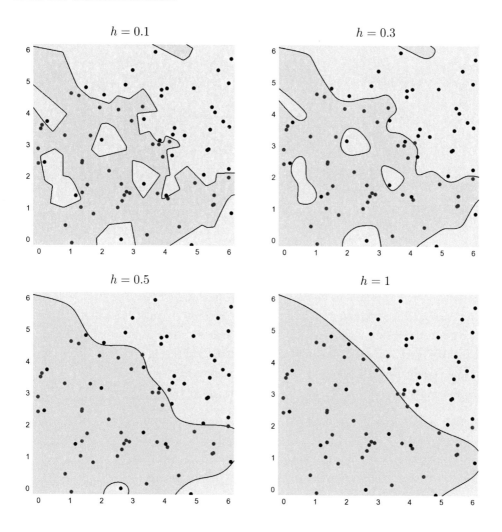

Figure 5.5: Classifiers produced by the Gaussian-RBF kernel classification rule on the training data in Figure 5.2 and 5.3 (plots generated by c05_kernel.py).

take negative values (see Figure 5.6 for a univariate example), but can be used for kernel classification. In fact, there are examples of problems where the Hermite kernel outperforms all nonnegative kernels (see Exercise 5.7). This illustrates the fact that classification and distribution estimation are different problems. For example, the statement that one needs good density estimation to design good classifiers is mistaken, as the former requires more conditions (and more data) than the latter.

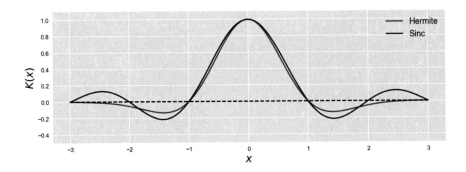

Figure 5.6: Univariate Hermite and sinc kernels (plot generated by c05_kern_neg.py).

5.5   Cover-Hart Theorem

In this section we present a famous and influential result, due to Cover and Hart [1967], about the asymptotic behavior of the expected error of the kNN classification rule. It is often expressed as "the error of the nearest-neighbor classifier with a large sample size cannot be worse than two times the Bayes error." We investigate below the rigorous formulation of this idea. For convenience, we divide the result into three separate theorems (but still call them collectively "The Cover-Hart Theorem"). One of the theorems is proved in this section, while the other two are proved in Appendix A4.

The first theorem shows that the NN rule, in the limit as sample size goes to infinity, cannot err more than twice as the Bayes classifier itself. This is a distribution-free result, i.e., it applies to any feature-label distribution. See Appendix A4 for a proof.

Theorem 5.1. *(Cover-Hart Theorem) The expected error of the NN classification rule satisfies*

$$\varepsilon_{\mathrm{NN}} = \lim_{n \to \infty} E[\varepsilon_n] = E[2\eta(\mathbf{X})(1 - \eta(\mathbf{X}))], \tag{5.13}$$

from which it follows that

$$\varepsilon_{\mathrm{NN}} \le 2\varepsilon^*(1 - \varepsilon^*) \le 2\varepsilon^*. \tag{5.14}$$

Recall that $\varepsilon_{\mathrm{NN}}$ was previously defined in (2.68), where it was called the nearest-neighbor distance (an example of F-error). The first part of Thm. 5.1 shows that this is in fact the asymptotic error of the NN classification rule.

Thm. 5.1 places the asymptotic classification error of the NN rule in the interval

$$\varepsilon^* \le \varepsilon_{\mathrm{NN}} \le 2\varepsilon^*(1 - \varepsilon^*). \tag{5.15}$$

Clearly, the interval shrinks to zero and $\varepsilon_{\mathrm{NN}} = \varepsilon^*$ (so the NN rule is consistent) if $\varepsilon^* = 0$. The next theorem shows when the lower and the upper bounds in (5.15) are achieved in the general case.

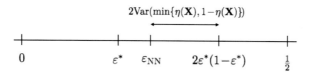

Figure 5.7: Relationship between asymptotic NN classification error ε_{NN} and the Bayes error ε^*.

Theorem 5.2. *The asymptotic classification error of the NN rule satisfies:*

(a) $\varepsilon_{\text{NN}} = 2\varepsilon^*(1 - \varepsilon^*)$ *if and only if* $\eta(\mathbf{X}) \in \{\varepsilon^*, 1 - \varepsilon^*\}$ *with probability 1.*

(b) $\varepsilon_{\text{NN}} = \varepsilon^*$ *if and only if* $\eta(\mathbf{X}) \in \{0, \frac{1}{2}, 1\}$ *with probability 1.*

Proof. From the proof of Theorem 5.1, we have

$$\varepsilon_{\text{NN}} = 2\varepsilon^*(1-\varepsilon^*) - 2\text{Var}(\min\{\eta(\mathbf{X}), 1-\eta(\mathbf{X})\}) . \tag{5.16}$$

See Figure 5.7 for an illustration. Hence, $\varepsilon_{\text{NN}} = 2\varepsilon^*(1-\varepsilon^*)$ if and only if $\text{Var}(\min\{\eta(\mathbf{X}), 1-\eta(\mathbf{X})\}) = 0$, which occurs if and only if $\min\{\eta(\mathbf{X}), 1 - \eta(\mathbf{X})\}$ is constant with probability 1, i.e., $\eta(\mathbf{X}) = a$ or $1 - a$, with probability 1, for some $0 \leq a \leq 1$. But then $\varepsilon^* = a$, showing part (a). Now, rewrite (5.16) as

$$\varepsilon_{\text{NN}} = 2\varepsilon^* - 2E[\min\{\eta(\mathbf{X}), 1 - \eta(\mathbf{X})\}^2] . \tag{5.17}$$

Let $r(\mathbf{X}) = \min\{\eta(\mathbf{X}), 1 - \eta(\mathbf{X})\}$. To obtain $\varepsilon_{\text{NN}} = \varepsilon^*$ we need $2E[r^2(\mathbf{X})] = \varepsilon^* = E[r(\mathbf{X})]$, i.e., $E[2r^2(\mathbf{X}) - r(\mathbf{X})] = 0$. But since $2r^2(\mathbf{X}) - r(\mathbf{X}) \geq 0$, this requires that $2r^2(\mathbf{X}) - r(\mathbf{X}) = 0$, i.e., $r(\mathbf{X}) \in \{0, \frac{1}{2}\}$, all with probability one, which shows part (b). $\diamond$

The previous theorem shows that the lower bound ε^* and upper bound $2\varepsilon^*(1 - \varepsilon^*)$ are achieved only in very special cases, but they are achieved; this shows that the inequality in the Cover-Hart Theorem is tight (it cannot be improved). The theorem also shows that the NN rule is consistent if and only if the class-conditional densities do not overlap ($\eta(\mathbf{X}) = 0$ or 1) or they are equal to each other if they overlap ($\eta(\mathbf{X}) = \frac{1}{2}$).

Theorem 5.3 can be generalized to the kNN classification rule, with odd $k > 1$, as the next theorem shows. See Appendix A4 for a proof.

Theorem 5.3. *The expected error of the kNN classification rule, with odd $k > 1$, satisfies*

$$\varepsilon_{\text{kNN}} = \lim_{n \to \infty} E[\varepsilon_n] = E[\alpha_k(\eta(\mathbf{X}))] , \tag{5.18}$$

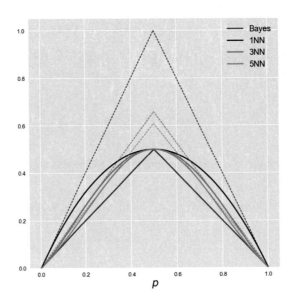

Figure 5.8: Plot of polynomial $\alpha_k(p)$ (solid line) and the function $a_k \min\{p, 1 - p\}$ (dashed line) in the interval $p \in [0,1]$, where $a_k > 1$ is the smallest possible. It can be observed that a_k is the slope of the tangent to $\alpha_k(p)$ through the origin. Also plotted is the function $\min\{p, 1 - p\}$ (red), corresponding to the Bayes error (plot generated by `c05_knn_thm.py`).

where $\alpha_k(p)$ is a polynomial of order $k + 1$ given by

$$\alpha_k(p) = \sum_{i=0}^{(k-1)/2} \binom{k}{i} p^{i+1}(1-p)^{k-i} + \sum_{i=(k+1)/2}^{k} \binom{k}{i} p^i (1-p)^{k+1-i}, \tag{5.19}$$

from which it follows that

$$\varepsilon_{\text{kNN}} \leq a_k \varepsilon^*, \tag{5.20}$$

where the constant $a_k > 1$ is the slope of the tangent line to $\alpha_k(p)$ through the origin, that is, a_k satisfies

$$a_k = \alpha_k'(p_0) = \frac{\alpha_k(p_0)}{p_0}, \tag{5.21}$$

for some $p_0 \in [0, \frac{1}{2}]$.

Notice that with $k = 1$, we have $\alpha_1(p) = 2p(1 - p)$, and

$$\alpha_1'(p_0) = 2 - 4p_0 = \frac{2p_0(1 - p_0)}{p_0} = \frac{\alpha_1(p_0)}{p_0} \Rightarrow p_0 = 0, \tag{5.22}$$

so that $a_1 = \alpha_1'(0) = 2$, which leads to (5.13) and (5.14) in the original Cover-Hart Theorem.

For $k = 3$, we have $\alpha_3(p) = p(1 - p)^3 + 6p^2(1 - p)^2 + p^3(1 - p)$, hence

$$\varepsilon_{3\text{NN}} = E[\eta(\mathbf{X})(1 - \eta(\mathbf{X}))^3] + 6E[\eta(\mathbf{X})^2(1 - \eta(\mathbf{X}))^2] + E[\eta(\mathbf{X})^3(1 - \eta(\mathbf{X}))] . \tag{5.23}$$

In addition, it can be checked that (5.21) requires p_0 to be a solution of

$$(p_0)^3 - \frac{4}{3}(p_0)^2 + \frac{1}{4}p_0 = 0 . \tag{5.24}$$

This has three solutions:

$$p_0^1 = 0 ,$$
$$p_0^2 = \frac{4 + \sqrt{7}}{6} , \tag{5.25}$$
$$p_0^3 = \frac{4 - \sqrt{7}}{6} .$$

The first candidate p_0^1 is invalid because the tangent is not above $\alpha_k(p)$, the second is invalid because $p_0^2 > 1$, while the last one is valid and gives:

$$a_3 = \alpha'_3\left(\frac{4 - \sqrt{7}}{6}\right) = \frac{17 + 7\sqrt{7}}{27} \approx 1.3156 . \tag{5.26}$$

Hence, $\varepsilon_{3\text{NN}} \leq 1.316\,\varepsilon^*$, which is better than the ε_{NN} bound: $a_1 = 2 > a_3 = 1.316$.

Notice that $\varepsilon_{k\text{NN}} = E[\alpha_k(\eta(\mathbf{X}))]$ has the general form of an F-error, defined in Section 2.6.2. However, as suggested by Figure 5.8, the function $\alpha_k(p)$ is concave in $p \in [0, 1]$ only for $k = 1$, so that *only ε_{NN} is an F-error*.

From Figure 5.8 it can also be deduced that $\alpha_1(p) \geq \alpha_3(p) \geq \alpha_5(p) \geq \cdots \geq \min\{p, 1 - p\}$, for all $p \in [0, 1]$, from which it follows that

$$\varepsilon_{\text{NN}} \geq \varepsilon_{3\text{NN}} \geq \varepsilon_{5\text{NN}} \geq \cdots \geq \varepsilon^* . \tag{5.27}$$

In fact, it can be shown that $\varepsilon_{k\text{NN}} \to \varepsilon^*$ as $k \to \infty$.

*5.6　Stone's Theorem

Given the distribution-free nature of nonparametric classification rules, it is not surprising that they make ideal candidates for universally consistent rules. However, if the amount of smoothing is fixed and not a function of sample size, then the rule cannot be universally consistent, as it was proved in Section 5.3 with regards to a kNN classification rule with a fixed number of nearest neighbors k.

Stone's theorem was the first result to prove that universally consistent rules exist, by showing that nonparametric classification rules of the form (5.3) can achieve consistency under any feature-label distribution, provided that the weights are carefully adjusted as sample size increases. We state Stone's theorem (see Appendix A5 for a proof) and give some of its corollaries, which provide results on the universal consistency of nonparametric classification rules examined previously.

Stone's theorem is in fact a result about consistency of *nonparametric regression*, not nonparametric classification. However, it turns out that it can be applied to nonparametric classification through the following lemma, the proof of which mirrors the steps followed in Exercise 4.6.

Lemma 5.1. *The nonparametric classification rule specified by (5.1)–(5.3) is consistent if* $E[|\eta_n(\mathbf{X}) - \eta(\mathbf{X})|] \to 0$, *as* $n \to \infty$.

Since $E[X]^2 \leq E[X^2]$ (as $\mathrm{Var}(X) = E[(X - E[X])^2] = E[X^2] - E[X]^2 \geq 0$), it turns out that a sufficient condition for consistency is that $E[(\eta_n(\mathbf{X}) - \eta(\mathbf{X}))^2] \to 0$, as $n \to \infty$. In other words, consistent regression estimation of $\eta(x)$, in the L^1 or L^2 norms, leads to consistent classification.

Theorem 5.4. *(Stone's Theorem) The classification rule specified by (5.1)–(5.3) is universally consistent, provided that the weights have the following properties, for all distributions of* $(\mathbf{X}, Y)$:

(i) $\sum_{i=1}^{n} W_{n,i}(\mathbf{X}) I_{||\mathbf{X}_i - \mathbf{X}|| > \delta} \xrightarrow{P} 0$, *as* $n \to \infty$, *for all* $\delta > 0$,

(ii) $\max_{i=1,\ldots,n} W_{n,i}(\mathbf{X}) \xrightarrow{P} 0$, *as* $n \to \infty$,

(iii) *there is a constant* $c \geq 1$ *such that, for every nonnegative integrable function* $f : R^d \to R$ *and all* $n \geq 1$, $E\left[\sum_{i=1}^{n} W_{n,i}(\mathbf{X}) f(\mathbf{X}_i)\right] \leq cf(\mathbf{X})$.

Condition (i) in Stone's Theorem says that the weights outside a ball of radius δ around the test point $\mathbf{X}$ vanish as $n \to \infty$, for all $\delta > 0$. This means that estimation becomes more and more local as sample size increases. Condition (ii) says that the weights go uniformly to zero so that no individual training point can dominate the inference. Finally, condition (iii) is a technical assumption that is required in the proof of the theorem.

The following result about the histogram classification rule can be proved by checking the conditions in Stone's theorem. See Devroye et al. [1996] for a direct proof.

Theorem 5.5. *(Universal consistence of the Histogram Classification Rule.) Let* A_n *be a sequence of partitions and let* $N_n(\mathbf{x})$ *be the number of training points in the zone* $A(\mathbf{x})$. *If*

(i) $\mathrm{diam}[A_n(\mathbf{X})] = \sup_{\mathbf{x}, \mathbf{y} \in A_n(\mathbf{X})} ||\mathbf{x} - \mathbf{y}|| \to 0$ *in probability,*

(ii) $N_n(\mathbf{X}) \to \infty$ *in probability,*

then $E[\varepsilon_n] \to \varepsilon^*$ *and the histogram classification rule is universally consistent.*

Conditions (i) and (ii) of the previous theorem correspond to conditions (i) and (ii) of Stone's Theorem, respectively. This application of Stone's Theorem illuminates the issue: condition (i) says that the zone of influence of the training data around the test point must shrink to zero, while condition (ii) says that this must happen slow enough that an infinite number of training points is allowed to accumulate inside the zone. These two conditions are natural conditions to obtain an accurate and universal estimator of the posterior-probability function. Condition (iii) of Stone's Theorem, necessary for its proof, can also be shown to hold.

As a corollary of the previous theorem, we get universal consistency of the cubic histogram rule, in which case the partitions are indexed by the hypercube side h_n. The proof of the result consists of checking conditions (i) and (ii) of Theorem 5.5.

Theorem 5.6. *(Universal consistence of the Cubic Histogram Rule.) Let $V_n = h_n^d$ be the common volume of all cells. If $h_n \to 0$ (so $V_n \to 0$) but $nV_n \to \infty$ as $n \to \infty$, then $E[\varepsilon_n] \to \varepsilon^*$ and the cubic histogram rule is universally consistent.*

One of the main achievements in Stone's paper was proving the universal consistency of the kNN classification rule, for a number of different weight families, including the uniform weights of the standard kNN rule.

Theorem 5.7. *(Universal consistence of the kNN Rule.) If $K \to \infty$ while $K/n \to 0$ as $n \to \infty$, then for all distributions $E[\varepsilon_n] \to \varepsilon^*$ and the kNN classification rule is universally consistent.*

The proof of the previous theorem is based again on checking the conditions of Theorem 5.5, though this turns out to be a nontrivial task. Detailed proofs can be found in Stone [1977] and in Devroye et al. [1996].

In the case of kernel rules, direct application of Stone's Theorem is problematic. If the class-conditional densities exist then any consistent kernel density estimator leads to a consistent rule. The following result is proved in Devroye et al. [1996] and does not require any assumptions on the distributions (only on the kernel).

Theorem 5.8. *(Universal consistence of Kernel Rules.) If the kernel k is nonnegative, uniformly continuous, integrable, and bounded away from zero in a neighborhood of the origin, and $h_n \to 0$ with $nh_n^d \to \infty$ as $n \to \infty$, then the kernel rule is (strongly) universally consistent.*

5.7 Bibliographical Notes

Nonparametric classification methods comprise some of the most classical work done in supervised learning. The area was inaugurated by the pioneering work by Fix and Hodges [1951] on the k-

nearest-neighbor classification rule, which stated: "There seems to be a need for discrimination procedures whose validity does not require the amount of knowledge implied by the normality assumption, the homoscedastic assumption, or any assumption of parametric form." Other milestones include the paper by Cover and Hart [1967] on the asymptotic error rate of the kNN rule (where Theorem 5.1 appears), and the paper by Stone [1977] on universal consistency of nonparametric rules. Nonparametric classification is covered in detail in Devroye et al. [1996].

5.8   Exercises

5.1. Consider that an experimenter wants to use A 2-D cubic histogram classification rule, with square cells with side length h_n, and achieve consistency as the sample size n increases, for any possible distribution of the data. If the experimenter lets h_n decrease as $h_n = \frac{1}{\sqrt{n}}$, would they be guaranteed to achieve consistency and why? If not, how would they need to modify the rate of decrease of h_n to achieve consistency?

5.2. Consider that an experimenter wants to use the kNN classification rule and achieve consistency as the sample size n increases. In each of the following alternatives, answer whether the experimenter is successful and why.

   (a) The experimenter does not know the distribution of (X, Y) and lets k increase as $k = \sqrt{n}$.

   (b) The experimenter does not know the distribution but knows that $\epsilon^* = 0$ and keeps k fixed, $k = 3$.

5.3. Show by example that the Hermite kernel can lead to a kernel classification rule that has a strictly smaller classification error than any positive kernel.
   Hint: Consider a simple discrete univariate distribution with equal masses at $x = 0$ and $x = 2$, with $\eta(0) = 0$ and $\eta(1) = 1$.

5.4. Extend the Cover-Hart theorem to M classes, proving that

$$\varepsilon^* < \varepsilon_{\mathrm{NN}} < \varepsilon^* \left(2 - \frac{M}{M-1} \varepsilon^* \right). \tag{5.28}$$

Hint: The main difference with respect to the proof of Theorem 5.1 is the way that the Bayes error ε^* and the conditional error rate $P(\psi_n(\mathbf{X}) \neq Y \mid \mathbf{X}, \mathbf{X}_1, \ldots, \mathbf{X}_n)$ are written in terms of the posterior-probability functions $\eta_i(\mathbf{X}) = P(Y = i \mid \mathbf{X})$, for $i = 1, \ldots, M$. Use the inequality

$$L \sum_{l=1}^{L} a_l^2 \geq \left(\sum_{l=1}^{L} a_l \right)^2. \tag{5.29}$$

5.5. Show that $\varepsilon_{kNN} \leq \tilde{\alpha}_k(\varepsilon^*)$, where $\tilde{\alpha}_k$ is the *least concave majorant* of α_k in the interval $[0,1]$, that is, $\tilde{\alpha}_k$ is the smallest concave function such that $\tilde{\alpha}_k(p) \geq \alpha_k(p)$, for $0 \leq p \leq 1$.
Note: It is clear that $\tilde{\alpha}_1 = \alpha_1$, giving the usual inequality $\varepsilon_{NN} \leq 2\varepsilon^*(1 - \varepsilon^*)$. For odd $k \geq 3$, $\tilde{\alpha}_k \neq \alpha_k$.

5.6. Assume that the feature X in a classification problem is a real number in the interval $[0,1]$. Assume that the classes are equally likely, with $p(x|Y = 0) = 2x I_{\{0 \leq x \leq 1\}}$ and $p(x|Y = 1) = 2(1 - x) I_{\{0 \leq x \leq 1\}}$.

  (a) Find the Bayes error ε^*.

  (b) Find the asymptotic error rate ε_{NN} for the NN classification rule.

  (c) Find $E[\varepsilon_n]$ for the NN classification rule for finite n.

  (d) Show that indeed $E[\varepsilon_n] \to \varepsilon_{NN}$ and that $\varepsilon^* < \varepsilon_{NN} < 2\varepsilon^*(1 - \varepsilon^*)$.

5.7. Consider a problem in R^2 with equally-likely classes, such that $p(\mathbf{x} \mid Y = 0)$ is uniform over a unit-radius disk centered at $(-3,0)$ and $p(\mathbf{x} \mid Y = 1)$ is uniform over a unit-radius disk centered at $(3,0)$. Since the class-conditional densities do not overlap, $\varepsilon^* = 0$.

  (a) Show that the expected error of the 1NN classification rule is strictly smaller than that of any kNN classification rule with $k > 1$. This shows that, despite its often poor performance in practice, the 1NN classification rule is not universally dominated by nearest neighbor rules with larger k.
  Hint: Remember that the expected classification error is just the unconditional probability $P(\psi_n(\mathbf{X}) \neq Y)$.

  (b) Show that, for fixed k, $\lim_{n \to \infty} E[\varepsilon_n] = 0$.
  Hint: See Theorem 5.3.

  (c) Show that, for fixed n, $\lim_{k \to \infty} E[\varepsilon_n] = 1/2$. This shows overfitting as k becomes much larger than n.

  (d) Show that, if k_n is allowed to vary, $\lim_{n \to \infty} E[\varepsilon_n] = 0$, if $k_n < n$ for all n.

5.9 Python Assignments

5.8. This assignment concerns Example 5.2.

  (a) Modify the code in `c05_cubic.py` to obtain plots for $l = 0.3, 0.5, 1, 2, 3, 6$, and $n = 50, 100, 250, 500$ per class. In addition, obtain and plot the classifiers over the range $[-6, 9] \times [-6, 9]$ in order to visualize the entire data. To facilitate visualization, you may want to reduce the marker size from 12 to 8. Which classifiers are closest to the optimal classifier? How do you explain this in terms of underfitting/overfitting?

Coding hint: In order to generate the same training data as in Figure 5.2 (with $n = 50$), the data for class 1 has to be simulated immediately following that for class 0. This can be accomplished, using python's list comprehension feature, by means of the code snippet

```
N = [50,100,250,500]
X = [[mvn.rvs(mm0,Sig,n),mvn.rvs(mm1,Sig,n)] for n in N]
```

The various data sets can be accessed by looping through this list.

(b) Under the assumption of equally-likely classes, obtain an estimate of the error rate of each of the 24 classifiers in part (a), by testing them on a large independent data set of size $M = 500$ per class (you may use the same test set for all classifiers). The estimate is the total number of errors made by each classifier on the test set divided by $2M$ (this test-set error estimate is close to the true classification error due to the large test sample size, as will be seen in Chapter 7). Also, compute the Bayes error using (2.50) and (2.53) (the test set is not necessary in this case and should not be used). Generate a table containing each classifier plot in part (a) with its test set error rate. Which combinations of sample size and grid size produce the top 5 smallest error rates?

(c) If plotted as a function of sample size, the classification errors in part (b) produce jagged curves, since they correspond to a single realization of the random error rate. The proper way of studying the effect of sample size is to consider the expected error rate instead, which can be estimated by repeating the experiment for each value of n and l for a total of R times and averaging the error rates. Plot these estimates of the expected classification error rates, in the same plot, as a function of sample size n for each value of l, with $R = 50$. Do the expected error rates approach the Bayes error and how do they compare to each other? Now, plot the expected error rates, in the same plot, as a function of l for each value of sample size. What value of l would you select for each sample size? (This is a model selection problem.) How do you interpret your conclusions in light of Theorem 5.6?

(d) Repeat parts (a)–(c) using a histogram classification rule that breaks ties randomly with equal probabilities to class 0 and 1. Does this significantly change the results?

Coding hint: The command `rnd.binomial(1,0.5)` generates a Bernoulli random variable with equal probabilities. In order to generate the same training and testing data sets as in parts (a)–(c), set the same random seed and generate the data sets at the top of the code.

5.9. This assignment concerns Example 5.3.

(a) Modify the code in `c05_knn.py` to obtain plots for $k = 1, 3, 5, 7, 9, 11$ and $n = 50, 100, 250, 500$ per class. Plot the classifiers over the range $[-3, 9] \times [-3, 9]$ in order to visualize the entire data and reduce the marker size from 12 to 8 to facilitate visualization. Which classifiers are closest to the optimal classifier? How do you explain this in terms of underfitting/overfitting? See the coding hint in part (a) of Problem 5.8.

(b) Compute test set errors for each classifier in part (a), using the same procedure as in part (b) of Problem 5.8. Generate a table containing each classifier plot in part (a) with its test set error rate. Which combinations of sample size and number of neighbors produce the top 5 smallest error rates?

Coding hint: the `score` method of `sklearn.neighbors.KNeighborsClassifier` returns the accuracy of the fitted classifier on an input data set.

(c) Compute expected error rates for the kNN classification rules in part (a), using the same procedure as in part (c) of Problem 5.8. Since error computation is faster here, a larger value $R = 200$ can be used, for better estimation of the expected error rates. Which number of neighbors should be used for each sample size?

(d) Repeat parts (a)–(c) using the L^1 distance $d(\mathbf{x}_0, \mathbf{x}_1) = \sum_{i=1}^{d} |x_{1i} - x_{0i}|$ to calculate nearest neighbors. Do the results change significantly?

Coding hint: change the `metric` attribute of `sklearn.neighbors.KNeighborsClassifier` to "manhattan" (the L^1 distance).

5.10. This assignment concerns Example 5.4.

(a) Modify the code in `c05_kernel.py` to obtain plots for $k = 1, 3, 5, 7, 9, 11$ and $n = 50, 100, 250, 500$ per class. Plot the classifiers over the range $[-3, 9] \times [-3, 9]$ in order to visualize the entire data and reduce the marker size from 12 to 8 to facilitate visualization. Which classifiers are closest to the optimal classifier? How do you explain this in terms of underfitting/overfitting? See the coding hint in part (a) of Problem 5.8.

(b) Compute test set errors for each classifier in part (a), using the same procedure as in part (b) of Problem 5.8. Generate a table containing each classifier plot in part (a) with its test set error rate. Which combinations of sample size and kernel bandwidth produce the top 5 smallest error rates?

(c) Compute expected error rates for the Gaussian kernel classification rule in part (a), using the same procedure as in part (c) of Problem 5.8. Since error computation is faster here, a larger value $R = 200$ can be used, for better estimation of the expected error rates. Which kernel bandwidth should be used for each sample size?

(d) Repeat parts (a)–(c) for the Epanechnikov kernel, which, unlike the Gaussian kernel, is a bounded support kernel. How do the results change?

Coding hint: change the `kernel` attribute of `sklearn.neighbors.KernelDensity` to "epanechnikov".

5.11. (Nonlinearly separable data.) The optimal decision boundary in the previous coding assignments is linear. However, nonparametric rules are most useful in nonlinear problems. Repeat Problems 5.8–5.10 with data generated from mixtures of Gaussians, where the mixture for class 0 has centers at $(2, 2)$ and $(4, 4)$, and the mixture for class 1 has centers at $(2, 4)$ and $(4, 2)$. Assume that all covariance matrices are spherical with variance equal to 4.

Chapter 6

Function-Approximation Classification

"That is why no reasonable scientist has ever claimed to know the ultimate cause of any natural process, or to show clearly and in detail what goes into the causing of any single effect in the universe These ultimate sources and principles are totally hidden from human enquiry."
–David Hume, *Enquiry Concerning Human Understanding*, 1748.

All the classification rules seen so far were plug-in rules, that is, they could be viewed as distribution estimation using training data. We consider now a different idea: iteratively adjusting a discriminant (decision boundary) to the training data by optimizing an error criterion. This is reminiscent in some ways to the process of human *learning*. Not surprisingly, this is the basic idea behind many popular classification rules: support vector machines, neural networks, decision trees, and rank-based classifiers. We examine each of these classification rules in this chapter.

6.1 Support Vector Machines

Rosenblatt's Perceptron, invented in the late 50's, was the first function-approximation classification rule. It assumes a linear discriminant function

$$g_n(x) = a_0 + \sum_{i=1}^{d} a_i x_i, \tag{6.1}$$

© Springer Nature Switzerland AG 2020
U. Braga-Neto, *Fundamentals of Pattern Recognition and Machine Learning*,
https://doi.org/10.1007/978-3-030-27656-0_6

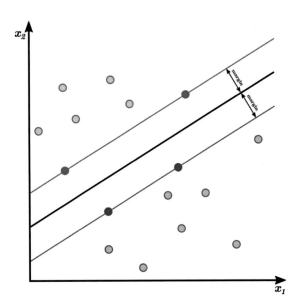

Figure 6.1: Sample data with indicated maximum-margin hyperplane, margin hyperplanes, and support vectors, in the case $d = 2$. The decision boundary is a function only of the support vectors.

so that the designed classifier is given by

$$\psi_n(\mathbf{x}) = \begin{cases} 1\,, & \text{if } a_0 + \sum a_i x_i \geq 0, \\ 0\,, & \text{otherwise,} \end{cases} \qquad (6.2)$$

for $\mathbf{x} \in R^d$. The *perceptron algorithm* iteratively adjusts ("learns") the parameters $a_0, a_1, \ldots, a_d$ in order to maximize an empirical criterion related to the performance of ψ_n on the training data.

Now, consider a version of the perceptron algorithm that attempts to adjust the linear discriminant so that the *margin*, i.e. the distance between the separating hyperplane and the closest data points, is maximal. The separating hyperplane in this case is called a *maximum-margin hyperplane* and the closest data points, which sit on the *margin hyperplanes*, are the *support vectors*. The distance between the maximum-margin hyperplane and each margin hyperplane is called the *margin*. See Figure 6.1 for an illustration in the case $d = 2$. Notice that the maximum-margin hyperplane is entirely determined by the support vectors (e.g., moving other data points does not alter the hyperplane, as long as they are further away from it than the margin). The resulting classification rule is called a linear *Support Vector Machine* (SVM). This section discusses a few different SVM classification rules, both linear and nonlinear, which are all based on this simple idea. We will see that nonlinear SVMs are nonparametric kernel classifiers, albeit of a peculiar sort.

6.1.1 Linear SVMs for Separable Data

In this section, we describe an algorithm for determining the maximum-margin hyperplane. We will assume, for now, that the data is linearly separable, as in Figure 6.1 — this assumption will be relaxed in Section 6.1.2.

Consider the linear classifier in (6.2). If we change our convention and assign labels $y_i = -1$ and $y_i = 1$ to the classes, then it is clear that each training point is correctly classified if:

$$y_i(\mathbf{a}^T \mathbf{x}_i + a_0) > 0, \quad i = 1, \ldots, n, \tag{6.3}$$

where $\mathbf{a} = (a_1, \ldots, a_d)$. In other to impose a margin, we change the constraint to

$$y_i(\mathbf{a}^T \mathbf{x}_i + a_0) \geq b, \quad i = 1, \ldots, n, \tag{6.4}$$

where $b > 0$ is a parameter to be determined. It is possible to show (see Exercise 6.1) that the constraint (6.4) is satisfied if and only if all training points points at a distance at least $b/||\mathbf{a}||$ from the decision hyperplane. Since the parameters $\mathbf{a}$, a_0 and b can be freely scaled without altering the constraint, they are not identifiable. In order to fix this, we can arbitrarily set $b = 1$, and the constraint becomes:

$$y_i(\mathbf{a}^T \mathbf{x}_i + a_0) \geq 1, \quad i = 1, \ldots, n. \tag{6.5}$$

With $b = 1$, the margin becomes $1/||\mathbf{a}||$, and the points that are at this exact distance from the hyperplane are the margin vectors. (We will determine in the sequel the condition that needs to be satisfied for them to be support vectors).

In order to obtain the maximal-margin hyperplane, we need to maximize $1/||\mathbf{a}||$, which is equivalent to minimizing $||\mathbf{a}||^2 = \frac{1}{2}\mathbf{a}^T\mathbf{a}$. The optimization problem to be solved is therefore:

$$\begin{aligned} \min \quad & \frac{1}{2}\mathbf{a}^T\mathbf{a} \\ \text{s.t.} \quad & y_i(\mathbf{a}^T\mathbf{x}_i + a_0) \geq 1, \quad i = 1, \ldots, n. \end{aligned} \tag{6.6}$$

This is a *convex optimization problem* in $(\mathbf{a}, a_0)$, with a convex cost function affine inequality constraints. The KKT conditions are necessary and sufficient for an optimal solution in this case. See Section A3 for a review of the required optimization results.

The *primal Lagrangian functional* codes the constraints into the cost functional as follows:

$$L_P(\mathbf{a}, a_0, \boldsymbol{\lambda}) = \frac{1}{2}\mathbf{a}^T\mathbf{a} - \sum_{i=1}^{n} \lambda_i \left(y_i(\mathbf{a}^T\mathbf{x}_i + a_0) - 1\right), \tag{6.7}$$

where $\lambda_i \geq 0$ is the Langrage multiplier for the ith constraint (data point), $i = 1, \ldots, n$, and $\boldsymbol{\lambda} = (\lambda_1, \ldots, \lambda_n)$. As shown in Section A3, a solution $(\mathbf{a}^*, a_0^*, \boldsymbol{\lambda}^*)$ of the previous constrained problem simultaneously minimizes L_p with respect to $(\mathbf{a}, a_0)$ and maximizes it with respect to $\boldsymbol{\lambda}$, i.e., we search for a *saddle point* in L_p.

The KKT stationarity condition demands that the gradient of L_p with respect to $(\mathbf{a}, a_0)$ be zero, which yields the equations:

$$\mathbf{a} = \sum_{i=1}^{n} \lambda_i y_i \mathbf{x}_i \,, \tag{6.8}$$

and

$$\sum_{i=1}^{n} \lambda_i y_i = 0 \,. \tag{6.9}$$

Substituting these back into L_p eliminates $\mathbf{a}$ and a_0, yielding the *dual Lagrangian functional*:

$$L_D(\boldsymbol{\lambda}) = \sum_{i=1}^{n} \lambda_i - \frac{1}{2} \sum_{i=1}^{n} \sum_{j=1}^{n} \lambda_i \lambda_j y_i y_j \mathbf{x}_i^T \mathbf{x}_j \,. \tag{6.10}$$

The functional $L_D(\boldsymbol{\lambda})$ must be maximized with respect to λ_i subject to the constraints:

$$\lambda_i \geq 0 \quad \text{and} \quad \sum_{i=1}^{n} \lambda_i y_i = 0 \,. \tag{6.11}$$

This is a quadratic programming problem that can be efficiently solved numerically.

According to the KKT complementary slackness condition, $\lambda_i^* = 0$ whenever

$$y_i((\mathbf{a}^*)^T \mathbf{x}_i + a_0^*) > 1 \text{ (inactive or slack constraint)}. \tag{6.12}$$

A *support vector* is a training point $(\mathbf{x}_i, y_i)$ such that $\lambda_i^* > 0$, i.e., for which

$$y_i((\mathbf{a}^*)^T \mathbf{x}_i + a_0^*) = 1 \text{ (active or tight constraint)}. \tag{6.13}$$

The degenerate case where the constraint is active but $\lambda_i^* = 0$ does not violate the complementary slackness condition and can happen; this would be a point that is just touching the margin but is not constraining the solution.

Once $\boldsymbol{\lambda}^*$ is found, the solution vector $\mathbf{a}^*$ is determined by (6.8):

$$\mathbf{a}^* = \sum_{i=1}^{n} \lambda_i^* y_i \mathbf{x}_i = \sum_{i \in \mathcal{S}} \lambda_i^* y_i \mathbf{x}_i \,, \tag{6.14}$$

where $\mathcal{S} = \{i \mid \lambda_i^* > 0\}$ is the set of support vector indices. Notice that the larger λ_i^* is, the larger the influence of the corresponding support vector on the direction vector $\mathbf{a}^*$ is. The intercept a_0^* can be determined from any of the support vectors, since the constraint $(\mathbf{a}^*)^T \mathbf{x}_i + a_0^* = y_i$ is active, or, for better numerical accuracy, from their sum:

$$|\mathcal{S}| a_0^* + (\mathbf{a}^*)^T \sum_{i \in \mathcal{S}} \mathbf{x}_i = \sum_{i \in \mathcal{S}} y_i \,, \tag{6.15}$$

which yields:

$$a_0^* = -\frac{1}{|\mathcal{S}|} \sum_{i \in \mathcal{S}} \sum_{j \in \mathcal{S}} \lambda_i^* y_i \mathbf{x}_i^T \mathbf{x}_j + \frac{1}{|\mathcal{S}|} \sum_{i \in \mathcal{S}} y_i \,. \tag{6.16}$$

Collecting all the results, the maximum-margin hyperplane classifier is given by

$$\psi_n(\mathbf{x}) = \begin{cases} 1, & \text{if } \sum_{i \in \mathcal{S}} \lambda_i^* y_i \mathbf{x}_i^T \mathbf{x} - \frac{1}{|\mathcal{S}|} \sum_{i \in \mathcal{S}} \sum_{j \in \mathcal{S}} \lambda_i^* y_i \mathbf{x}_i^T \mathbf{x}_j + \frac{1}{|\mathcal{S}|} \sum_{i \in \mathcal{S}} y_i > 0\,, \\ 0, & \text{otherwise.} \end{cases} \tag{6.17}$$

Despite the apparent complexity of the expression, this is a linear classifier, with a hyperplane decision boundary. Notice that the classifier is a function only of the support vectors and no other training points. Furthermore, it is a function of inner products of the form $\mathbf{x}^T \mathbf{x}'$, a fact that will be important later.

6.1.2 General Linear SVMs

The algorithm given in the previous section is simple, but tends to be unusable in practice, since it is not possible to guarantee that the training data will be linearly separable (unless the Bayes error is zero and the class-conditional densities are linearly separable). Luckily, it is possible to modify the basic algorithm to allow for nonlinearly separable data. This leads to the general linear SVM classification rule described in this section.

To handle nonlinearly separable data, one introduces a vector of nonnegative *slack variables* $\boldsymbol{\xi} = (\xi_1, \ldots, \xi_n)$, one slack for each of the constraints, resulting in a new set of $2n$ constraints:

$$y_i(\mathbf{a}^T \mathbf{x}_i + a_0) \geq 1 - \xi_i \text{ and } \xi_i \geq 0, \quad i = 1, \ldots, n\,. \tag{6.18}$$

If $\xi_i > 0$, the corresponding training point is an *outlier*, i.e., it can lie closer to the hyperplane than the margin, or even be misclassified. In order to keep slackness under control, one introduces a penalty term $C \sum_{i=1}^n \xi_i$ in the functional, which then becomes:

$$\frac{1}{2} \mathbf{a}^T \mathbf{a} + C \sum_{i=1}^n \xi_i\,. \tag{6.19}$$

The constant C modulates how large the penalty for the presence of outliers are. If C is small, the penalty is small and a solution is more likely to incorporate outliers. If C is large, the penalty is large and therefore a solution is unlikely to incorporate many outliers. This means that small C favors a *soft margin* and therefore less overfitting, whereas large C leads to a *hard margin* and more overfitting. In summary, the amount of overfitting is directly proportional to the magnitude of C. Too small a C on the other hand may lead to *underfitting*, that is, too much slackness is allowed and the classifier does not fit the data at all.

The optimization problem is now

$$
\begin{aligned}
\min \quad & \frac{1}{2}\mathbf{a}^T\mathbf{a} + C\sum_{i=1}^{n}\xi_i \\
\text{s.t.} \quad & y_i(\mathbf{a}^T\mathbf{x}_i + a_0) \geq 1 - \xi_i, \quad i = 1, \ldots, n \\
& \xi_i \geq 0, \quad i = 1, \ldots, n.
\end{aligned}
\tag{6.20}
$$

This is a convex problem in $(\mathbf{a}, a_0, \boldsymbol{\xi})$ with affine constraints. We follow a similar solution method as in the separable case. This time there are $2n$ constraints, so there is a total of $2n$ Lagrange multipliers: $\boldsymbol{\lambda} = (\lambda_1, \ldots, \lambda_n)$ and $\boldsymbol{\rho} = (\rho_1, \ldots, \rho_n)$. The primal functional can then be written as:

$$
\begin{aligned}
L_P(\mathbf{a}, a_0, \boldsymbol{\xi}, \boldsymbol{\lambda}, \boldsymbol{\rho}) &= \frac{1}{2}\mathbf{a}^T\mathbf{a} + C\sum_{i=1}^{n}\xi_i - \sum_{i=1}^{n}\lambda_i\left(y_i(\mathbf{a}^T\mathbf{x}_i + a_0) - 1 + \xi_i\right) - \sum_{i=1}^{n}\rho_i\xi_i \\
&= \frac{1}{2}\mathbf{a}^T\mathbf{a} - \sum_{i=1}^{n}\lambda_i\left(y_i(\mathbf{a}^T\mathbf{x}_i + a_0) - 1\right) + \sum_{i=1}^{n}(C - \lambda_i - \rho_i)\xi_i.
\end{aligned}
\tag{6.21}
$$

This is equal to the previous primal functional (6.7) plus an extra term $\sum_{i=1}^{n}(C - \lambda_i - \rho_i)\xi_i$. Setting the derivatives of L_p with respect to $\mathbf{a}$ and a_0 to zero yields therefore the same equations (6.8) and (6.9) as in the separable case. Setting the derivatives of L_p with respect to ξ_i to zero yields the additional equations

$$
C - \lambda_i - \rho_i = 0, \quad i = 1, \ldots, n.
\tag{6.22}
$$

Substituting these equations back into L_P clearly leads to the same quadratic dual Lagrangian functional (6.10), which must be maximized with respect to $\boldsymbol{\lambda}$ under the same constraints (6.11), plus the additional constraints:

$$
\lambda_i \leq C, \quad i = 1, \ldots, n,
\tag{6.23}
$$

which follow from (6.22) and the nonnegativity condition $\rho_i \geq 0$, for $i = 1, \ldots, n$. The solution $(\boldsymbol{\lambda}^*, \boldsymbol{\rho}^*)$ can be obtained by quadratic programming methods, as before.

An outlier is a support vector $(\mathbf{x}_i, y_i)$ for which

$$
\xi_i^* > 0 \Rightarrow \rho_i^* = 0 \Rightarrow \lambda_i^* = C.
\tag{6.24}
$$

If $0 < \lambda_i^* < C$, then $(\mathbf{x}_i, y_i)$ is a regular support vector, i.e., it lies on one of the margin hyperplanes. We refer to these support vectors as *margin vectors*.

Notice that the separable case corresponds to $C = \infty$. In that case, no outlier is possible and all support vectors are margin vectors (the case $\lambda_i = C$ is not possible). For finite C, outliers are possible. If C is small, more of the constraints $\lambda_i \leq C$ may be active (i.e., $\lambda_i = C$), so that there can be more outliers. If C is large, the opposite happens and there are fewer outliers. This is in agreement with the conclusion that C controls overfitting.

The solution vector a^* is given, as before, by

$$a^* = \sum_{i=1}^{n} \lambda_i^* y_i \mathbf{x}_i = \sum_{i \in \mathcal{S}} \lambda_i^* y_i \mathbf{x}_i . \tag{6.25}$$

The intercept a_0^* can be determined from any of the active constraints $(a^*)^T \mathbf{x}_i + a_0^* = y_i$ with $\xi_i^* = 0$, that is, the constraints for which $0 < \lambda_i^* < C$ (the margin vectors), or from their sum:

$$|\mathcal{S}_m| a_0^* + (a^*)^T \sum_{i \in \mathcal{S}_m} \mathbf{x}_i = \sum_{i \in \mathcal{S}_m} y_i , \tag{6.26}$$

where $\mathcal{S}_m = \{ i \,| 0 < \lambda_i^* < C \} \subseteq \mathcal{S}$ is the margin vector index set, resulting in

$$a_0^* = -\frac{1}{|\mathcal{S}_m|} \sum_{i \in \mathcal{S}} \sum_{j \in \mathcal{S}_m} \lambda_i^* y_i \mathbf{x}_i^T \mathbf{x}_j + \frac{1}{|\mathcal{S}_m|} \sum_{i \in \mathcal{S}_m} y_i . \tag{6.27}$$

The general linear SVM classifier is thus given by:

$$\psi_n(\mathbf{x}) = \begin{cases} 1, & \text{if } \sum_{i \in \mathcal{S}} \lambda_i^* y_i \mathbf{x}_i^T \mathbf{x} - \frac{1}{|\mathcal{S}_m|} \sum_{i \in \mathcal{S}} \sum_{j \in \mathcal{S}_m} \lambda_i^* y_i \mathbf{x}_i^T \mathbf{x}_j + \frac{1}{|\mathcal{S}_m|} \sum_{i \in \mathcal{S}_m} y_i > 0, \\ 0, & \text{otherwise.} \end{cases} \tag{6.28}$$

Once again, this classifier is only a function of the support vectors and inner products of the form $\mathbf{x}^T \mathbf{x}'$.

6.1.3   Nonlinear SVMs

The idea behind nonlinear SVMs is to apply the general algorithm in the previous section in a transformed space R^p, with $p > d$. If p is sufficiently large, then the data can be made linearly separable (or close to) in the high-dimensional space. If $\phi : R^d \to R^p$ denotes the transformation, then the previous derivation goes through unchanged with $\phi(\mathbf{x})$ in place of $\mathbf{x}$ everywhere. Thus, the nonlinear SVM classifier in the original space R^d is given by

$$\psi_n(\mathbf{x}) = \begin{cases} 1, & \text{if } \sum_{i \in \mathcal{S}} \lambda_i^* y_i \phi(\mathbf{x}_i)^T \phi(\mathbf{x}) - \frac{1}{|\mathcal{S}_m|} \sum_{i \in \mathcal{S}} \sum_{j \in \mathcal{S}_m} \lambda_i^* y_i \phi(\mathbf{x}_i)^T \phi(\mathbf{x}_j) + \frac{1}{|\mathcal{S}_m|} \sum_{i \in \mathcal{S}_m} y_i > 0, \\ 0, & \text{otherwise.} \end{cases}$$
$$\tag{6.29}$$

This produces in general a nonlinear decision boundary.

This classification rule would not be that useful if one had to compute the high-dimensional transformation explicitly. Luckily, this is not the case, due to the so-called *Kernel Trick*. Let us introduce a kernel function:

$$k(\mathbf{x}, \mathbf{x}') = \phi^T(\mathbf{x}) \phi(\mathbf{x}'), \quad \mathbf{x}, \mathbf{x}' \in R^d . \tag{6.30}$$

Such kernels differ from the ones in Chapter 5 by being functions of two points in feature space. (These kernels will appear again in Chapter 11.) The kernel trick consists in avoiding computing $\phi(\mathbf{x})$ completely by using $k(\mathbf{x}, \mathbf{x}')$. This is possible because the linear SVM classifier (6.28) is only a function of inner products of the form $\mathbf{x}^T \mathbf{x}'$ and hence the nonlinear SVM classifier (6.29) is only a function of quantities $k(\mathbf{x}, \mathbf{x}') = \phi(\mathbf{x})^T \phi(\mathbf{x}')$. In order to determine the support vectors and associated Lagrange multipliers, one needs to maximize the dual functional:

$$L_D(\lambda) = \sum_{i=1}^{n} \lambda_i - \frac{1}{2} \sum_{i=1}^{n} \sum_{j=1}^{n} \lambda_i \lambda_j y_i y_j k(\mathbf{x}_i, \mathbf{x}_j), \tag{6.31}$$

with respect to $\boldsymbol{\lambda}$, subject to the constraints

$$0 \le \lambda_i \le C \quad \text{and} \quad \sum_{i=1}^{n} \lambda_i y_i = 0. \tag{6.32}$$

Once again, this is possible because the original dual functional (6.10) is only a function of terms $\mathbf{x}_i^T \mathbf{x}_j$.

Once $\boldsymbol{\lambda}^*$ is found, the nonlinear SVM classifier is given by:

$$\psi_n(\mathbf{x}) = \begin{cases} 1, & \text{if } \sum_{i \in \mathcal{S}} \lambda_i^* y_i k(\mathbf{x}_i, \mathbf{x}) - \frac{1}{|\mathcal{S}_m|} \sum_{i \in \mathcal{S}} \sum_{j \in \mathcal{S}_m} \lambda_i^* y_i k(\mathbf{x}_i, \mathbf{x}_j) + \frac{1}{|\mathcal{S}_m|} \sum_{i \in \mathcal{S}_m} y_i > 0, \\ 0, & \text{otherwise.} \end{cases}$$
$$\tag{6.33}$$

This corresponds to *sparse* kernel classification, where only a subset of the training set, the support vectors, have any influence on the decision. Furthermore, the kernel influence is weighted by the corresponding support vector sensitivity parameter λ_i^*.

The next question is whether it is necessary to construct a kernel by specifying $\phi(\mathbf{x})$ and using the definition (6.30). As we will see next, not even that is necessary, so that $\phi(\mathbf{x})$ is entirely superfluous. One can specify directly a function $k(\mathbf{x}, \mathbf{x}')$, as long as certain conditions are met. To be admissible, $k(\mathbf{x}, \mathbf{x}')$ must be expressible as an inner product $\phi^T(\mathbf{x})\phi(\mathbf{x}')$ in some space (which can be an infinite-dimensional space). It can be shown that this is true if and only if

- k is symmetric: $k(\mathbf{x}, \mathbf{x}') = k(\mathbf{x}', \mathbf{x})$, for all $\mathbf{x}, \mathbf{x}' \in R^d$.

- k is positive semi-definite:
$$\int k(\mathbf{x}, \mathbf{x}')g(\mathbf{x})g(\mathbf{x}') \, d\mathbf{x} \, d\mathbf{x}' \ge 0, \tag{6.34}$$

for any function *square-integrable function* $g : R^d \to R$, that is, any function g satisfying $\int g^2(\mathbf{x}) \, d\mathbf{x} < \infty$.

This result is called *Mercer's Theorem* and the conditions above are *Mercer's conditions*.

Some examples of kernels used in applications are:

- Polynomial kernel: $k(\mathbf{x}, \mathbf{x}') = (1 + \mathbf{x}^T \mathbf{x}')^p$, for $p = 1, 2, \ldots$.

- Gaussian kernel: $k(\mathbf{x}, \mathbf{x}') = \exp(-\gamma||\mathbf{x} - \mathbf{x}'||^2)$, for $\gamma > 0$.

- Sigmoid kernel: $k(\mathbf{x}, \mathbf{x}') = \tanh(\gamma \mathbf{x}^T \mathbf{x}' - \delta)$.

The Gaussian kernel is also called the *radial basis function* (RBF) kernel in the context of SVMs. It can be shown that all examples of kernels above satisfy Mercer's condition (for the sigmoid kernel, this will be true for certain values of γ and δ), so that $k(\mathbf{x}, \mathbf{x}') = \phi^T(\mathbf{x})\phi(\mathbf{x}')$ for a suitable mapping ϕ. In the case of the Gaussian kernel, the mapping ϕ is into an infinite-dimensional space, so it cannot be computed exactly (but we do not need to). The choice of the kernel parameters, as was the case with the smoothing parameters in Chapter 6, is a model selection problem. In some cases, they can be chosen automatically based on the data, as is the case of the parameter γ in scikit-learn's implementation of the RBF SVM.

Example 6.1. (XOR data set.)[1] The XOR (read "X-or") data set is the simplest (i.e., with the minimal number of points) nonlinearly separable data set in two dimensions (more on this topic in Chapter 8). The data set is $S_n = \{((-1,1), -1), ((1,-1), -1), ((-1,-1), 1), ((1,1), 1)\}$. See the left plot in Figure 6.2(a) for an illustration. We let $C = 1$ and consider the polynomial kernel of order $p = 2$:

$$k(\mathbf{x}, \mathbf{x}') = (1 + \mathbf{x}^T \mathbf{x}')^2 = (1 + x_1 x_1' + x_2 x_2')^2. \tag{6.35}$$

Note that the dual functional (6.10) can be written compactly as:

$$L_D(\boldsymbol{\lambda}) = \boldsymbol{\lambda}^T \mathbf{1} - \frac{1}{2} \boldsymbol{\lambda}^T H \boldsymbol{\lambda}, \tag{6.36}$$

where $H_{ij} = y_i y_j k(\mathbf{x}_i, \mathbf{x}_j)$. In the present case,

$$H = \begin{bmatrix} 9 & -1 & -1 & 1 \\ -1 & 9 & 1 & -1 \\ -1 & 1 & 9 & -1 \\ 1 & -1 & -1 & 9 \end{bmatrix}. \tag{6.37}$$

In this example, it is possible to maximize $L_D(\boldsymbol{\lambda})$ analytically. First we obtain the gradient of $L_D(\boldsymbol{\lambda})$ in (6.36) and set it to zero:

$$\frac{\partial L_D}{\partial \boldsymbol{\lambda}} = H\boldsymbol{\lambda} - \mathbf{1} = 0 \Rightarrow H\boldsymbol{\lambda} = \mathbf{1}. \tag{6.38}$$

---

[1]This example is adapted from Example 2 of Chapter 5 in Duda et al. [2001].

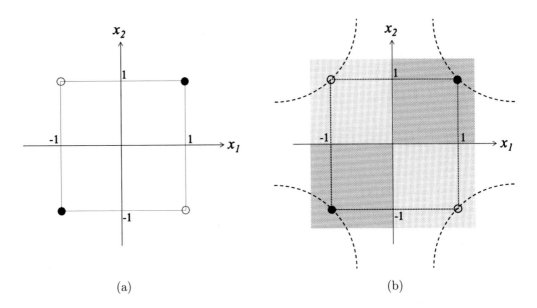

Figure 6.2: Nonlinear SVM with the XOR data set. (a) Training data, where the solid and hollow points have labels 1 and -1, respectively. (b) SVM classifier, where the red and green decision regions correspond to labels 1 and -1, respectively. Dashed lines indicated the hyperbolic margin boundaries.

Hence, we need to solve the following system of equations:

$$9\lambda_1 - \lambda_2 - \lambda_3 + \lambda_4 = 1$$
$$-\lambda_1 + 9\lambda_2 + \lambda_3 - \lambda_4 = 1$$
$$-\lambda_1 + \lambda_2 + 9\lambda_3 - \lambda_4 = 1 \tag{6.39}$$
$$\lambda_1 - \lambda_2 - \lambda_3 + 9\lambda_4 = 1$$

The solution to this system of equations is $\lambda_1^* = \lambda_2^* = \lambda_3^* = \lambda_4^* = \frac{1}{8}$. Because matrix H is positive definite, the functional $L_D(\boldsymbol{\lambda})$ is strictly concave and the condition $\frac{\partial L_D}{\partial \boldsymbol{\lambda}} = 0$ is necessary and sufficient for an unconstrained global maximum. But since the vector $\boldsymbol{\lambda}^* = (\frac{1}{8}, \frac{1}{8}, \frac{1}{8}, \frac{1}{8})^T$ is an interior point of the feasible region defined by the conditions (6.32), this is in fact the solution $\boldsymbol{\lambda}^*$ to the dual problem (note that all 4 training points are support vectors and margin vectors, an usual occurrence). Plugging these values in (6.33) yields the following simple expression for the nonlinear SVM classifier

$$\psi_n(\mathbf{x}) = \begin{cases} 1\,, & \text{if } x_1 x_2 > 0\,, \\ 0\,, & \text{otherwise.} \end{cases} \tag{6.40}$$

as can be easily verified. The decision boundary is $x_1 x_2 = 0$, which is the union of the lines $x_1 = 0$ and $x_2 = 0$. The margin boundaries are the loci of points where the discriminant is equal to ± 1. In this case, we have $x_1 x_2 = \pm 1$, which describe hyperbolas passing through the training data points. The classifier is depicted in Figure 6.2(b). It can be seen that this highly nonlinear classifier is able to separate the XOR data set, i.e., obtain zero error on the training data.

The high-dimensional space where the data is mapped to was never invoked in the previous derivation. Let us investigate it, though it is not a necessary step for SVM classification. Expanding (6.35) reveals that

$$
\begin{aligned}
k(\mathbf{x}, \mathbf{x}') &= 1 + 2x_1 x_1' + 2x_2 x_2' + 2x_1 x_2 x_1' x_2' + x_1^2 (x_1')^2 + x_2^2 (x_2')^2 \\
&= \phi^T(\mathbf{x}) \phi(\mathbf{x}'),
\end{aligned}
\tag{6.41}
$$

where

$$
\mathbf{z} = \phi(\mathbf{x}) = (z_1, z_2, z_3, z_4, z_5, z_6) = (1, \sqrt{2}x_1, \sqrt{2}x_2, \sqrt{2}x_1 x_2, x_1^2, x_2^2).
\tag{6.42}
$$

Hence, the transformation is to a six-dimensional space. The original data points are projected to

$$
\begin{aligned}
\mathbf{z}_1 &= \phi((-1, 1)) = (1, -\sqrt{2}, \sqrt{2}, -\sqrt{2}, 1, 1), \\
\mathbf{z}_2 &= \phi((1, -1)) = (1, \sqrt{2}, -\sqrt{2}, -\sqrt{2}, 1, 1), \\
\mathbf{z}_3 &= \phi((-1, -1)) = (1, -\sqrt{2}, -\sqrt{2}, \sqrt{2}, 1, 1), \\
\mathbf{z}_4 &= \phi((1, 1)) = (1, \sqrt{2}, \sqrt{2}, \sqrt{2}, 1, 1).
\end{aligned}
\tag{6.43}
$$

The hyperplane decision boundary designed by the SVM in this six-dimensional space has parameters:

$$
\mathbf{a}^* = \sum_{i=1}^{4} \lambda_i^* y_i \mathbf{z}_i = (0, 0, 0, 1/\sqrt{2}, 0, 0)^T
\tag{6.44}
$$

and

$$
a_0^* = -\frac{1}{4} \sum_{i=1}^{4} \sum_{j=1}^{4} \lambda_i^* y_i \mathbf{z}_i^T \mathbf{z}_j + \frac{1}{4} \sum_{i=1}^{4} y_i = 0
\tag{6.45}
$$

The decision boundary is therefore determined by

$$
(\mathbf{a}^*)^T \mathbf{z} + a_0^* = z_4 / \sqrt{2} = 0 \Rightarrow z_4 = 0,
\tag{6.46}
$$

with margin is $1/\|\mathbf{a}^*\| = \sqrt{2}$, while the margin hyperplanes are determined by

$$
(a^*)^T \mathbf{z} + a_0^* = z_4 / \sqrt{2} = \pm 1 \Rightarrow z_4 = \pm \sqrt{2}.
\tag{6.47}
$$

We cannot visualize the data set or these boundaries in six-dimensional space. However, we observe that z_1, z_5, and z_6 are constant in the transformed data (6.43). Hence, the transformation is essentially to the three-dimensional subspace defined by z_2, z_3, and z_4. In this subspace, the data lie on four vertices of a cube of size $\sqrt{2}$ centered at the origin, and the decision boundary $z_4 = 0$ is a plane going through the origin and perpendicular to z_4, while the the margin boundaries are the

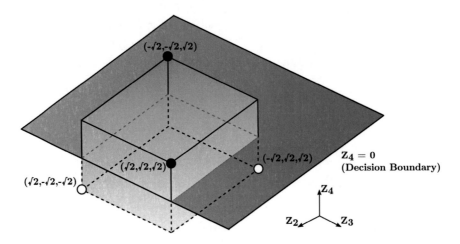

Figure 6.3: Projection of the XOR data set into a three-dimensional subspace, where it is linearly separable. The SVM decision boundary is a plane passing through the origin.

planes $z_4 = \pm\sqrt{2}$. We can see that the points from class -1 are on the plane $z_4 = -\sqrt{2}$ while the ones from class 1 are on the plane $z_4 = \sqrt{2}$. See Figure 6.3 for an illustration. Notice that the data are now linearly separable (three is the minimum number of dimensions necessary to linearly separate a general set of four points; more on this in Chapter 8). Notice also that, with $z_4 = \sqrt{2}x_1x_2$, (6.46) and (6.47) yield the same decision and margin boundaries in the original space as before. ◇

Example 6.2. Figure 6.4 displays classifiers produced by a nonlinear SVM on the synthetic Gaussian training data used in Figures 5.2, 5.3, and 5.5 of the previous chapter. The Gaussian-RBF kernel with $\gamma = 1/2$ and varying C is used. The plots show that the decision boundaries in all cases are complex, but this is true especially for large C, in which case outliers are not allowed and there is overfitting, as was discussed previously. With $C = 1$, the nonlinear SVM produces a decision boundary that is not too different from the optimal one (a line with slope -45° going through the center of the plot). The choice of C is a model selection problem (see Chapter 8). ◇

6.2 Neural Networks

Neural Networks (NNs) combine linear functions and univariate nonlinearities to produce complex discriminants with arbitrary approximation capability (as discussed in Section 6.2.2). A neural network consists of units called *neurons*, which are organized in layers. Each neuron produces a univariate *output* that is a nonlinear function of the neuron *activation*, which consists of a linear

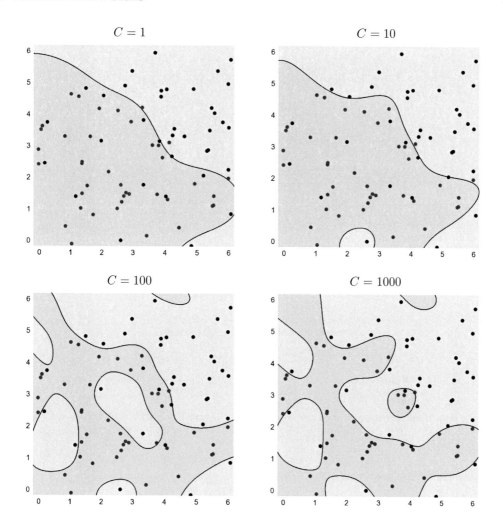

Figure 6.4: Classifiers produced by a Gaussian-RBF nonlinear SVM rule on the training data in Figures 5.2, 5.3, and 5.5 of Chapter 5 (plots generated by c06_svm.py).

combination of the neuron univariate inputs, where the coefficients of the linear combination are the neural network *weights*. Figure 6.5 displays three consecutive layers of a neural network. The neuron activations are the functions $\alpha(\mathbf{x})$, the neuron outputs are the functions $\beta(\mathbf{x})$, and the weights are the scalars w, where $\mathbf{x}$ is the input feature vector.

Notice that Rosenblatt's Perceptron, introduced at the beginning of Section 6.1, can be understood as a neural network classifier with a single layer. For this reason, neural networks with more than one neuron are also called *multilayer perceptrons*.

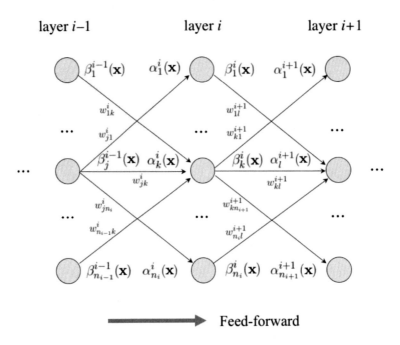

Figure 6.5: Three consecutive layers of a neural network.

The univariate nonlinearities are often *sigmoids*, i.e., nondecreasing functions $\sigma(x)$ such that $\sigma(-\infty) = -1$ and $\sigma(\infty) = 1$. A few examples are:

- Threshold sigmoid:

$$\sigma(x) = \begin{cases} 1, & \text{if } x > 0, \\ 0, & \text{otherwise.} \end{cases} \tag{6.48}$$

- Logistic sigmoid:

$$\sigma(x) = \frac{1}{1 + e^{-x}}. \tag{6.49}$$

- Arctan sigmoid:

$$\sigma(x) = \frac{1}{2} + \frac{1}{\pi} \arctan(x). \tag{6.50}$$

- Gaussian sigmoid:

$$\sigma(x) = \Phi(x) = \frac{1}{2\pi} \int_{-\infty}^{x} e^{-\frac{u^2}{2}} du. \tag{6.51}$$

These nonlinearities differ by their slope (derivative), from extremely sharp (threshold) to very smooth (arctan). This distinction matters in backpropagation training, as we will see.

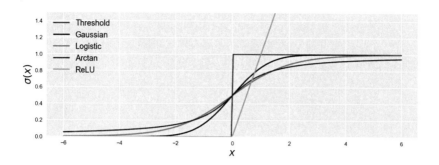

Figure 6.6: Neural network nonlinearities (plot generated by c06_nonlin.py).

In addition, a popular nonlinearity that is not a sigmoid is the *Rectifier Linear Unit (ReLU)*:

$$\sigma(x) = \max(x, 0) = \begin{cases} x, & \text{if } x > 0, \\ 0, & \text{otherwise.} \end{cases} \tag{6.52}$$

See Figure 6.6 for an illustration. Although the ReLU stands out by not being bounded, its derivative is just the threshold nonlinearity. So it can be seen as a kind of smoothed threshold nonlinearity. (Similarly, one could define nonlinearities whose derivatives are logistic, arctan, and so on.)

Consider a two-layer neural network with k neurons in one *hidden layer* and one neuron in an *output layer*. The activation function of the output neuron in such a network corresponds to the simple discriminant:

$$\zeta(\mathbf{x}) = c_0 + \sum_{i=1}^{k} c_i \xi_i(\mathbf{x}), \tag{6.53}$$

where $\xi_i(\mathbf{x}) = \sigma(\phi(\mathbf{x}))$ is the output of the i-th neuron in the hidden layer, which in turn is the result of passing the linear activation

$$\phi_i(\mathbf{x}) = b_i + \sum_{j=1}^{d} a_{ij} x_j, \quad i = 1, \dots, k, \tag{6.54}$$

through the nonlinearity σ. (For simplicity, we will assume here that all neurons use the same nonlinearity.) The vector of parameters, or weights, for this neural network is

$$\mathbf{w} = (c_0, \dots, c_k, a_{10}, \dots, a_{1d}, \dots, a_{k1}, \dots, a_{kd}) \tag{6.55}$$

for a total of $(k+1) + k(d+1) = k(d+2) + 1$ weights. Thus, as k and d increase, the number of weights in this two-layer network is roughly equal to $k \times d$.

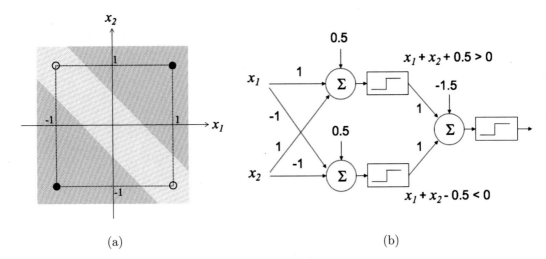

(a)   (b)

Figure 6.7: Neural network with the XOR data set. (a) Original data and neural network classifier (same color conventions as in Figure 6.2 are used). (b) Neural network graph.

A neural network classifier can be obtained by thresholding the discriminant at zero, as usual:

$$\psi_n(\mathbf{x}) = \begin{cases} 1, & \zeta(\mathbf{x}) > 0, \\ 0, & \text{otherwise.} \end{cases} \tag{6.56}$$

This is equal to the output of the neural network if the nonlinearity of the output neuron is a threshold sigmoid.

Neural networks are similar to nonlinear SVMs in that any hidden layers nonlinearly map the original feature space into a different space, and the output layer acts on the transformed features by means of a linear decision (hyperplane). The decision in the original feature space is nonlinear.

Example 6.3.[2] Here we separate the XOR data set in Example 6.1 using a neural network with two neurons in one hidden layer. Here the $k(d+2)+1 = 9$ weights are set manually to achieve zero empirical error on the data. The neural network classifier and corresponding graph are displayed in Figure 6.7. The upper and lower linear boundaries of the class 1 decision region are implemented by the two perceptrons in the hidden layer, while the output perceptron generates the labels of the decision regions. ◇

Example 6.4. Figure 6.8 displays classifiers produced by a neural network on the training data in Figure 6.4. One hidden layer containing a varying number of neurons with logistic sigmoids is considered. The plots show that overfitting quickly occurs with an increasing number of neurons.

---

[2]This example is adapted from the example in Figure 6.1 in Duda et al. [2001].

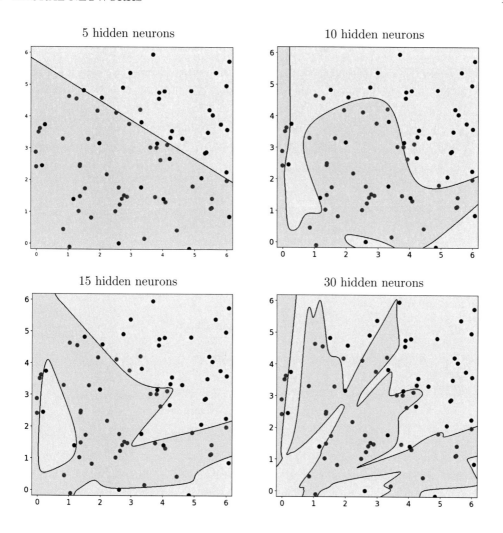

Figure 6.8: Classifiers produced by a neural network with one hidden layer and logistic sigmoids on the training data in Figure 6.4 (plots generated by c06_nnet.py).

With 5 hidden neurons, the neural network produces a nearly linear decision boundary, which is close to the optimal one (a line with slope -45° going through the center of the plot). Choosing the number of neurons, and the architecture itself, is a model selection problem (see Chapter 8). ◇

6.2.1 Backpropagation Training

All neural network training methods seek to minimize a empirical score $J(\mathbf{w})$, which is a function of how well the NN predicts the labels of the training data. This minimization is necessarily iterative due to the complexity of the problem. It is this process of iteratively adjusting the weights based on "rewards and punishments" dictated by the labeled training data that draws a comparison with learning in biological neural networks. Given data $S_n = \{(\mathbf{x}_1, y_1), \ldots, (\mathbf{x}_n, y_n)\}$ (since we threshold discriminants at zero, it is convenient here to assume $y_i = \pm 1$), examples of empirical scores include:

- The empirical classification error:

$$J(\mathbf{w}) = \frac{1}{n} \sum_{i=1}^{n} I_{y_i \neq \psi_n(\mathbf{x}_i; \mathbf{w})} \,, \tag{6.57}$$

  where $\psi_n(\cdot; \mathbf{w})$ is the NN classifier with weights $\mathbf{w}$.

- The mean absolute error:

$$J(\mathbf{w}) = \sum_{i=1}^{n} |y_i - \zeta(\mathbf{x}_i; \mathbf{w})| \,, \tag{6.58}$$

  where $\zeta_n(\cdot; \mathbf{w})$ is the NN discriminant with weights $\mathbf{w}$.

- The mean-square error:

$$J(\mathbf{w}) = \frac{1}{2} \sum_{i=1}^{n} (y_i - \zeta(\mathbf{x}_i; \mathbf{w}))^2. \tag{6.59}$$

Using the empirical classification error in training is not practical, since it is not differentiable, and thus not amenable for gradient descent methods (it is still of theoretical interest). The mean absolute error is differentiable but not twice differentiable. The most widespread criterion used in practice is the mean-square error. The classical *backpropagation algorithm* arises from the application of gradient descent to minimize the mean-square error criterion. Backpropagation is therefore a least-squares fitting procedure. We remark that gradient descent is not by any means the only procedure used in NN training. Other popular nonlinear minimization procedures include the Gauss-Newton Algorithm and the Levenberg-Marquardt Algorithm (see the Bibliographical Notes).

In *online backpropagation training*, the score is evaluated at each training point $(\mathbf{x}_i, y_i)$ separately:

$$J_i(\mathbf{w}) = \frac{1}{2} \left(y_i - \zeta_n(\mathbf{x}_i; \mathbf{w}) \right)^2. \tag{6.60}$$

By contrast, in *batch backpropagation training*, the total error over the entire training set is used, which is the score $J(\mathbf{w})$ in (6.59). Notice that

$$J(\mathbf{w}) = \sum_{i=1}^{n} J_i(\mathbf{w}) \,. \tag{6.61}$$

In the sequel, we will examine in detail online backpropagation. The basic gradient descent step in this case is:

$$\Delta \mathbf{w} = -\ell \, \nabla_{\mathbf{w}} J_i(\mathbf{w}) \,, \tag{6.62}$$

that is, upon presentation of the training point $(\mathbf{x}_i, y_i)$, the vector of weights $\mathbf{w}$ is updated in the direction that decreases the error criterion, according to the step length ℓ. Writing the above in component form, we get the updates for each weight separately:

$$\Delta w_i = -\ell \, \frac{\partial J}{\partial w_i} \,. \tag{6.63}$$

The backpropagation algorithm applies the chain rule of calculus to compute the partial derivatives in (6.63).

A single presentation of the entire training set is called an *epoch*. The amount of training is typically measured by the number of epochs. Each update in (6.62) is guaranteed to lower the individual error $J_i(\mathbf{w})$ for the pattern, but it could increase the total error $J(\mathbf{w})$. However, in the long run (over several epochs), $J(\mathbf{w})$ will generally decrease as well. It has been claimed that online training, in addition to leading to simpler updates, can avoid flat patches in the batch score $J(\mathbf{w})$ that slow down convergence.

Let us examine how to compute the backpropagation updates (6.63) for a depth-2, one-hidden layer neural network, in which case the weight vector $\mathbf{w}$ is given by (6.55), upon presentation of a training point $(\mathbf{x}, y)$. Using artificial bias units, we can write (we omit the dependencies on the weights):

$$\zeta(\mathbf{x}) = \sum_{i=0}^{k} c_i \xi_i(\mathbf{x}) \,, \tag{6.64}$$

where $\xi_i(\mathbf{x}) = \sigma(\phi(\mathbf{x})$, with

$$\phi_i(\mathbf{x}) = \sum_{j=0}^{d} a_{ij} x_j \,, \quad i = 1, \ldots, k. \tag{6.65}$$

When computing the activations and outputs $\phi_i(\mathbf{x})$, $\xi_i(\mathbf{x})$, and $\zeta(\mathbf{x})$ from the input $\mathbf{x}$, the network is said to be in *feed-forward* mode.

To find the update of the *hidden-to-output* weights c_i, notice that

$$\frac{\partial J}{\partial c_i} = \frac{\partial J}{\partial \zeta} \frac{\partial \zeta}{\partial c_i} = -[y - \zeta(\mathbf{x})] \, \xi_i(\mathbf{x}) \,. \tag{6.66}$$

Define the output error

$$\delta^o = -\frac{\partial J}{\partial \zeta} = y - \zeta(\mathbf{x}) \,. \tag{6.67}$$

According to (6.63), the update is

$$\Delta c_i = \ell \, \delta^o \xi_i(\mathbf{x}) \,. \tag{6.68}$$

For the *input-to-hidden* weights a_{ij}, we have

$$\frac{\partial J}{\partial a_{ij}} = \frac{\partial J}{\partial \phi_i} \frac{\partial \phi_i}{\partial a_{ij}} = \frac{\partial J}{\partial \phi_i} x_j \,, \tag{6.69}$$

where

$$\frac{\partial J}{\partial \phi_i} = \frac{\partial J}{\partial \zeta} \frac{\partial \zeta}{\partial \xi_i} \frac{\partial \xi_i}{\partial \phi_i} = -\delta^o \, c_i \, \sigma'(\phi_i(\mathbf{x})) \,. \tag{6.70}$$

Define the errors

$$\delta_i^H = -\frac{\partial J}{\partial \phi_i} = \delta^o \, c_i \, \sigma'(\phi_i(\mathbf{x})) \,. \tag{6.71}$$

This is the *backpropagation equation.* According to (6.63), the update is

$$\Delta a_{ij} = \ell \, \delta_i^H x_j \,. \tag{6.72}$$

Notice that the slope of the nonlinearity affects δ_i^H and the magnitude of the update to a_{ij}.

In summary, one iteration of backpropagation for this simple network proceeds as follows. At the start of each iteration, the new training point $\mathbf{x}$ is passed through the network in feed-forward mode using the old weights, which updates all network activations and outputs. The output error δ_0 is then computed using the label y and (6.67), and the weights c_i are updated using (6.68). Next, the error δ_o is *backpropagated* using (6.71), to obtain the errors δ_i^H at the hidden layer, and the weights a_{ij} are updated using (6.72). When computing the errors δ_i^H from the error δ^o, the network is in *backpropagation* mode.

To obtain the general update equations for a multilayer network, consider the graph in Figure 6.5. Here, $\beta_k^i(\mathbf{x}) = \sigma(\alpha_k^i(\mathbf{x}))$, where

$$\alpha_k^i(\mathbf{x}) = \sum_{j=0}^{n_{i-1}} w_{jk}^i \beta_j^{i-1}(\mathbf{x}) \,. \tag{6.73}$$

Now,

$$\frac{\partial J}{\partial w_{jk}^i} = \frac{\partial J}{\partial \alpha_k^i} \frac{\partial \alpha_k^i}{\partial w_{jk}^i} = -\delta_k^i \, \beta_j^{i-1}(\mathbf{x}) \,, \tag{6.74}$$

where the error is defined as

$$\delta_k^i = -\frac{\partial J}{\partial \alpha_k^i} \,. \tag{6.75}$$

Hence, the update rule for weight w_{jk}^i can be written as

$$\Delta w_{jk}^i = -\ell \frac{\partial J}{\partial w_{jk}^i} = \ell \, \delta_k^i \, \beta_j^{i-1}(\mathbf{x}) \,. \tag{6.76}$$

To determine δ_k^i one uses the chain rule:

$$\delta_k^i = -\frac{\partial J}{\partial \alpha_k^i} = -\sum_{l=1}^{n_{i+1}} \frac{\partial J}{\partial \alpha_l^{i+1}} \frac{\partial \alpha_l^{i+1}}{\partial \alpha_k^i} = \sum_{l=1}^{n_{i+1}} \delta_l^{i+1} \frac{\partial \alpha_l^{i+1}}{\partial \alpha_k^i} \,, \tag{6.77}$$

while

$$\frac{\partial \alpha_l^{i+1}}{\partial \alpha_k^i} = \frac{\partial \alpha_l^{i+1}}{\partial \beta_k^i} \frac{\partial \beta_k^i}{\partial \alpha_k^i} = w_{kl}^{i+1} \sigma'(\alpha_k^i(\mathbf{x})). \tag{6.78}$$

This results in the backpropagation equation:

$$\delta_k^i = \sigma'(\alpha_k^i(\mathbf{x})) \sum_{l=1}^{n_{i+1}} w_{kl}^{i+1} \delta_l^{i+1}. \tag{6.79}$$

Hence, the deltas at one hidden layer are backpropagated from the deltas in the next layer. The slope of the nonlinearity affects δ_k^i and thus the magnitude of the update to w_{jk}^i in (6.76). SS Online backpropagation training proceeds as follows. First, all weights are initialized with random numbers (they cannot be all zeros). At each backpropagation iteration, a new training point is presented, and the network is run in feed-forward mode to update all activations and outputs. Then weight updates proceed from the output layer backwards to the first hidden layer. Weight updates are based on errors backpropagated from the later layers to earlier ones. This process is repeated for all points in the training data, which consists in one epoch of training. The process can continue, and the training data reused, for any desired number of epochs. Training stops when there is no significant improvement to the empirical score function.

In practice, training needs to be stopped early to avoid overfitting. This can be done by selecting ahead of time a fixed number of training epochs, or by checking performance on a separate validation data set. More will be said on this in Chapter 8. In addition, gradient descent can get stuck in local minima of the empirical score function. The simplest method to deal with this is to train the network multiple times with different random initialization of weights and compare the results.

6.2.2 Convolutional Neural Networks

Convolutional neural networks (CNN) have a special architecture that makes them very effective in computer vision and image analysis applications. Even though they were invented in the 1980's, interest in them has surged recently due to the ability of training deep versions of these networks and their impressive performance in various tasks (see the Bibliographical Notes).

In a CNN, there are special layers, called *convolutional layers*, where the activation function $g(i)$ of each neuron i is a very sparse function of the outputs $f(j)$ of the neurons j of the previous convolutional layer. In fact, by arranging the neurons in layer i in a 2-dimensional array of size $n \times m$, called a *feature map*, then the relation between activation and outputs is given by

$$g(i, j) = \sum_{(k,l) \in N} \sum w(k, l) f(i + k, j + l), \tag{6.80}$$

for $(i, j) \in \{(0, 0), \ldots, (n, m)\}$, where the *filter* $w(k, l)$ is a (typically small) array defined on a square domain N. It is common to *zero-pad* f, i.e., add a proper number of rows and columns of zeros around it, to avoid indices $(i + k, j + l)$ out of bounds in (6.80) and allow g to be of the same dimensions as f. Recall that each activation $g(i, j)$ needs to pass through a nonlinearity to obtain the output to the next layer, so that nonlinear discriminants can be implemented (in CNNs, the ReLU nonlinearity is usually employed).

A common example of domain is an origin-centered 3×3 array $N = \{(-1, -1), (-1, 0), \ldots, (0, 0), \ldots, (1, 0), (1, 1)\}$, in which case (6.80) can be expanded as

$$
\begin{aligned}
g(i, j) = w(-1, -1)f(i - 1, j - 1) + w(-1, 0)f(i - 1, j) + \ldots \\
+ w(0, 0)f(i, j) + \ldots + w(1, 0)f(i + 1, j) + w(1, 1)f(i + 1, j + 1)
\end{aligned}
\tag{6.81}
$$

for $(i, j) \in \{(0, 0), \ldots, (n, m)\}$. Notice that in this case each neuron is connected to only nine neurons in the previous layer. The sum in (6.80 is known as the *convolution* of f by w in signal and image processing (strictly speaking, this sum is a *correlation*, since convolution requires flipping the domain N, but the two operations are closely related).

The values of the filter $w(k, l)$ over the domain N specify the operation being performed. Two example of filters are specified by the arrays

$\frac{1}{9}$	$\frac{1}{9}$	$\frac{1}{9}$
$\frac{1}{9}$	$\frac{1}{9}$	$\frac{1}{9}$
$\frac{1}{9}$	$\frac{1}{9}$	$\frac{1}{9}$

-1	-1	-1
0	0	0
1	1	1

The first one performs averaging, which has the effect of smoothing or blurring, while the second one performs differentiation in the vertical direction and thus enhances horizontal edges. The main idea behind CNNs is that, instead of being manually specified for a particular application, the filter coefficients become weights of the neural network and are learned from the data as part of the training process.

In practice, the convolutional layers of a CNN are a little more complex than the previous discussion indicates, as the neurons in each layer are actually arranged in a volume $n \times m \times r$, rather than a 2-dimensional array. The filters are of dimension $p \times q \times r$, where the depth r must exactly match that of the previous layer, in order to make the result of the convolution a 2-dimensional feature map as before (no zero padding is performed in the depth direction). Other than that, p and q can be freely selected (though they are usually very small, between 3×3 and 7×7). However, several different filters w_k, $k = 1, \ldots, s$, all of the same dimensions, are employed, which means that the activation g is again a volume, of dimensions $n \times m \times s$ (with the appropriate zero-padding). There is no required relationship between r and s: the first can be smaller, equal, or larger than the second, but in many architectures in practice, $s \geq r$.

Example 6.5. Suppose that the output volume from the previous layer f has dimensions $8 \times 8 \times 3$ (e.g., this would be the case if f were a 8×8 color image input to the first layer, in which case the 3 feature maps are the red, green and blue components). Assume that there are a total of six filters of dimensions $3 \times 3 \times 3$ (the last dimension *must* be 3, in order to match the depth of f). With zero-padding of size 1 in both height and width, the result is a $8 \times 8 \times 6$ activation volume g for the current layer.

In the previous example, g has the same height n and width m as f. In practice, dimensionality reduction (i.e., reduction in n and m) from one layer to the next is accomplished in two different ways. First, dimensionality reduction can be accomplished by *striding*, i.e., skipping some of the indices (i, j) in (6.80). For example, a stride of 2 in each direction means that only every other point of the convolution is computed, resulting in a reduction by 2 in the dimensions of the activation array g. In the previous example, this would result in a $4 \times 4 \times 6$ activation volume g. Secondly, a special layer, called a *max-pooling layer*, can be interposed between two consecutive convolutional layers. The max-pooling layer applies a maximum filter (typically of size 2×2 with a stride of 2) to each feature map, producing an activation volume to the next layer of the same depth, but reduced height and width. (In the first CNNs, average-pooling was also common, but now max-pooling is almost exclusively used).

Figure 6.9 displays the architecture of the *VGG16* (Visual Geometry Group 16-layer) CNN for image classification, which is typical representative of the class of deep CNNs in current use. The VGG16 CNN distinguishes itself by the use of very small filters, of height and width 3×3 (or smaller), in all layers. The idea is that, through the use of max-pooling and a large number of convolutional layers, the local nature of small filters in the early layers become nonlocal in the later layers. The ReLU nonlinearity is used in all non-pooling layers. We can see in Figure 6.9 that there are 16 non-pooling hidden layers (plus the output layer), hence the name VGG16, and 5 max-pooling layers. Notice that the height and width of the volumes are progressively reduced from input to output, while the depth (i.e., the number of feature maps, or filters) is increased. The three last layers, before the softmax, are not convolutional, but are *fully-connected layers*, i.e., ordinary layers of hidden neurons.

The VGG16 CNN was originally an entry in the ImageNet Large-Scale Visual Recognition Challenge (ILSVRC), which classifies images in a very large database into one of 1000 classes. To handle the multiple classes, the output layer has 1000 neurons (compare to the one-neuron output layer for binary classification in previous sections), which uses as nonlinearity the *softmax function*. The softmax nonlinearity is a multivariate function $S : R^c \to R^c$, where c is the number of classes (it thus differs from the univariate nonlinearities in the previous sections). Each component S_i of the softmax function is given by

$$S_i(\mathbf{z}) = \frac{e^{z_i}}{\sum_{j=1}^{c} e^{z_j}}, \tag{6.82}$$

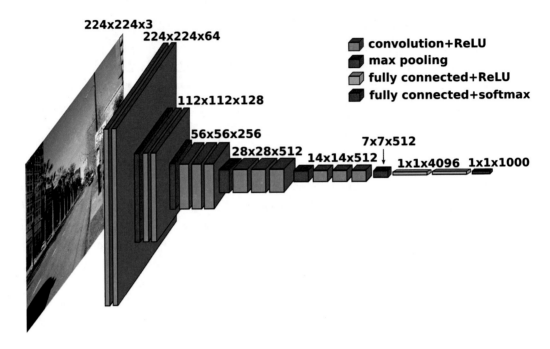

Figure 6.9: Architecture of the VGG16 network. Adapted from Hassan [2018].

for $i = 1, \ldots, c$, where $\mathbf{z} = (z_1, \ldots, z_c)$ is the activation of the output layer ($\mathbf{z}$ generalizes the discriminant ζ of previous sections). We can see that the softmax function "squashes" all activations into the interval $[0, 1]$ and makes them add up to one. Hence, the output can be interpreted as a vector of discrete probabilities, which can be trained to indicate the likelihood of each class.

The total number of weights in the VGG16 network is 138 million, an astounding number, nearly 90% of which are in the last three fully-connected layers (see Exercise 6.5). Such a large network requires an inordinately large amount of data and is very difficult to train. According to the authors, it originally took 2-3 weeks of 2014 state-of-the-art high-performance computer time to train on the ImageNet database, which consists of roughly 14M images (as of 2010) and 1000 classes. Regularization techniques, known as *dropout*, where a fraction, usually 50%, of the neurons are randomly ignored during each backpropagation iteration, and *weight decay*, which constrains the L^2-norm of the weight vector to be small by adding a penalty term to the score (this is similar to ridge regression, discussed in Chapter 11), are used in an effort to avoid overfitting. Due to this training difficulty, the VGG16, and other similar deep networks, are often used with the pre-trained weights in other imaging applications, in what is called a *transfer learning* approach (see Python Assignment 6.12).

*6.2.3 Universal Approximation Property of Neural Networks

In this section, we review classical and recent results on the expressive power of neural networks. These results do not take into account training, so they are not directly relatable to universal consistency (see Section 6.2.2 for such results). All the theorems in the section are given without proof; see the Bibliographical Notes for the sources.

The idea behind neural networks was to some extent anticipated by the following result. Let $C(I^d)$ be the set of all continuous functions defined on the closed hypercube I^d.

Theorem 6.1. *(Kolmogorov-Arnold Theorem.) Every $f \in C(I^d)$ can be written as:*

$$f(\mathbf{x}) = \sum_{i=1}^{2d+1} F_i \left(\sum_{j=1}^{d} G_{ij}(x_j) \right), \tag{6.83}$$

where $F_i : R \to R$ and $G_{ij} : R \to R$ are continuous functions.

The previous result guarantees that a multivariate function can be computed as a finite sum of univariate functions of sums of univariate functions of the coordinates. For example, for $d = 2$, the function $f \in C(I^2)$ given by $f(x,y) = xy$ can be written as:

$$f(x,y) = \frac{1}{4} \left((x+y)^2 - (x-y)^2 \right). \tag{6.84}$$

However, the Kolmogorov-Arnold theorem does not say how to find the functions F_i and G_{ij} required to compute exactly a general function f. Furthermore, the functions F_i and G_{ij} could be quite complicated. Neural networks give up exact representation and instead use a combination of simple functions: linear functions and univariate nonlinearities.

The discriminant of a neural network with k neurons in one hidden layer is specified by (6.53) and (6.54). This discriminant can be written in a single equation as:

$$\zeta(\mathbf{x}) = \sum_{i=0}^{k} c_i \sigma \left(\sum_{j=0}^{d} a_{ij} x_j \right). \tag{6.85}$$

where we use *artificial bias units* with constant unit output in order to include the coefficients a_{i0} and c_0 in the summations. Comparing (6.85) with (6.83) reveals the similarity with the Kolmogorov-Arnold result.

Even though exact representation is lost, neural networks are universal approximators in the sense of the following classical theorem by Cybenko.

Theorem 6.2. *(Cybenko's Theorem.) Let $f \in C(I^d)$. For any $\tau > 0$, there is a one-hidden layer neural network discriminant ζ in (6.85) and sigmoid nonlinearity σ, such that*

$$|f(\mathbf{x}) - \zeta(\mathbf{x})| < \tau , \quad \text{for all } \mathbf{x} \in I^d . \tag{6.86}$$

An equivalent way to state this theorem is as follows. A subset A of a metric space X is *dense* if given every point $x \in X$, there is a point $a \in A$ arbitrarily close to it, i.e., given any $x \in X$ and $\tau > 0$, the ball $B(x,\tau)$ in the metric of X contains at least one point of A. Hence, any $x \in X$ can be obtained as the limit of a sequence $\{a_n\}$ in A. The classic example is the set of rational numbers, which is dense in the real line. If $Z_k(\sigma)$ denotes the set set of neural network discriminants with k neurons in one hidden layer, Theorem 6.2 states that $Z(\sigma) = \bigcup_{k=1}^{\infty} Z_k(\sigma)$ is dense in the metric space $C(I^d)$, with the metric $\rho(f,g) = \sup_{\mathbf{x} \in R^d} |f(\mathbf{x}) - g(\mathbf{x})|$, provided that σ is a continuous sigmoid. Any continuous function on I^d is the limit of a sequence of neural network discriminants, and so can be approximated arbitrarily well by such discriminants. We remark that the result can be modified to replace I^d by any bounded domain D in R^d.

Since classifiers can be defined from discriminants, it is not surprising that denseness of discriminants translate to universal approximation of the optimal classifier by neural networks. This is shown by the following result, which is a corollary of Theorem 6.2.

Theorem 6.3. *If C_k is the class of one-hidden layer neural networks with k hidden nodes and sigmoid nonlinearity σ, then*

$$\lim_{k \to \inf} \inf_{\psi \in C_k} \varepsilon[\psi] = \varepsilon^* , \tag{6.87}$$

for any distribution of $(\mathbf{X}, Y)$.

The ReLU nonlinearity is not included in Theorem 6.2, since it is not a sigmoid. In particular, it is not bounded. Before we give a universal approximation result for the ReLU, we point out that Theorem 6.2 is predicated on the fact that the number of neurons k in the hidden layer, i.e., the *width* of the neural network, must be allowed to increase without bound. The only bound is on the *depth* of the neural network, which in this case is two (one hidden layer and one output layer). These theorems apply therefore to *depth-bound* networks.

The next recent result by Lu and collaborators applies to *width-bound* networks, where the maximum number of neurons per layer is fixed, but the number of layers themselves is allowed to increase freely. Let $\xi : R^d \to R$ be the discriminant implemented by such a network (we do not give an explicit expression of it here).

Theorem 6.4. *Let f be an integrable function defined on R^n. Given any $\tau > 0$, there is a neural network discriminant ξ with ReLU nonlinearities and width $\leq d + 4$, such that*

$$\int_{R^n} |f(\mathbf{x}) - \xi(\mathbf{x})| \, d\mathbf{x} < \tau . \tag{6.88}$$

The important point here is that to obtain universal approximation capability with a width-bound neural network, an arbitrary number of layers may be necessary. These are known as *deep neural networks*. This represents a paradigm shift with respect to the previous practice, where wide networks with small depth (perhaps only one hidden layer) were widely preferred.

Interestingly, for approximation of arbitrary accuracy, a deep network has to be sufficiently (but not arbitrarily) wide as well, as shown by the next result by the same authors.

Theorem 6.5. *Given any $f \in C(I^n)$ that is not constant along any direction, there exists a $\tau^* > 0$ such that*

$$\int_{R^n} |f(\mathbf{x}) - \xi(\mathbf{x})| \, d\mathbf{x} \geq \tau^*, \tag{6.89}$$

for all neural network discriminants ξ with ReLU nonlinearities and width $\leq d - 1$.

*6.2.4 Universal Consistency Theorems

The results in Section 6.2.2 do not consider data. In particular, they do not weigh directly on the issue of consistency of neural network classification. In this section, we state without proof (see Bibliographical Notes) two strong universal consistency for neural network classification rules. The first one is based on minimization of the empirical classification error (6.57), while the second assumes minimization of the mean absolute error (6.58). Though these results are mostly of theoretical interest, given the difficulty of minimizing these scores in practice, they do illustrate the expressive power of depth-2 neural networks, in agreement with Theorems 6.2 and 6.3 (which do not consider training).

The first result applies to to threshold sigmoids and empirical error minimization.

Theorem 6.6. *Let Ψ_n be the classification rule that uses minimization of the empirical error to design a neural network with k hidden nodes and threshold sigmoids. If $k \to \infty$ such that $k \ln n / n \to 0$ as $n \to \infty$, then Ψ_n is strongly universally consistent.*

As usual for these kinds of results, Theorem 6.6 requires that the complexity of the classification rule, in terms of the number of neurons k in the hidden layer, be allowed to grow, slowly, with the sample size n.

The next result applies to arbitrary sigmoids and absolute error minimization, but regularization must be applied in the form of a constraint on the magnitude of the output weights.

Theorem 6.7. *Let Ψ_n be the classification rule that uses minimization of the absolute error to design a neural network with k_n hidden nodes and arbitrary sigmoid, with the constraint that the* *output weights satisfy*

$$\sum_{i=0}^{k_n} |c_i| \leq \beta_n \, . \tag{6.90}$$

If $k_n \to \infty$ and $\beta_n \to \infty$ and $\frac{k_n \beta_n^2 \ln(k_n \beta_n)}{n} \to 0$ as $n \to \infty$, then Ψ_n is universally consistent.

Notice that this result requires that complexity be allowed to grow, slowly, with sample size, both in terms of the number of neurons k in the hidden layer and the bound on the output weights.

6.3 Decision Trees

The main idea behind decision tree classification rules is to cut up the space recursively, as in a game of "20 questions." This means that decision trees can handle *categorical features* as well as numeric ones.

One of the strong points of decision trees is that they provide nonlinear, complex classifiers that also have a high degree of *interpretability*, i.e., the inferred classifier specifies simple logical rules that can be validated by domain experts and help generate new hypotheses.

A tree classifier consists of a hierarchy of *nodes* where data splitting occurs, except for terminal *leaf* nodes where label assignment are made. All tree classifiers have an initial *splitting node*, known as the *root node*. Each splitting node has at least two *descendants*. The *depth* of a tree classifier is the maximum number of splits between the root node and any leaf node. In a *binary tree*, all splitting nodes have exactly two descendants. It is easy to see that the leaf nodes partition the feature space, i.e., they are do not overlap and cover the entire space. The following example illustrates these concepts.

Example 6.6. Here we separate the XOR data set in Example 6.1 using a binary tree with three splitting nodes and four leaf nodes. The tree classifier and node diagram are displayed in Figure 6.10. (We are back to employing the usual labels 0 and 1 for the classes.) The decision boundary and regions are the same as for the nonlinear SVM classifier in Example 6.1. In the diagram, the root node is depicted at the top, and the leaf nodes at the bottom. Each leaf in this case corresponds to a different rectangular connected piece of the decision regions. A general fact about decision trees is that each leaf node can be represented by a logical AND of the intermediate decisions. In the present example, the left most leaf node corresponds to the clause "$[x_1 \leq 0]$ AND $[x_2 \leq 0]$." Different leaf nodes can be combined with the OR logic. For example, for the present tree classifier we have

$$\begin{aligned} \text{Class 0:} \quad & ([x_1 \leq 0] \text{ AND } [x_2 > 0]) \text{ OR } ([x_1 > 0] \text{ AND } [x_2 \leq 0]) \, . \\ \text{Class 1:} \quad & ([x_1 \leq 0] \text{ AND } [x_2 \leq 0]) \text{ OR } ([x_1 > 0] \text{ AND } [x_2 \leq 0]) \, . \end{aligned} \tag{6.91}$$

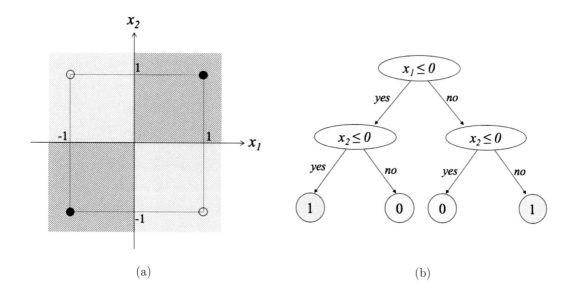

(a)   (b)

Figure 6.10: Decision tree with the XOR data set. (a) Original data and tree classifier (same color conventions as in Figure 6.2 are used). (b) Tree diagram.

Logical simplifications should be carried out if possible to obtain the smallest expressions. This illustrates the interpretability capabilities of decision tree classifiers. ◇

Next we examiner how to train tree classifiers from sample data. We consider the well-known *Classification and Regression Tree* (CART) rule, where the decision at each node is of the form $x^j \leq \alpha$ where x^j is one of the coordinates of the point $\mathbf{x}$, just as in the case of Example 6.6. It is clear that this always partitions the feature space into unions of rectangles. Here we consider CART for classification, while the regression case will be considered in Section 11.5.

For each node, the coordinate to split on j and the split threshold α are determined in training by using the concept of an *impurity* criterion. Given a node R (a hyper-rectangle in feature space, in the case of CART), let $N_i(R)$ be the number of training points from class i in R, for $i = 0, 1$. Hence, $N(R) = N_0(R) + N_1(R)$ is the total number of points in R. The *impurity* of R is defined by

$$\kappa(R) = \xi(p, 1 - p), \qquad (6.92)$$

where $p = N_1(R)/N(R)$ and $1 - p = N_0(R)/N(R)$. The *impurity function* $\xi(p, 1 - p)$ is nonnegative and satisfies the following intuitive conditions:

(1) $\xi(0.5, 0.5) \geq \xi(p, 1 - p)$ for any $p \in [0, 1]$ (so that $\kappa(R)$ is maximum when $p = 1 - p = 0.5$, i.e. $N_1(R) = N_0(R)$, corresponding to maximum impurity).

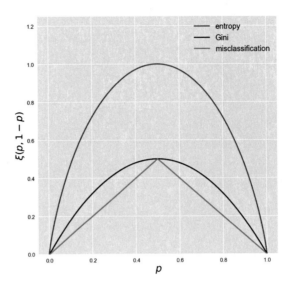

Figure 6.11: Common impurity functions used in CART (plot generated by `c06_impurity.py`).

(2) $\xi(0,1) = \xi(1,0) = 0$ (so that $\kappa(R) = 0$ if either $N_1(R) = 0$ or $N_0(R) = 0$; i.e. R is *pure*).

(3) As a function of p, $\xi(p, 1 - p)$ increases for $p \in [0, 0.5]$ and decreases for $p \in [0.5, 1]$ (so that $\kappa(R)$ increases as the ratio $N_1(R)/N_0(R)$ approaches 1).

Three common choices for ξ are:

1. The entropy impurity: $\xi(p, 1 - p) = -p \ln p - (1-p) \ln(1-p)$.

2. The *Gini* or variance impurity: $\xi(p, 1 - p) = 2p(1 - p)$.

3. The misclassification impurity: $\xi(p, 1 - p) = \min(p, 1 - p)$.

See Figure 6.11 for an illustration. These functions should look familiar, as they are just the same as some of the criteria used in F-errors in Chapter 2. They are related to, respectively, the entropy of a binary source in (2.69), the function used in the nearest-neighbor distance in (2.68), and the function used in the Bayes error in (2.67).

For a node R, the coordinate j to split on and the split threshold α are determined as follows. Let

$$
\begin{aligned}
R_{\alpha,-}^{j} &= \text{sub-rectangle resulting from split} x^j \le \alpha \\
R_{\alpha,+}^{j} &= \text{sub-rectangle resulting from split} x^j > \alpha
\end{aligned}
\tag{6.93}
$$

The *impurity drop* is defined by

$$\Delta_R(j,\alpha) = \kappa(R) - \frac{N(R^j_{\alpha,-})}{N(R)}\kappa(R^j_{\alpha,-}) - \frac{N(R^j_{\alpha,+})}{N(R)}\kappa(R^j_{\alpha,+}) \qquad (6.94)$$

The strategy is to search for the j and α that maximize the impurity drop. Since there are only a finite number of data points and coordinates, there is only a finite number of distinct candidate splits (j,α), so the search can be exhaustive (e.g., consider as candidate α only the mid-points between successive data points along the given coordinate j). In a fully-grown tree, splitting stops when a pure node is encountered, which is then declared a leaf node and given the label of the points contained in it. Training terminates when all current nodes are pure.

As a general rule, however, fully-grown trees should never be employed, as they are almost certain to produce overfitting, even in simple problems. Instead, regularization techniques should be used to avoid overfitting. Simple examples of such techniques include:

- Stopped Splitting: Call a node a leaf and assign the majority label if (1) there are fewer than a specified number of points in the node, (2) the best impurity drop is below a specified threshold, or (3) a maximum tree depth has been reached, or (4) a maximum number of leaf nodes has been obtained.

- Pruning: Fully grow the tree, then successively merge pairs of neighboring leafs that increase impurity the least until a specified maximum level of impurity is achieved.

A more sophisticated form of stopped splitting rules out candidate splits that leave fewer than a specified number of points in either of the child nodes. Splitting stops when no remaining legal splits are left.

Another regularization strategy to reduce overfitting of decision trees is to use ensemble classification, described in Section 3.5.1. For example, in the *random forest* family of classification rules, a number of fully-grown trees are trained on randomly perturbed data (e.g. through bagging), and the decisions are combined by majority voting.

Example 6.7. Figure 6.12 displays classifiers produced by a CART classification rule with Gini impurity on the training data in Figure 6.4. Regularization is employed, whereby no splits that leave fewer than s points in either child node are allowed. The plots show that overfitting quickly occurs with decreasing s. The case $s = 1$, which corresponds to a fully-grown tree, displays gross overfitting, in agreement with the remark made previously that fully-grown trees should not be used. The case $s = 20$ is interesting: there is only one split and the decision tree consists of a single node, i.e., the root node. Such one-split decision trees are known as *stumps* and can be surprisingly effective (see the Bibliographical Notes). Stumps can always be obtained by making s large enough

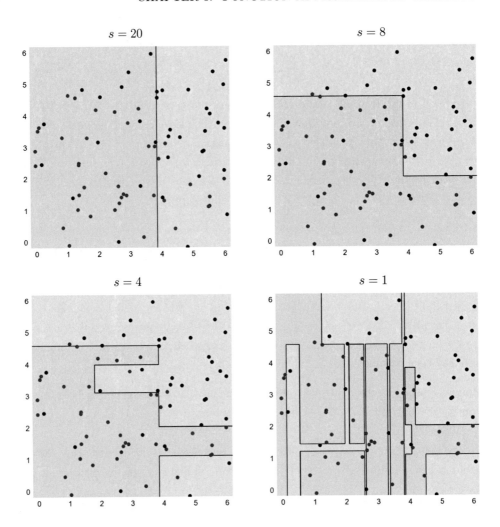

Figure 6.12: Classifiers produced by a CART rule with Gini impurity on the training data in Figures 6.4 and 6.8. No splits that leave fewer than s points in either child node are allowed. The case $s = 1$ corresponds to a fully-grown tree (plots generated by c06_tree.py).

(of course they can be also obtained more simply by imposing a maximum depth of 1 or a maximum leaf node number of 2). The choice of the regularization parameter s is a model selection problem (see Chapter 8). ◇

Finally, we remark that CART with impurity-based splitting is, surprisingly, *not* universally consistent in general, meaning that, for every impurity criterion, one can find a distribution of the data for which things go wrong. There are however other universally consistent decision tree rules. For example, if splitting depends only on $\mathbf{X}_1, \ldots, \mathbf{X}_n$, and the labels are used only for majority voting at the leaves, one can apply Theorem 5.5 to obtain universal consistency, provided that the partitions satisfy conditions (i) and (ii) of that theorem. In addition, there are examples of universally consistent tree rules that use the labels in splitting (see the Bibliographical notes).

6.4 Rank-Based Classifiers

In this section, we discuss briefly the family of *rank-based* classification rules, which are based only on the relative order between feature values and not their numerical magnitude. This makes these rules resistant to noisy and unnormalized data. Rank-based rules generally produce simple classifiers that are less prone to overfit and likely to be interpretable.

The original and most well-known example of rank-based classifier is the *Top Scoring Pair* (TSP) classification rule. Given two feature indices $1 \leq i, j \leq d$, with $i \neq j$, the TSP classifier is given by

$$\psi(\mathbf{x}) = I_{x_i < x_j} = \begin{cases} 1, & x_i < x_j, \\ 0, & \text{otherwise.} \end{cases} \tag{6.95}$$

Therefore, the TSP classifier uses a fixed decision boundary (a 45-degree line) and depends only on the feature ranks, and not on their magnitudes.

Training of a TSP classifier involves only finding a pair (i^*, j^*) that maximizes an empirical score given the data:

$$\begin{aligned} (i^*, j^*) &= \arg \max_{1 \leq i, j \leq d} \left[\hat{P}(X_i < X_j \mid Y = 1) - \hat{P}(X_i < X_j \mid Y = 0) \right] \\ &= \arg \max_{1 \leq i, j \leq d} \left[\frac{1}{N_0} \sum_{k=1}^{n} I_{X_{ki} < X_{kj}} I_{Y_k=1} - \frac{1}{N_1} \sum_{k=1}^{n} I_{X_{ki} < X_{kj}} I_{Y_k=0} \right] . \end{aligned} \tag{6.96}$$

There are efficient greedy search procedures to perform this maximization in case d is large (see the Bibliographical Notes).

We classify TSP as a function-approximation classification rule because it, like all preceding examples of classification rules in this chapter, seeks to adjust a discriminant, in this case, the simple discriminant $g(\mathbf{x}) = x_i - x_j$, to the training data by optimizing an empirical score. Notice that the amount of adjusting is small since, given the selected pair (i^*, j^*), the TSP classifier is independent of the sample data.

The TSP approach can be extended to any number k of pairs of features, by means of majority voting between k TSP classifiers. This could be done by a perturbation approach, as in the ensemble classification rules in Section 3.5.1, but the published k-TSP classification rule does this by taking the top TSPs according to the score in (6.96). Another example of rank-based classification rule is the *Top-Scoring Median* (TSM) algorithm, which is based on comparing the median ranks between two groups of features.

6.5   Bibliographical Notes

The Support Vector Machine was introduced in Boser et al. [1992]. Our discussion of SVMs follows for the most part that in Webb [2002].

The model of the neuron as a linear network followed by a trigger nonlinearity was originally proposed by McCulloch and Pitts [1943]. Rosenblatt's Perceptron algorithm appeared in Rosenblatt [1957]; see Duda et al. [2001] for a thorough description. A history of the Neural Networks is provided in Section 16.5.3 of Murphy [2012a], where the invention of the backpropagation algorithm is credited to Bryson and Ho [1969]. It seems to have been rediscovered at least two other times [Werbos, 1974; Rumelhart et al., 1985].

Convolutional Neural Networks go back at least to the *Neocognitron* architecture proposed by Fukushima [1980]. CNNs became well-known after the publication of the LeNet-5 architecture for recognition of signatures in bank checks in the 1990's [LeCun et al., 1998]. "AlexNet," as it became eventually known, was an innovative CNN architecture that used ReLU nonlinearities and dropout, winning the ImageNet competition in 2012 by a large margin [Krizhevsky et al., 2012]. The VGG16 architecture was one of a few architectures proposed in Simonyan and Zisserman [2014], which introduced the use of very large CNNs using small 3×3 filters. The ImageNet database was described in Deng et al. [2009]. The statistics about ImageNet are from Fei-Fei et al. [2010]. A recent compact treatment of CNNs and other deep neural network architectures is found in Buduma and Locascio [2017].

CART was introduced in Breiman et al. [1984]. A counter-example that shows that CART is not universally consistent is given in Section 20.9 of Devroye et al. [1996]. On the other hand, Theorem 21.2 in Devroye et al. [1996] gives conditions that a decision tree rule that uses the labels in splitting must satisfy to achieve strong consistency. The same authors go on to exhibit some specific cases where such rules are universally strongly consistent. The application of bagging to CART was explored by Breiman [2001]. There has been considerable interest in the application of random forests in different areas of Bioinformatics; e.g. see Alvarez et al. [2005]; Izmirlian [2004]; Knights et al. [2011].

The TSP classification rule was introduced in Geman et al. [2004], while the $kTSP$ extension of it appeared in Tan et al. [2005]. The TSM classification rule was introduced in Afsari et al. [2014]. The latter reference defines the family of *rank-in-context* (RIC) classification rules, which includes all the previous examples as special cases. This reference also discusses greedy search procedures to find the top scoring pairs.

Theorem 6.1 appears in Lorentz [1976]; see also Devroye et al. [1996]. Girosi and Poggio [1989] cite results from the Russian literature [Vitushkin, 1954] to argue that Theorem 6.1 is not useful in practice, given the high irregularity of the functions involved. Theorem 6.2 is Theorem 2 of Cybenko [1989]. The continuity of the sigmoid assumed in that theorem is shown to be unnecessary by Theorem 30.4 in Devroye et al. [1996]. See also Hornik et al. [1989]; Funahashi [1989]. Theorem 6.3 is Corollary 30.1 of Devroye et al. [1996]. Theorems 6.4 and 6.5 are from Lu et al. [2017]. For a description of nonlinear least-squares algorithms, including the Gauss-Newton and Levenberg-Marquardt Algorithms, see Chapter 9 of Nocedal and Wright [2006]. Theorem 6.6 is due to is due to Faragó and Lugosi, while Theorem 6.7 is due to Lugosi and Zeger; they are Theorems 30.7 and 30.9 in Devroye et al. [1996], respectively.

6.6 Exercises

6.1. Consider a linear discriminant $g(\mathbf{x}) = a^T \mathbf{x} + b$.

   (a) Use the method of Lagrange multipliers to show that the distance of a point $\mathbf{x}_0$ to the hyperplane $g(\mathbf{x}) = 0$ is given by $|g(\mathbf{x}_0)|/\|\mathbf{a}\|$.
   Hint: Set up a minimization problem with an equality constraint. (The theory is similar to the case with inequality constraints, except that the Lagrange multiplier is unconstrained and there is no complementary slackness condition).

   (b) Use the result in item (a) to show that the margin in a linear SVM is equal to $1/\|\mathbf{a}\|$.

6.2. Show that the polynomial kernel $K(x, y) = (1 + x^T y)^p$ satisfies Mercer's condition.

6.3. Show that the decision regions produced by a neural network with k threshold sigmoids in the *first* hidden layer, no matter what nonlinearities are used in succeeding layers, are equal to the intersection of k half-spaces, i.e., the decision boundary is piecewise linear.
   Hint: All neurons in the first hidden layer are perceptrons and the output of the layer is a binary vector.

6.4. A neural network with l and m neurons in two hidden layers implements the discriminant:

$$\zeta(\mathbf{x}) = c_0 + \sum_{i=1}^{l} c_i \xi_i(x), \qquad (6.97)$$

where $\xi_i(\mathbf{x}) = \sigma(\phi_i(x))$, for $i = 1, \ldots, l$,

$$\phi_i(\mathbf{x}) = b_{i0} + \sum_{j=1}^{m} b_{ij} v_j(x), \quad i = 1, \ldots, l, \tag{6.98}$$

where $v_j(x) = \sigma(\chi_j(x))$, for $j = 1, \ldots, m$, and

$$\chi_j(\mathbf{x}) = a_{j0} + \sum_{k=1}^{d} a_{jk} x_k, \quad j = 1, \ldots, m. \tag{6.99}$$

(a) Determine the backpropagation algorithm updates for the coefficients c_i, b_{ij}, and a_{jk}.

(b) Find the backpropagation equations for this problem.

6.5. For the VGG16 CNN architecture (see Figure 6.9):

(a) Determine the number of filters used in each convolution layer.

(b) Based on the fact that all filters are of size $3 \times 3 \times r$, where r is the depth of the previous layer, determine the total number of convolution weights in the entire network.

(c) Add the weights used in the fully-connected layers to obtain the total number of weights used by VGG16.

6.6. Consider the simple CART classifier in R^2 depicted below, consisting of three splitting nodes and four leaf nodes.

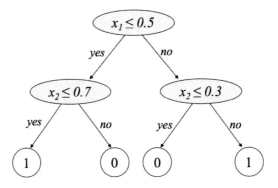

Figure 6.13: Diagram for Problem 6.6

Find the weights of a two-hidden-layer neural network with threshold sigmoids, with three neurons in the first hidden layer and four neurons in the second hidden layer, which implements the same classifier.

Hint: Note the correspondence between the number of neurons in the first and second hidden layers and the numbers of splitting nodes and leaf nodes, respectively.

6.7. Consider the training data set given in the figure below.

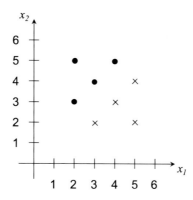

Figure 6.14: Diagram for Problem 6.7

(a) By inspection, find the coefficients of the linear SVM hyperplane $a_1x_1 + a_2x_2 + a_0 = 0$ and plot it. What is the value of the margin? Say as much as you can about the values of the Lagrange multipliers associated with each of the points.

(b) Apply the CART rule, using the misclassification impurity, and stop after finding one splitting node (this is the "1R" or "stump" rule). If there is a tie between best splits, pick one that makes at most one error in each class. Plot this classifier as a decision boundary superimposed on the training data and also as a binary decision tree showing the splitting and leaf nodes.

(c) How do you compare the classifiers in (a) and (b)? Which one is more likely to have a smaller classification error in this problem?

6.8. Consider the training data set given in the figure below.

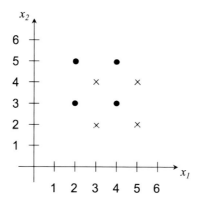

Figure 6.15: Diagram for Problem 6.8

(a) There are exactly two CART classifiers with two splitting nodes (root node plus another node) that produce an apparent error of 0.125. Plot these classifiers in the form of a decision boundary superimposed on the training data and also as a binary decision tree showing the splitting and leaf nodes.

(b) Find the architecture and weights of neural network implementations of the two classifiers in item (a).

Hint: See Exercise 6.6.

6.7 Python Assignments

6.9. This assignment concerns Example 6.2.

(a) Modify the code in c06_svm.py to obtain plots for $C = 0.001, 0.01, 0.1, 1, 10, 100, 1000$ and $n = 50, 100, 250, 500$ per class. Plot the classifiers over the range $[-3, 9] \times [-3, 9]$ in order to visualize the entire data and reduce the marker size from 12 to 8 to facilitate visualization. Which classifiers are closest to the optimal classifier? How do you explain this in terms of underfitting/overfitting? See the coding hint in part (a) of Problem 5.8.

(b) Compute test set errors for each classifier in part (a), using the same procedure as in part (b) of Problem 5.8. Generate a table containing each classifier plot in part (a) with its test set error rate. Which combinations of sample size and parameter C produce the top 5 smallest error rates?

(c) Compute expected error rates for the SVM classification rules in part (a), using the same procedure as in part (c) of Problem 5.8, with $R = 200$. Which parameter C should be used for each sample size?

(d) Repeat parts (a)–(c) using the polynomial kernel with order $p = 2$. Do the results change significantly?

6.10. This assignment concerns Example 6.4.

(a) Modify the code in c06_nnet.py to obtain plots for $H = 2, 3, 5, 8, 10, 15, 30$ neurons in the hidden layer and $n = 50, 100, 250, 500$ per class. Plot the classifiers over the range $[-3, 9] \times [-3, 9]$ in order to visualize the entire data and reduce the marker size from 12 to 8 to facilitate visualization. Which classifiers are closest to the optimal classifier? How do you explain this in terms of underfitting/overfitting? See the coding hint in part (a) of Problem 5.8.

(b) Compute test set errors for each classifier in part (a), using the same procedure as in part (b) of Problem 5.8. Generate a table containing each classifier plot in part (a) with its test set error rate. Which combinations of sample size and number of hidden neurons produce the top 5 smallest error rates?

(c) Compute expected error rates for the neural network classification rules in part (a), using the same procedure as in part (c) of Problem 5.8, with $R = 200$. Which number of hidden neurons should be used for each sample size?

(d) Repeat parts (a)–(c) using ReLU nonlinearities. How do the results change?

6.11. This assignment concerns Example 6.7.

(a) Modify the code in `c06_tree.py` to obtain plots for $s = 20, 16, 8, 5, 4, 2, 120, 16, 8, 5, 4, 2, 1$ and $n = 50, 100, 250, 500$ per class. Plot the classifiers over the range $[-3, 9] \times [-3, 9]$ in order to visualize the entire data and reduce the marker size from 12 to 8 to facilitate visualization. Which classifiers are closest to the optimal classifier? How do you explain this in terms of underfitting/overfitting? See the coding hint in part (a) of Problem 5.8.

(b) Compute test set errors for each classifier in part (a), using the same procedure as in part (b) of Problem 5.8. Generate a table containing each classifier plot in part (a) with its test set error rate. Which combinations of sample size and parameter s produce the top 5 smallest error rates?

(c) Compute expected error rates for the tree classification rules in part (a), using the same procedure as in part (c) of Problem 5.8, with $R = 200$. Which parameter s should be used for each sample size?

(d) Repeat parts (a)–(c) by limiting the number of leaf nodes to $l = 10, 6, 4, 2$ (notice that $l = 2$ must produce a stump). How do you compare this to the previous method as a regularization technique? Is it as effective in producing accurate tree classifiers?

6.12. In this assignment, we apply the VGG16 convolutional neural networks and a Gaussian Radial-Basis Function (RBF) nonlinear SVM to the Carnegie Mellon University ultrahigh carbon steel (UHCS) dataset (see Section A8.6). We will apply a *transfer learning* approach, where the VGG16 network with pre-trained weights is used to generate features to train the SVM.

We will classify the micrographs according to `primary microconstituent`. As explained in Section A8.6, there are a total of seven different labels, corresponding to different phases of steel resulting from different thermal processing. The training data will be the first 100 data points in the spheroidite, carbide network, pearlite categories and the first 60 points in the spheroidite+Widmanstätten category. The remaining data points will compose the various test sets (more on this below). Figure 6.16 displays sample micrographs in each of the four categories. These materials are sufficiently different that classifying their micrographs should be easy, *provided that* one has the right features. In this assignment, we use the pre-trained convolutional layers of the VGG16 to provide the featurization.

Spheroidite   Pearlite

Carbide Network   Spheroidite+Widmanstätten

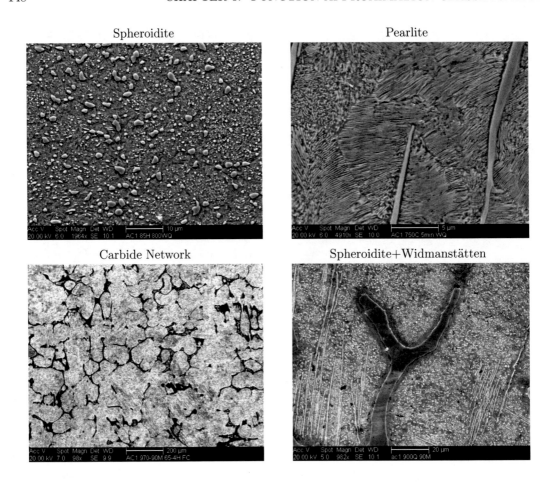

Figure 6.16: Sample micrographs from the CMU-UHCS database. From the top left clockwise, these are micrographs 2, 5, 9, and 58 in the database. Source: CMU-UHCS database [Hecht et al., 2017; DeCost et al., 2017]. See Section A8.6. Images used with permission.

The classification rule to be used is a Radial Basis Function (RBF) nonlinear SVM. We will use the one-vs-one approach (see Exercise 4.2), to deal with the multiple labels, where each of 4 choose 2 = 6 classification problems for each pair of labels are carried out. Given a new image, each of the six classifiers is applied and then a vote is taken to achieve a consensus for the most often predicted label.

To featurize the images, we will use the pre-trained VGG16 deep convolutional neural network (CNN), discussed in Section 6.2.2. We will ignore the fully connected layers, and take the features from the input to the max-pooling layers only, using the "channels" mean value as the feature vector (each channel is a 2D image corresponding to the output of a different filter). This results in feature vectors of length 64, 128, 256, 512, 512, respectively (these lengths correspond to the number of filters in each layer and are fixed, having nothing to do with the image size). In each pairwise classification experiment, we will select one of the five layers according to the best 10-fold cross-validation error estimate (cross-validation estimators will be discussed in detail in Chapter 7).

You are supposed to record the following:

(a) The convolution layer used and the cross-validated error estimate for each of the six pairwise two-label classifiers.

(b) Separate test error rates on the unused micrographs of each of the four categories, for the pairwise two-label classifiers and the multilabel one-vs-one voting classifier described previously. For the pairwise classifiers use only the test micrographs with the two labels used to train the classifier. For the multilabel classifier, use the test micrographs with the corresponding four labels.

(c) For the mixed pearlite + spheroidite test micrographs, apply the trained pairwise classifier for pearlite vs. spheroidite and the multilabel voting classifier. Print the predicted labels by these two classifiers side by side (one row for each test micrograph). Comment your results.

(d) Now apply the multilabel classifier on the pearlite + Widmanstätten and martensite micrographs and print the predicted labels. Compare to the results in part (c).

In each case above, interpret your results. Implementation should use the scikit-learn and Keras python libraries.

Coding hints:

1. The first step is to read in and preprocess each micrograph and featurize it. This will take most of the computation time. First read in the images using the Keras `image` utility:
   `img = image.load_img('image file name')`
   `x = image.img_to_array(img)`
   Next, crop the images to remove the subtitles:
   `x = x[0:484,:,:]`
   add an artificial dimension to specify a batch of one image (since Keras works with batches of images):
   `x = np.expand_dims(x,axis=0)`
   and use the Keras function `preprocess_input` to remove the image mean and perform other conditioning:
   `x = preprocess_input(x)`

Notice that no image size reduction is necessary, as Keras can accept any input image size when in featurization mode.

2. The features are computed by first specifying as base model VGG16 with the pretrained ImageNet weights in featurization mode (do not include the top fully-connected layers):

   `base_model = VGG16(weights='imagenet', include_top=False)`

   extracting the desired feature map (e.g. for the first layer):

   `model = Model(inputs=base_model.input,`
   `                outputs=base_model.get_layer('block1_pool').output)`

   `xb = model.predict(x)`

   and computing its mean

   `F = np.mean(xb,axis=(0,1,2))`

   Notice that steps 1 and 2 have to be repeated for each of the micrographs.

3. The next step is to separate the generated feature vectors by label and by training/testing status, using standard python code. You should save the features to disk since they are expensive to compute.

4. You should use the scikit-learn functions `svm.SVC` and `cross_val_score` to train the SVM classifiers and calculate the cross-validation error rates, respectively, and record which layer produces the best results. When calling `svm.SVC`, set the kernel option to `'rbf'` (this corresponds to the Gaussian RBF) and the C and gamma parameters to 1 and `'auto'`, respectively.

5. Obtain test sets using standard python code. Compute the test set errors for each pairwise classifier and for the multiclass classifier, using the correspond best featurization (layer) for each pairwise classifier, obtained in the previous item. For the multiclass classifier, you should use the scikit-learn function `OnevsOneClassifier`, which has its own internal tie-breaking procedure.

6. The remaining classifiers and error rates for parts (c) and (d) can be obtained similarly.

Chapter 7

Error Estimation for Classification

"The game of science is, in principle, without end. He who
decides one day that scientific statements do not call for
any further test, and that they can be regarded as finally
verified, retires from the game."
– Sir Karl Popper, *The Logic of Scientific Discovery*, 1935.

If one knew the feature-label distribution of a problem, then one could in principle compute the exact error of a classifier. In many practical cases, such knowledge is not available, and it is necessary to estimate the error of a classifier using sample data. If the sample size is large enough, it is possible to divide the available data into training and testing samples, design a classifier on the training set, and evaluate it on the test set. In this chapter, it is shown that this produces a very accurate error estimator, provided that there is an abundance of test points. However, large test samples are not possible if the overall sample size is small, in which case training and testing has to be performed on the same data. This chapter provides a comprehensive survey of different classification error estimation procedures and how to assess their performance.

7.1 Error Estimation Rules

Good classification rules produce classifiers with small average error rates. But is a classification rule good if its error rate cannot be estimated accurately? In other words, is a classifier with a small error rate useful if this error rate cannot be stated with confidence? It can hardly be disputed that the answer to the question is negative. Hence, at a fundamental level, one can only speak of the goodness of a classification rule together with an *error estimation rule* that complements it. We

© Springer Nature Switzerland AG 2020
U. Braga-Neto, *Fundamentals of Pattern Recognition and Machine Learning*,
https://doi.org/10.1007/978-3-030-27656-0_7

will focus on estimation of the error ε_n of a sample-based classifier, since this is the most important error rate in practice. Estimation of other error rates, such as the expected error rate μ_n or the Bayes error ε^*, are comparatively rare in current practice (however, we shall have opportunity to comment on those as well).

Formally, given a classification rule Ψ_n and sample data $S_n = \{(\mathbf{X}_1, Y_1), \ldots, (\mathbf{X}_n, Y_n)\}$, an *error estimation rule* is a mapping $\Xi_n : (\Psi_n, S_n, \xi) \mapsto \hat{\varepsilon}_n$, where $0 \leq \hat{\varepsilon}_n \leq 1$ and ξ denotes internal random factors (if any) of Ξ_n that do not depend on the random sample data. If there are no such internal random factors, the error estimation rule is said to be *nonrandomized*, otherwise, it is said to be *randomized*.

If Ψ_n, S_n, and ξ are all fixed, then $\hat{\varepsilon}_n$ is called an error *estimate* of the fixed classifier $\psi_n = \Psi_n(S_n)$. If only Ψ_n is specified, $\hat{\varepsilon}_n$ is called an *error estimator*. This emphasizes the distinction that, while an error estimation rule is a general procedure, an error estimator is associated with a classification rule, and therefore can have different properties and perform differently with different classification rules. The goodness of a given classification procedure relative to a specific feature-label distribution involves both how well the classification error ε_n approximates the optimal error ε^* as well as how well the error estimator $\hat{\varepsilon}_n$ approximates the classification error ε_n. The latter depends on the joint distribution between the random variables ε_n and $\hat{\varepsilon}_n$.

Example 7.1. (*Resubstitution Error Estimation Rule.*) This rule is based on testing the classifier directly on the training data:

$$\Xi_n^r(\Psi_n, S_n) = \frac{1}{n} \sum_{i=1}^{n} |Y_i - \Psi_n(S_n)(\mathbf{X}_i)| . \tag{7.1}$$

The resulting resubstitution estimate is also variously known as the *apparent error*, the *training error*, or the *empirical error*. It is simply the fraction of errors committed on S_n by the classifier designed by Ψ_n on S_n itself. For example, in Figure 3.1, the resubstitution error estimate is 13/40 = 32.5%. On the other hand, all the classifiers in Examples 4.2 and 4.3 of Chapter 4 have zero apparent error. Note that ξ is omitted in (7.1), as this is a nonrandomized error estimation rule. ◇

Example 7.2. (*Cross-Validation Error Estimation Rule.*) This rule is based on training on a subset of the data, testing on the remaining data, and averaging the results over a number of repetitions. In the basic version of this rule, known as *k-fold cross-validation*, the sample S_n is randomly partitioned into k equal-size *folds* $S_{(i)} = \{(\mathbf{X}_j^{(i)}, Y_j^{(i)}); j = 1, \ldots, n/k\}$, for $i = 1, \ldots, k$ (assume that k divides n), each fold $S_{(i)}$ is left out of the design process to obtain a deleted sample $S_n - S_{(i)}$, a classification rule $\Psi_{n-n/k}$ is applied to $S_n - S_{(i)}$, and the error of the resulting classifier is estimated as the proportion of errors it makes on $S_{(i)}$. The average of these k error rates is the cross-validated error estimate.

The *k-fold cross-validation error estimation rule* can thus be formulated as

$$\Xi_n^{\mathrm{cv}(k)}(\Psi_n, S_n, \xi) = \frac{1}{n}\sum_{i=1}^{k}\sum_{j=1}^{n/k}|Y_j^{(i)} - \Psi_{n-n/k}(S_n - S_{(i)})(\mathbf{X}_j^{(i)})|.\tag{7.2}$$

The random factors ξ specify the random partition of S_n into the folds; therefore, k-fold cross-validation is, in general, a randomized error estimation rule. However, if $k = n$, then each fold contains a single point, and randomness is removed as there is only one possible partition of the sample. As a result, the error estimation rule is nonrandomized. This is called the *leave-one-out error estimation* rule. ◇

7.2 Error Estimation Performance

The performance of error estimators depends on the the classification rule, feature-label distribution, and sample size. From a *frequentist* perspective, performance is determined by the joint sampling distribution of the random variables ε_n and $\hat{\varepsilon}_n$ as they vary with respect to the sample data S_n. Roughly speaking, we are interested in the "average" performance over all possible sample data (and random factors, if any). This section examines performance criteria for error estimators, including the deviation with respect to the true error, bias, variance, RMS, and consistency.

7.2.1 Deviation Distribution

As the purpose of an error estimator is to approximate the true error, the distribution of the *deviation* $\hat{\varepsilon}_n - \varepsilon_n$ is central to accuracy characterization; see Figure 7.1 for an illustration. In contrast to the joint distribution between ε_n and $\hat{\varepsilon}_n$, the deviation distribution has the advantage of being a univariate distribution; it contains a useful subset of the larger information contained in the joint distribution. For a specified feature-label distribution and classification rule, the deviation distribution can be estimated by designing classifiers over a large number of simulated training data sets, computing the estimated and true errors $\hat{\varepsilon}_n$ and ε_n (the latter can be computed analytically if formulas are available, or estimated using large separate test sets) and then fitting a smooth finite-support density to the differences $\hat{\varepsilon}_n - \varepsilon_n$.

7.2.2 Bias, Variance, RMS, and Tail Probabilities

Certain moments and probabilities associated with the deviation distribution play key roles as performance metrics for error estimation:

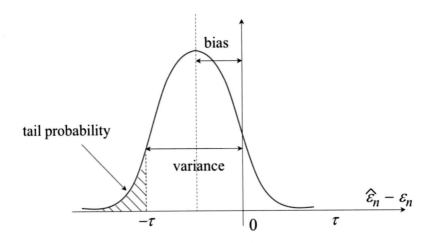

Figure 7.1: Deviation distribution and a few performance metrics derived from it.

1. The *bias*,
$$\text{Bias}(\hat{\varepsilon}_n) \; = \; E[\hat{\varepsilon}_n - \varepsilon_n] \; = \; E[\hat{\varepsilon}_n] - E[\varepsilon_n]. \tag{7.3}$$

The error estimator $\hat{\varepsilon}_n$ is said to be *optimistically biased* if $\text{Bias}(\hat{\varepsilon}_n) < 0$ and *pessimistically biased* if $\text{Bias}(\hat{\varepsilon}_n) > 0$. It is *unbiased* if $\text{Bias}(\hat{\varepsilon}_n) = 0$.

2. The *deviation variance*,
$$\text{Var}_{\text{dev}}(\hat{\varepsilon}_n) \; = \; \text{Var}(\hat{\varepsilon}_n - \varepsilon_n) \; = \; \text{Var}(\hat{\varepsilon}_n) + \text{Var}(\varepsilon_n) - 2\text{Cov}(\varepsilon_n, \hat{\varepsilon}_n). \tag{7.4}$$

3. The *root mean-square error*,
$$\text{RMS}(\hat{\varepsilon}_n) \; = \; \sqrt{E[(\hat{\varepsilon}_n - \varepsilon_n)^2]} \; = \; \sqrt{\text{Bias}(\hat{\varepsilon}_n)^2 + \text{Var}_{\text{dev}}(\hat{\varepsilon}_n)}. \tag{7.5}$$

The square of the RMS is called the *mean-square error* (MSE).

4. The *tail probabilities*,
$$P(|\hat{\varepsilon}_n - \varepsilon_n| \geq \tau) \; = \; P(\hat{\varepsilon}_n - \varepsilon_n \geq \tau) + P(\hat{\varepsilon}_n - \varepsilon_n \leq -\tau), \quad \text{for } \tau > 0. \tag{7.6}$$

Good error estimation performance requires that the bias (magnitude), deviation variance, RMS/MSE, and the tail probabilities be as close as possible to zero. Note that the bias, the deviation variance, and the MSE are respectively the first moment, central second moment, and second moment of the deviation distribution (see Figure 7.1). Therefore, the deviation distribution should be centered at zero and be as thin and tall as possible, i.e., it should approach a point mass centered at zero.

The RMS is generally considered the most important error estimation performance metric. The other performance metrics appear within the computation of the RMS; indeed, all of the five basic moments — the expectations $E[\varepsilon_n]$ and $E[\hat{\varepsilon}_n]$, the variances $\mathrm{Var}(\varepsilon_n)$ and $\mathrm{Var}(\hat{\varepsilon}_n)$, and the covariance $\mathrm{Cov}(\varepsilon_n, \hat{\varepsilon}_n)$ — appear within the RMS. In addition, applying Markov's Inequality (A.70) with $X = |\hat{\varepsilon}_n - \varepsilon_n|^2$ and $a = \tau^2$ yields

$$P(|\hat{\varepsilon}_n - \varepsilon_n| \geq \tau) = P(|\hat{\varepsilon}_n - \varepsilon_n|^2 \geq \tau^2) \leq \frac{E[|\hat{\varepsilon}_n - \varepsilon_n|^2]}{\tau^2} = \left(\frac{\mathrm{RMS}(\hat{\varepsilon}_n)}{\tau} \right)^2 , \quad \text{for } \tau > 0. \quad (7.7)$$

Therefore, a small RMS implies small tail probabilities.

A further comment about the deviation variance is in order. In classical statistics, one considers only the variance of the estimator, in this case, $\mathrm{Var}(\hat{\varepsilon}_n)$. However, unlike in classical statistics, here the quantity being estimated, ε_n, is a "moving target," as it is random itself. This is why it is appropriate to consider the variance of the difference, $\mathrm{Var}(\hat{\varepsilon}_n - \varepsilon_n)$. However, if the classification rule is competent with respect to the sample size and complexity of the problem, then the classification error should not change much with varying training data, i.e., $\mathrm{Var}(\varepsilon_n) \approx 0$ — in fact, overfitting could be defined as present if $\mathrm{Var}(\varepsilon_n)$ is large, since in that case the classification rule is learning the changing data and not the fixed underlying feature-label distribution (more on this in Chapter 8). It follows, from the Cauchy-Schwarz Inequality (A.68) with $X = \varepsilon_n - E[\varepsilon_n]$ and $Y = \hat{\varepsilon}_n - E[\hat{\varepsilon}_n]$ that $\mathrm{Cov}(\varepsilon_n, \hat{\varepsilon}_n) \leq \sqrt{\mathrm{Var}(\varepsilon_n)\mathrm{Var}(\hat{\varepsilon}_n)} \approx 0$, and thus, from (7.4), $\mathrm{Var}(\hat{\varepsilon}_n - \varepsilon_n) \approx \mathrm{Var}(\hat{\varepsilon}_n)$. In other words, as $\mathrm{Var}(\varepsilon_n)$ becomes small, the estimation problem becomes closer to the one in classical statistics.

The effect of randomization of the error estimation rule on $\mathrm{Var}(\hat{\varepsilon}_n)$ can be investigated as follows. If the error estimation rule is randomized, then the error estimator $\hat{\varepsilon}_n$ is still random even after S_n is specified. Accordingly, we define the *internal variance* of $\hat{\varepsilon}_n$ as

$$V_{\mathrm{int}} = \mathrm{Var}(\hat{\varepsilon}_n | S_n). \quad (7.8)$$

The internal variance measures the variability due only to the internal random factors, while the full variance $\mathrm{Var}(\hat{\varepsilon}_n)$ measures the variability due to both the sample S_n and the internal random factors ξ. Letting $X = \hat{\varepsilon}_n$ and $Y = S_n$ in the Conditional Variance Formula (A.84), one obtains

$$\mathrm{Var}(\hat{\varepsilon}_n) = E[V_{\mathrm{int}}] + \mathrm{Var}(E[\hat{\varepsilon}_n | S_n]). \quad (7.9)$$

This variance decomposition equation illuminates the issue. The first term on the right-hand side contains the contribution of the internal variance to the total variance. For nonrandomized $\hat{\varepsilon}_n$, $V_{\mathrm{int}} = 0$; for randomized $\hat{\varepsilon}_n$, $E[V_{\mathrm{int}}] > 0$. Randomized error estimation rules, such as cross-validation, attempt to make the internal variance small by averaging and intensive computation.

*7.2.3   Consistency

One can also consider asymptotic performance as sample size increases. In particular, an error estimator is said to be *consistent* if $\hat{\varepsilon}_n \to \varepsilon_n$ in probability as $n \to \infty$, and *strongly consistent* if convergence is with probability 1. It turns out that consistency is related to the tail probabilities of the deviation distribution: by definition of convergence in probability, $\hat{\varepsilon}_n$ is consistent if and only if, for all $\tau > 0$,

$$\lim_{n \to \infty} P(|\hat{\varepsilon}_n - \varepsilon_n| \geq \tau) = 0. \tag{7.10}$$

By Theorem A.8, we obtain that $\hat{\varepsilon}_n$ is, in addition, strongly consistent if the following stronger condition holds:

$$\sum_{n=1}^{\infty} P(|\hat{\varepsilon}_n - \varepsilon_n| \geq \tau) < \infty, \tag{7.11}$$

for all $\tau > 0$, i.e., the tail probabilities vanish sufficiently fast for the summation to converge. If an estimator is consistent regardless of the underlying feature-label distribution, one says it is *universally consistent*.

It will be seen in Chapter 8 that, if the classification rule has finite *VC dimension*, then the tail probabilities of the resubstitution estimator satisfy, regardless of the feature-label distribution,

$$P\left(|\hat{\varepsilon}_n^r - \varepsilon_n| > \tau\right) \leq 8(n+1)^{V_C} e^{-n\tau^2/32}, \tag{7.12}$$

for all $\tau > 0$, where V_C is the VC dimension. Since $\sum_{n=1}^{\infty} P(|\hat{\varepsilon}_n^r - \varepsilon_n| \geq \tau) < \infty$, for all $\tau > 0$, the resubstitution estimator is universally strongly consistent if $V_C < \infty$. The finite VC dimension ensures that the classification rule is well-behaved in the sense that, as the sample size goes to infinity, the classifier produced is "stable," so that the empirical error is allowed to converge to the true error, in a similar fashion as sample means converge to true means. For example, linear classification rules, such as LDA, linear SVM, and the perceptron, produce simple hyperplane decision boundaries and have finite VC dimension, but decision boundaries produced by nearest-neighbor classification rules are too complex and do not. Incidentally, it will also be shown in Chapter 8 that $|E[\hat{\varepsilon}_n^r - \varepsilon_n]|$ is $O(\sqrt{\ln(n)/n})$ as $n \to \infty$,[1] regardless of the feature-label distribution, so that the resubstitution error estimator is not only universally strongly consistent but also universally asymptotically unbiased, provided that the VC dimension is finite.

If the given classification rule is consistent and the error estimator is consistent, then $\varepsilon_n \to \varepsilon^*$ and $\hat{\varepsilon}_n \to \varepsilon_n$, implying $\hat{\varepsilon}_n \to \varepsilon^*$. Hence, $\hat{\varepsilon}_n$ provides a good estimate of the Bayes error, provided one has a large sample S_n. The question of course is how large the sample size needs to be. While the rate of convergence of $\hat{\varepsilon}_n$ to ε_n can be bounded for some classification rules and some error estimators regardless of the distribution (more on this later), there will always be distributions for which ε_n converges to ε^* arbitrarily slowly, as demonstrated by the "no-free lunch" Theorem 3.3. Hence, one cannot guarantee that $\hat{\varepsilon}_n$ is close to ε^* for a given n, unless one has additional information about the distribution.

---

[1] The notation $f(n) = O(g(n))$ as $n \to \infty$ means that the ratio $|f(n)/g(n)|$ remains bounded as $n \to \infty$; in particular, $g(n) \to 0$ implies that $f(n) \to 0$ and that $f(n)$ goes to 0 at least as fast as $g(n)$ goes to zero.

As in the case of classification, an important caveat is that, in small-sample cases, i.e., when the number of training points is small for the dimensionality or complexity of the problem, consistency does not play a significant role in the choice of an error estimator, the key issue being accuracy as measured, for example, by the RMS.

7.3 Test-Set Error Estimation

We begin the examination of specific error estimation rules by considering the most natural one, namely, test-set error estimation. This error estimation rule assumes the availability of a second random sample $S_m = \{(\mathbf{X}_i^t, Y_i^t); i = 1, \ldots, m\}$, known as the *test data*, which is independent and identically distributed with the training data S_n, and is to be set aside and not used in classifier design. The classification rule is applied on the training data and the error of the resulting classifier is estimated to be the proportion of errors it makes on the test data. This *test-set* or *holdout error estimation rule* error estimation rule can thus be formulated as:

$$\Xi_{n,m}(\Psi_n, S_n, S_m) = \frac{1}{m} \sum_{i=1}^{m} |Y_i^t - \Psi_n(S_n)(\mathbf{X}_i^t)|. \tag{7.13}$$

This is a randomized error estimation rule, with the test data themselves as the internal random factors, $\xi = S_m$.

For any given classification rule Ψ_n, the *test-set error estimator* $\hat{\varepsilon}_{n,m} = \Xi_{n,m}(\Psi_n, S_n, S_m)$ has a few remarkable properties. First of all, it is clear that the test-set error estimator is unbiased in the sense that

$$E[\hat{\varepsilon}_{n,m} \mid S_n] = \varepsilon_n, \tag{7.14}$$

for $m = 1, 2, \ldots$ From this and the Law of Large Numbers (see Theorem A.12) it follows that, given the training data S_n, $\hat{\varepsilon}_{n,m} \to \varepsilon_n$ with probability 1 as $m \to \infty$, regardless of the feature-label distribution. In addition, it follows from (7.14) that $E[\hat{\varepsilon}_{n,m}] = E[\varepsilon_n]$. Hence, the test-set error estimator is also unbiased in the sense of Section 7.2.2:

$$\mathrm{Bias}(\hat{\varepsilon}_{n,m}) = E[\hat{\varepsilon}_{n,m}] - E[\varepsilon_n] = 0. \tag{7.15}$$

We now show that the internal variance of this randomized estimator vanishes as the number of testing samples increases to infinity. Using (7.14) allows one to obtain

$$V_{\mathrm{int}} = E[(\hat{\varepsilon}_{n,m} - E[\hat{\varepsilon}_{n,m} \mid S_n])^2 \mid S_n] = E[(\hat{\varepsilon}_{n,m} - \varepsilon_n)^2 \mid S_n]. \tag{7.16}$$

Moreover, given S_n, $m\hat{\varepsilon}_{n,m}$ is binomially distributed with parameters (m, ε_n):

$$P(m\hat{\varepsilon}_{n,m} = k \mid S_n) = \binom{m}{k} \varepsilon_n^k (1 - \varepsilon_n)^{m-k}, \tag{7.17}$$

for $k = 0, \ldots, m$. From the formula for the variance of a binomial random variable, it follows that

$$
\begin{aligned}
V_{\text{int}} &= E[(\hat{\varepsilon}_{n,m} - \varepsilon_n)^2 | S_n] = \frac{1}{m^2} E[(m\hat{\varepsilon}_{n,m} - m\varepsilon_n)^2 | S_n] \\
&= \frac{1}{m^2} \text{Var}(m\hat{\varepsilon}_{n,m} | S_n) = \frac{1}{m^2} m\varepsilon_n 1 - \varepsilon_n) = \frac{\varepsilon_n(1 - \varepsilon_n)}{m}.
\end{aligned}
\tag{7.18}
$$

The internal variance of this estimator can therefore be bounded as follows:

$$
V_{\text{int}} \leq \frac{1}{4m}. \tag{7.19}
$$

Hence, $V_{\text{int}} \to 0$ as $m \to \infty$. Furthermore, by using (7.9), we get that the full variance of the holdout estimator is simply

$$
\text{Var}(\hat{\varepsilon}_{n,m}) = E[V_{\text{int}}] + \text{Var}(\varepsilon_n). \tag{7.20}
$$

Thus, provided that m is large, so that V_{int} is small — this is guaranteed by (7.19) for large enough m — the variance of the holdout estimator is approximately equal to the variance of the true error itself, which is typically small if n is large.

It follows from (7.16) and (7.19) that

$$
\text{MSE}(\hat{\varepsilon}_{n,m}) = E[(\hat{\varepsilon}_{n,m} - \varepsilon_n)^2] = E[E[(\hat{\varepsilon}_{n,m} - \varepsilon_n)^2 | S_n]] = E[V_{\text{int}}] \leq \frac{1}{4m}. \tag{7.21}
$$

Thus, we have the distribution-free bound on the RMS:

$$
\text{RMS}(\hat{\varepsilon}_{n,m}) \leq \frac{1}{2\sqrt{m}}. \tag{7.22}
$$

With $m = 400$ test points, the RMS is guaranteed to be no larger than 2.5%, regardless of the classification rule and the feature-label distribution, which would be accurate enough for most applications.

Despite its many favorable properties, the holdout estimator has a considerable drawback. In practice, one has n total labeled sample points that must be split into training and testing sets S_{n-m} and S_m. For large n, $E[\varepsilon_{n-m}] \approx E[\varepsilon_n]$, so that $n - m$ is still large enough to train an accurate classifier, while m is still large enough to obtain a good (small-variance) test-set estimator $\hat{\varepsilon}_{n,m}$. However, in scientific applications data is often scarce, and $E[\varepsilon_{n-m}]$ may be too large with respect to $E[\varepsilon_n]$ (i.e., the loss of performance may be intolerable, see Figure 7.2 for an illustration), m may be too small (leading to a poor test-set error estimator), or both. Setting aside a part of the sample for testing means that there are fewer data available for design, thereby typically resulting in a poorer performing classifier, and if fewer points are set aside to reduce this undesirable effect, there are insufficient data for testing.

The variance can also be a problem if sample data are limited. If the number m of testing samples is small, then the variance of $\hat{\varepsilon}_{n,m}$ is usually large and this is reflected in the RMS bound of (7.22).

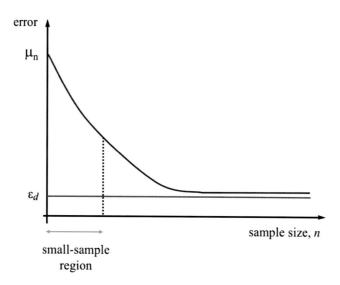

Figure 7.2: Expected error as a function of training sample size, demonstrating the problem of higher rate of classification performance deterioration with a reduced training sample size in the small-sample region.

For example, to get the RMS bound down to 0.05, it would be necessary to use 100 test points. Data sets containing fewer than 100 overall sample points are quite common. Moreover, the bound of (7.19) is tight, it being achieved by letting $\varepsilon_n = 0.5$ in (7.18).

In conclusion, splitting the available data into training and test data is problematic in practice due to the unavailability of enough data to be able to make both n and m large enough. Therefore, for a large class of practical problems, where data are at a premium, test-set error estimation is effectively ruled out. In such cases, one must use the same data for training and testing.

7.4 Resubstitution

The simplest and fastest error estimation rule based solely on the training data is the resubstitution estimation rule, defined previously in Example 7.1. Given the classification rule Ψ_n, and the classifier $\psi_n = \Psi_n(S_n)$ designed on S_n, the resubstitution error estimator is given by

$$\hat{\varepsilon}_n^r = \frac{1}{n} \sum_{i=1}^{n} |Y_i - \psi_n(\mathbf{X}_i)|. \tag{7.23}$$

An alternative way to look at resubstitution is as the classification error according to the *empirical feature-label distribution* given the training data, which is the probability mass function $p_n(\mathbf{X}, Y) =$

$P(\mathbf{X} = \mathbf{X}_i, Y = Y_i \mid S_n) = \frac{1}{n}$, for $i = 1, \ldots, n$. It can be shown that the resubstitution estimator is given by

$$\hat{\varepsilon}_n^r = E_{p_n}[|Y - \psi_n(\mathbf{X})|]. \tag{7.24}$$

The concern with resubstitution is that it is typically optimistically biased; that is, $\text{Bias}(\hat{\varepsilon}_n^r) < 0$ in most cases. The optimistic bias of resubstitution may not be of great concern if it is small; however, it tends to become unacceptably large for overfitting classification rules, especially in small-sample situations. An extreme example is given by the 1-nearest neighbor classification rule, where $\hat{\varepsilon}_n^r \equiv 0$ for all sample sizes, classification rules, and feature-label distributions. Although resubstitution may be optimistically biased, the bias will often disappear as sample size increases, as shown below. In addition, resubstitution is typically a low-variance error estimator. This often translates to good asymptotic RMS properties.

It will be seen in Chapter 8 that resubstitution is useful in a classification rule selection procedure known as *structural risk minimization*, provided that all classification rules considered have finite VC dimension.

7.5   Cross-Validation

The cross-validation error estimation rule improves upon the bias of resubstitution by using a re-sampling strategy in which classifiers are trained and tested on non-overlapping subsets of the data. Several variants of the basic k-fold cross-validation estimation rule defined in Example 7.2 are possible. In *stratified cross-validation*, the classes are represented in each fold in the same proportion as in the original data; it has been claimed that this can reduce variance. As with all randomized error estimation rules, there is concern about the internal variance of cross-validation; this can be reduced by repeating the process of random selection of the folds several times and averaging the results. This is called *repeated k-fold cross-validation*. In the limit, one will be using all possible folds of size n/k, and there will be no internal variance left, the estimation rule becoming nonrandomized. This has been called *complete k-fold cross-validation*.

The most well-known property of the k-fold cross-validation error estimator is its *near-unbiasedness property*

$$E[\hat{\varepsilon}_n^{\text{cv}(k)}] = E[\varepsilon_{n-n/k}]. \tag{7.25}$$

This is a distribution-free property, which holds as long as sampling is i.i.d. (However, see Exercise 7.10.) This property is often and mistakenly taken to imply that the k-fold cross-validation error estimator is unbiased, but this is not true, since $E[\varepsilon_{n-n/k}] \neq E[\varepsilon_n]$, in general. In fact, $E[\varepsilon_{n-n/k}]$ is typically larger than $E[\varepsilon_n]$, since the former error rate is based on a smaller sample size, and so the k-fold cross-validation error estimator tends to be pessimistically biased. To reduce the bias, it is advisable to increase the number of folds (which in turn will tend to increase estimator variance). The maximum value $k = n$ corresponds to the leave-one-out error estimator $\hat{\varepsilon}_n^l$, already introduced in Example 7.2. This is typically the least biased cross-validation estimator, with

$$E[\hat{\varepsilon}_n^l] \; = \; E[\varepsilon_{n-1}] \,. \tag{7.26}$$

The major concern regarding cross-validation is not bias but estimation variance. Despite the fact that, as mentioned previously, leave-one-out cross-validation is a nonrandomized error estimation rule, the leave-one-out error estimator can still exhibit large variance. Small internal variance (in this case, zero) by itself is not sufficient to guarantee a low-variance estimator. Following (7.9), the variance of the cross-validation error estimator also depends on the term $\mathrm{Var}(E[\hat{\varepsilon}_n^{\mathrm{cv}(k)}|S_n])$, which is the variance corresponding to the randomness of the sample S_n. This term tends to be large when sample size is small due to overfitting. In the case of leave-one-out, the situation is made worse by the large degree of overlap between the folds. In small-sample cases, cross-validation not only tends to display large variance, but can also display large bias, and even negative correlation with the true error; for an example, see Exercise 7.10.

One determining factor for the the variance of cross-validation is the fact that it attempts to estimate the error of a classifier $\psi_n = \Psi_n(S_n)$ via estimates of the error of *surrogate classifiers* $\psi_{n,i} = \Psi_{n-n/k}(S_n - S_{(i)})$, for $i = 1, \ldots, k$. If the classification rule is unstable (e.g. overfitting), then deleting different sets of points from the sample may produce wildly disparate surrogate classifiers. The issue is beautifully quantified by the following theorem, which we state without proof. A classification rule is *symmetric* if the designed classifier is not affected by permuting the points in the training sample (this includes the vast majority of classification rules encountered in practice).

Theorem 7.1. *For a symmetric classification rule* Ψ_n,

$$\mathrm{RMS}(\hat{\varepsilon}_n^l) \leq \sqrt{\frac{E[\varepsilon_{n-1}]}{n} + 6P(\Psi_n(S_n)(X) \neq \Psi_{n-1}(S_{n-1})(\mathbf{X}))} \,. \tag{7.27}$$

The previous theorem relates the RMS of leave-one-out with the stability of the classification rule: the smaller the term $P(\Psi_n(S_n)(\mathbf{X}) \neq \Psi_{n-1}(S_{n-1})(\mathbf{X}))$ is, i.e., the more stable the classification rule is (with respect to deletion of a point), the smaller the RMS of leave-one-out is guaranteed to be. For large n, $P(\Psi_n(S_n)(\mathbf{X}) \neq \Psi_{n-1}(S_{n-1})(\mathbf{X}))$ will be small for most classification rules; however, with small n, the situation is different. This is illustrated in Figure 7.3, which shows different classifiers resulting from a single point left out from the original sample. (These are the classifiers that most differ from the original one in each case.) We can observe in these plots that LDA is the most stable rule among the three and CART is the most unstable, with 3NN displaying moderate stability. Accordingly, it is expected that $P(\Psi_n(S_n)(\mathbf{X}) \neq \Psi_{n-1}(S_{n-1})(\mathbf{X}))$ will be smallest for LDA,

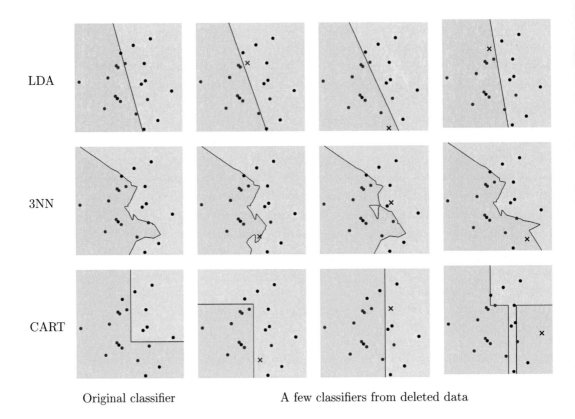

LDA

3NN

CART

Original classifier   A few classifiers from deleted data

Figure 7.3: Illustration of instability of different classification rules. Original classifier and a few classifiers obtained after one of the sample points is deleted. Deleted sample points are represented by "x" (plots generated by c07_deleted.py).

guaranteeing a smaller RMS for leave-one-out, and largest for CART. Instability of the classification rule is linked to its overfitting potential.

Theorem 7.1 proves very useful in generating distribution-free bounds on the RMS of the leave-one-out estimator for classification rules for which bounds for $P(\Psi_n(S_n)(\mathbf{X}) \neq \Psi_{n-1}(S_{n-1})(\mathbf{X}))$ are known. For example, for the kNN classification rule with randomized tie-breaking, $\Psi_n(S_n)(\mathbf{X}) \neq \Psi_{n-1}(S_{n-1})(\mathbf{X})$ is only possible if the deleted point is one of the k nearest neighbors of $\mathbf{X}$. Owing to symmetry, the probability of this occurring is k/n, as all sample points are equally likely to be among the k nearest neighbors of $\mathbf{X}$. Together with the fact that $E[\varepsilon_{n-1}]/n < 1/n$, Theorem 7.1 proves that

$$\mathrm{RMS}(\hat{\varepsilon}_n^l) \leq \sqrt{\frac{6k+1}{n}}. \tag{7.28}$$

Via (7.7), bounds on the RMS lead to corresponding bounds on the tail probabilities $P(|\hat{\varepsilon}_n^l - \hat{\varepsilon}_n| \geq \tau)$. For example, from (7.28) it follows that

$$P(|\hat{\varepsilon}_n^l - \hat{\varepsilon}_n| \geq \tau) \leq \frac{6k+1}{n\tau^2} . \tag{7.29}$$

Clearly, $P(|\hat{\varepsilon}_n^l - \varepsilon_n| \geq \tau) \to 0$, as $n \to \infty$, i.e., leave-one-out cross-validation is universally consistent (see Section 7.2.2) for kNN classification.

The previous bounds are "large-sample bounds." By being distribution-free, they are worst-case, and tend to be loose for small n. Failure to acknowledge this may lead to meaningless results. For example, with $k = 5$ and $n = 31$, the bound in (7.28) yields $\text{RMS}(\hat{\varepsilon}_n^l) \leq 1$. To be assured that the RMS is less than 0.1, one must have $n \geq 1900$.

Finally, we remark that, since k-fold cross-validation relies on an average of error rates of surrogate classifiers designed on samples of size $n - n/k$, it is more properly an estimator of $E[\varepsilon_{n-n/k}]$ than of ε_n. This is a reasonable approach if both the cross-validation and true error variances are small, since the rationale here follows the chain $\hat{\varepsilon}_n^{\text{cv}(k)} \approx E[\hat{\varepsilon}_n^{\text{cv}(k)}] = E[\varepsilon_{n-n/k}] \approx E[\varepsilon_n] \approx \varepsilon_n$. However, the large variance typically displayed by cross-validation under small sample sizes implies that the first approximation $\hat{\varepsilon}_n^{\text{cv}(k)} \approx E[\hat{\varepsilon}_n^{\text{cv}(k)}]$ in the chain is not valid, so that cross-validation cannot provide accurate error estimation in small-sample settings.

7.6 Bootstrap

The bootstrap methodology is a resampling strategy that can be viewed as a kind of smoothed cross-validation, with reduced variance The bootstrap methodology employs the notion of the feature-label empirical distribution $F_{\mathbf{X}Y}^*$ introduced at the beginning of Section 7.4. A *bootstrap sample* S_n^* from $F_{\mathbf{X}Y}^*$ consists of n equally-likely draws with replacement from the original sample S_n. Some sample points will appear multiple times, whereas others will not appear at all. The probability that any given sample point will not appear in S_n^* is $(1 - 1/n)^n \approx e^{-1}$. It follows that a bootstrap sample of size n contains on average $(1 - e^{-1})n \approx 0.632n$ of the original sample points.

The basic bootstrap error estimation procedure is as follows: given a sample S_n, independent bootstrap samples $S_n^{*,j}$, for $j = 1, \ldots, B$, are generated; B has been recommended to be between 25 and 200. Then a classification rule Ψ_n is applied to each bootstrap sample $S_n^{*,j}$, and the number of errors the resulting classifier makes on $S_n - S_n^{*,j}$ is recorded. The average number of errors made over all B bootstrap samples is the bootstrap estimate of the classification error. This *zero bootstrap error estimation rule* can thus be formulated as

$$\Xi_n^{\text{boot}}(\Psi_n, S_n, \xi) = \frac{\sum_{j=1}^{B} \sum_{i=1}^{n} |Y_i - \Psi_n(S_n^{*,j})(\mathbf{X}_i)| \, I_{(\mathbf{X}_i, Y_i) \notin S_n^{*,j}}}{\sum_{j=1}^{B} \sum_{i=1}^{n} I_{(\mathbf{X}_i, Y_i) \notin S_n^{*,j}}} . \tag{7.30}$$

The random factors ξ govern the resampling of S_n to create the bootstrap sample $S_n^{*,j}$, for $j = 1, \ldots, B$; hence, the zero bootstrap is a randomized error estimation rule. Given a classification rule Ψ_n, the *zero bootstrap error estimator* $\hat{\varepsilon}_n^{\text{boot}} = \Xi_n^{\text{boot}}(\Psi_n, S_n, \xi)$ tends to be a pessimistically biased estimator of $E[\varepsilon_n]$, since the number of points available for classifier design is on average only $0.632n$.

As in the case of cross-validation, several variants of the basic bootstrap scheme exist. In the *balanced bootstrap*, each sample point is made to appear exactly B times in the computation; this is supposed to reduce estimation variance, by reducing the internal variance associated with bootstrap sampling. This variance can also be reduced by increasing the value of B. In the limit as B increases, all bootstrap samples will be used up, and there will be no internal variance left, the estimation rule becoming nonrandomized. This is called a *complete zero bootstrap* error estimator, and corresponds to the expectation of the $\hat{\varepsilon}_n^{\text{boot}}$ over the bootstrap sampling mechanism:

$$\hat{\varepsilon}_n^{\text{zero}} = E[\hat{\varepsilon}_n^{\text{boot}} \mid S_n], \tag{7.31}$$

The ordinary zero bootstrap $\hat{\varepsilon}_n^{\text{boot}}$ is therefore a randomized Monte-Carlo approximation of the nonrandomized complete zero bootstrap $\hat{\varepsilon}_n^{\text{zero}}$. By the Law of Large Numbers (see Thm. A.12), $\hat{\varepsilon}_n^{\text{boot}}$ converges to $\hat{\varepsilon}_n^{\text{zero}}$ with probability 1 as the number of bootstrap samples B grows without bound. Note that $E[\hat{\varepsilon}_n^{\text{boot}}] = E[\hat{\varepsilon}_n^{\text{zero}}]$. The pessimistic bias of the zero bootstrap error estimator stems from the fact that, typically, $E[\hat{\varepsilon}_n^{\text{zero}}] > E[\varepsilon_n]$.

A practical way of implementing the complete zero bootstrap estimator is to generate beforehand all distinct bootstrap samples (this is possible if n is not too large) and form a weighted average of the error rates based on each bootstrap sample, the weight for each bootstrap sample being its probability. As in the case of cross-validation, for all bootstrap variants there is a trade-offbetween computational cost and estimation variance.

The *0.632 bootstrap error estimator* is a convex combination of the (typically optimistic) resubstitution and the (typically pessimistic) zero bootstrap estimators:

$$\hat{\varepsilon}_n^{632} = 0.368\,\hat{\varepsilon}_n^r + 0.632\,\hat{\varepsilon}_n^{\text{boot}}. \tag{7.32}$$

Alternatively, the complete zero bootstrap $\hat{\varepsilon}_n^{\text{zero}}$ can be substituted for $\hat{\varepsilon}_n^{\text{boot}}$. The weights in (7.32) are set heuristically; they can be far from optimal for a given feature-label distribution and classification rule. A simple example where the weights are offis afforded by the 1NN rule, for which $\hat{\varepsilon}_n^r \equiv 0$.

In an effort to overcome these issues, an adaptive approach to finding the weights has been proposed. The basic idea is to adjust the weight of the resubstitution estimator when resubstitution is exceptionally optimistically biased, which indicates strong overfitting. Given the sample

$S_n = \{(\mathbf{X}_1, Y_1), \ldots, (\mathbf{X}_n, Y_n)\}$, let

$$\hat{\gamma} = \frac{1}{n^2} \sum_{i=1}^{n} \sum_{j=1}^{n} I_{Y_i \neq \Psi_n(S_n)(\mathbf{X}_j)}, \tag{7.33}$$

and define the relative overfitting rate by

$$\hat{R} = \frac{\hat{\varepsilon}_n^{\text{boot}} - \hat{\varepsilon}_n^r}{\hat{\gamma} - \hat{\varepsilon}_n^r}. \tag{7.34}$$

To be certain that $\hat{R} \in [0, 1]$, one sets $\hat{R} = 0$ if $\hat{R} < 0$ and $\hat{R} = 1$ if $\hat{R} > 1$. The parameter $\hat{R}$ indicates the degree of overfitting via the relative difference between the zero bootstrap and resubstitution estimates, a larger difference indicating more overfitting. Although we shall not go into detail, $\hat{\gamma} - \hat{\varepsilon}_n^r$ approximately represents the largest degree to which resubstitution can be optimistically biased, so that usually $\hat{\varepsilon}_n^{\text{boot}} - \hat{\varepsilon}_n^r \leq \hat{\gamma} - \hat{\varepsilon}_n^r$. The weight

$$\hat{w} = \frac{0.632}{1 - 0.368\hat{R}} \tag{7.35}$$

replaces 0.632 in (7.32) to produce the *0.632+ bootstrap error estimator*,

$$\hat{\varepsilon}_n^{632+} = (1 - \hat{w})\hat{\varepsilon}_n^r + \hat{w}\hat{\varepsilon}_n^{\text{boot}}, \tag{7.36}$$

except when $\hat{R} = 1$, in which case $\hat{\gamma}$ replaces $\hat{\varepsilon}_n^{\text{boot}}$ in (7.36). If $R = 0$, then there is no overfitting, $\hat{w} = 0.632$, and the 0.632+ bootstrap estimate is equal to the plain 0.632 bootstrap estimate. If $R = 1$, then there is maximum overfit, $\hat{w} = 1$, the 0.632+ bootstrap estimate is equal to $\hat{\gamma}$, and the resubstitution estimate does not contribute, as in this case it is not to be trusted. As R changes between 0 and 1, the 0.632+ bootstrap estimate ranges between these two extremes. In particular, in all cases we have $\hat{\varepsilon}_n^{632+} \geq \hat{\varepsilon}_n^{632}$. The nonnegative difference $\hat{\varepsilon}_n^{632+} - \hat{\varepsilon}_n^{632}$ corresponds to the amount of "correction" introduced to compensate for too much resubstitution bias.

7.7   Bolstered Error Estimation

The resubstitution estimator is written in (7.24) in terms of the empirical distribution $p_n(\mathbf{X}, Y)$, which is confined to the original data points, so that no distinction is made between points near or far from the decision boundary. If one spreads out the probability mass put on each point by the empirical distribution, variation is reduced in (7.24) because points near the decision boundary will have more mass go to the other side than will points far from the decision boundary. Another way of looking at this is that more confidence is attributed to points far from the decision boundary than points near it. For $i = 1, \ldots, n$, consider a d-variate probability density function p_i°, called a *bolstering kernel*. Given the sample $S_n = \{(\mathbf{X}_1, Y_1), \ldots, (\mathbf{X}_n, Y_n)\}$, the *bolstered empirical distribution*

$p_n^\diamond(\mathbf{X}, Y)$ has probability density:[2]

$$p_n^\diamond(\mathbf{x}, y) = \frac{1}{n} \sum_{i=1}^{n} p_i^\diamond(\mathbf{x} - \mathbf{X}_i) I_{y=Y_i}. \tag{7.37}$$

Given a classification rule Ψ_n, and the designed classifier $\psi_n = \Psi_n(S_n)$, the *bolstered resubstitution error estimator* is obtained by replacing p_n by $p_n^\diamond$ in (7.24):

$$\hat{\varepsilon}_n^{\,\text{br}} = E_{p_n^\diamond}\left[|Y - \psi_n(\mathbf{X})|\right]. \tag{7.38}$$

The following result gives a computational expression for the bolstered resubstitution error estimate.

Theorem 7.2. *Let $A_j = \{\mathbf{x} \in R^d \mid \psi_n(\mathbf{x}) = j\}$, for $j = 0, 1$, be the decision regions for the designed classifier. Then the bolstered resubstitution error estimator can be written as*

$$\hat{\varepsilon}_n^{\,\text{br}} = \frac{1}{n} \sum_{i=1}^{n} \left(\int_{A_1} p_i^\diamond(\mathbf{x} - \mathbf{X}_i)\, d\mathbf{x}\, I_{Y_i=0} + \int_{A_0} p_i^\diamond(\mathbf{x} - \mathbf{X}_i)\, d\mathbf{x}\, I_{Y_i=1} \right). \tag{7.39}$$

Proof. From (7.38),

$$
\begin{aligned}
\hat{\varepsilon}_n^{\,\text{br}} &= \int |y - \psi_n(\mathbf{x})|\, dF^\diamond(\mathbf{x}, y) \\
&= \sum_{y=0}^{1} \int_{R^d} |y - \psi_n(\mathbf{x})|\, f^\diamond(\mathbf{x}, y)\, d\mathbf{x} \tag{7.40} \\
&= \frac{1}{n} \sum_{y=0}^{1} \sum_{i=1}^{n} \int_{R^d} |y - \psi_n(\mathbf{x})|\, f_i^\diamond(\mathbf{x} - \mathbf{X}_i) I_{y=Y_i}\, d\mathbf{x} \\
&= \frac{1}{n} \sum_{i=1}^{n} \int_{R^d} \psi_n(\mathbf{x})\, f_i^\diamond(\mathbf{x} - \mathbf{X}_i)\, d\mathbf{x}\, I_{Y_i=0} \\
&\quad + \int_{R^d} (1 - \psi_n(\mathbf{x}))\, f_i^\diamond(\mathbf{x} - \mathbf{X}_i)\, d\mathbf{x}\, I_{Y_i=1}.
\end{aligned}
$$

But $\psi_n(\mathbf{x}) = 0$ over A_0 and $\psi_n(\mathbf{x}) = 1$ over A_1, from which (7.39) follows. $\diamond$

The integrals in (7.39) are the error contributions made by the data points, according to the label $Y_i = 0$ or $Y_i = 1$. The bolstered resubstitution estimate is equal to the sum of all error contributions divided by the number of points (see Figure 7.4 for an illustration, where the bolstering kernels are given by uniform circular distributions). Notice that this allows counting partial errors, including errors for correctly classified points that are near the decision boundary.

---

[2]This is not a true density since Y is discrete; see Section 2.6.4.

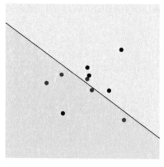 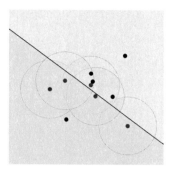 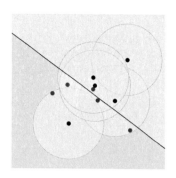

Original classifier   Bolstering kernels for class 0   Bolstering kernels for class 1

Figure 7.4: Bolstered resubstitution for LDA with uniform circular bolstering kernels. The error contribution made by each point is the area of the disk segment extending across the decision boundary (if any) divided by the area of the entire disk. The bolstered resubstitution error is the sum of all contributions divided by the number of points (plots generated by `c07_bolst.py`).

In general, the integrals in (7.39) are too complicated and must be computed numerically. For example, one may apply a simple Monte-Carlo estimate to obtain:

$$\hat{\varepsilon}_n^{\mathrm{br}} \approx \frac{1}{nM} \sum_{i=1}^{n} \left(\sum_{j=1}^{M} I_{X_{ij} \in A_1} I_{Y_i=0} + \sum_{j=1}^{M} I_{X_{ij} \in A_0} I_{Y_i=1} \right), \tag{7.41}$$

where $\{X_{ij}; j = 1, \ldots, M\}$ are random points drawn from the density $p_i^\diamond$. In this case, the estimation rule is randomized due to MC sampling.

However, in some cases the integrals can be computed in closed-form, rendering the estimator nonrandomized and very fast to compute. The example in Figure 7.4 is one such example. The next theorem shows that using zero-mean multivariate Gaussian densities as bolstering kernels

$$p_i^\diamond(\mathbf{x}) = \frac{1}{\sqrt{(2\pi)^d \det(C_i)}} \exp\left(-\frac{1}{2} \mathbf{x}^T C_i^{-1} \mathbf{x} \right), \tag{7.42}$$

where the kernel covariance matrix C_i can in principle be distinct at each training point $\mathbf{X}_i$, and the classification rule is produced by a linear discriminant (e.g. this includes LDA, linear SVMs, perceptrons), also leads to closed-form integration.

Theorem 7.3. *Let Ψ_n be a linear classification rule, defined by*

$$\Psi_n(S_n)(\mathbf{x}) = \begin{cases} 1, & \mathbf{a}_n^T x + b_n \leq 0, \\ 0, & otherwise, \end{cases} \tag{7.43}$$

where $\mathbf{a}_n \in R^d$ and $b_n \in R$ are sample-based coefficients. Then the Gaussian-bolstered resubstitution error estimator can be written as

$$\hat{\varepsilon}_n^{\mathrm{br}} = \frac{1}{n} \sum_{i=1}^{n} \left(\Phi\left(-\frac{\mathbf{a}_n^T \mathbf{X}_i + b_n}{\sqrt{\mathbf{a}_n^T C_i \mathbf{a}_n}} \right) I_{Y_i=0} + \Phi\left(\frac{\mathbf{a}_n^T \mathbf{X}_i + b_n}{\sqrt{\mathbf{a}_n^T C_i \mathbf{a}_n}} \right) I_{Y_i=1} \right), \qquad (7.44)$$

where $\Phi(x)$ is the cumulative distribution function of a standard $N(0,1)$ Gaussian random variable.

Proof. By comparing (7.39) and (7.44), it suffices to show that

$$\int_{A_j} p_i^\diamond(\mathbf{x} - \mathbf{X}_i) \, d\mathbf{x} = \Phi\left((-1)^j \frac{\mathbf{a}_n^T \mathbf{X}_i + b_n}{\sqrt{\mathbf{a}_n^T C_i \mathbf{a}_n}} \right), \quad j = 0,1. \qquad (7.45)$$

We show the case $j = 1$, the other case being entirely similar. Let $p_i^\diamond(\mathbf{a} - \mathbf{a}_i)$ be the density of a random vector Z. We have that

$$\int_{A_1} p_i^\diamond(\mathbf{x} - \mathbf{X}_i) \, d\mathbf{x} = P(Z \in A_1) = P(\mathbf{a}_n^T Z + b_n \leq 0). \qquad (7.46)$$

Now, by hypothesis, $Z \sim N_d(\mathbf{X}_i, C_i)$. From the properties of the multivariate Gaussian (see Section A1.7), it follows that $\mathbf{a}_n^T Z + b_n \sim N(\mathbf{a}_n^T \mathbf{X}_i + b_n, \mathbf{a}_n^T C_i \mathbf{a}_n)$. The result follows from the fact that $P(U \leq 0) = \Phi(-E[U]/\sqrt{\mathrm{Var}(U)})$ for a Gaussian random variable U. ⋄

Note that for any training point $\mathbf{X}_i$ exactly on the decision hyperplane, we have $\mathbf{a}_n^T \mathbf{X}_i + b_n = 0$, and thus its contribution to the bolstered resubstitution error estimate is exactly $\Phi(0) = 1/2$, regardless of the label Y_i.

Selecting the correct amount of bolstering, that is, the "size" of the bolstering kernels, is critical for estimator performance. Accordingly, we describe a simple non-parametric procedure for adjusting the kernel covariance matrices C_i, for $i = 1, \ldots, d$ to the sample data. The only (implicit) assumption is that the class-conditional densities are approximately unimodal; no other distributional assumption is made.

Under small sample sizes, restrictions have to be imposed on the kernel covariances. First, a natural assumption is to make all kernel densities, and thus covariance matrices, equal for training points with the same class label: $C_i = D_0$ if $Y_i = 0$ or $C_i = D_1$ if $Y_i = 1$. This reduces the number of parameters to be estimated to $2d(d+1)$. Furthermore, it could be assumed that the covariance matrices D_0 and D_1 are diagonal, with diagonal elements, or *kernel variances*, $\sigma_{01}^2, \ldots, \sigma_{0d}^2$ and $\sigma_{11}^2, \ldots, \sigma_{1d}^2$. This further reduces the number of parameters to $2d$. Finally, it could be assumed that $D_0 = \sigma_0^2 I_d$ and $D_1 = \sigma_1^2 I_d$, which corresponds to spherical kernels with variances σ_0^2 and σ_1^2, respectively, in which case there are only two parameters to be estimated.

In the diagonal and spherical cases, there is a simple interpretation of the relationship between the bolstered and plain resubstitution estimators. As the kernel variances all shrink to zero, it is easy to see that the bolstered resubstitution error estimator reduces to the plain resubstitution estimator. On the other hand, as the kernel variances increase, the bolstered estimator becomes more and more distinct from plain resubstitution. As resubstitution is usually optimistically biased, increasing the kernel variances will initially reduce the bias, but increasing them too much will eventually make the estimator pessimistic. Hence, from the point of view of bias, there is typically optimal values for the kernel variances that make the estimator unbiased. In addition, increasing the kernel variances generally decreases variance of the error estimator. Therefore, by properly selecting the kernel variances, one can simultaneously reduce estimator bias and variance.

If the kernel densities are multivariate Gaussian, then (9.17) and (9.18) apply in both the spherical and diagonal cases, which presents a computational advantage in the presence of feature selection.

Let us consider first a method to fit spherical kernels, in which case we need to determine the kernel variances σ_0 and σ_1 for the kernel covariance matrices $D_0 = \sigma_0^2 I_d$ and $D_1 = \sigma_1^2 I_d$. One reason why plain resubstitution is optimistically biased is that the test points in (7.23) are all at distance zero from the training points (i.e., the test points are the training points). Bolstered estimators compensate for that by spreading the test point probability mass over the training points. We propose to find the amount of spreading that makes the test points to be as close as possible to the true mean distance to the training data points. The true mean distance d_0 and d_1 among points from populations Π_0 and Π_1, respectively, are estimated here by the sample-based mean minimum distance among points from each population:

$$\hat{d}_j = \frac{1}{n_j} \sum_{i=1}^{n} \left(\min_{\substack{i'=1,\ldots,n \\ i'\neq i, Y_{i'}=j}} \{||\mathbf{X}_i - \mathbf{X}_{i'}||\} \right) I_{Y_i=j}, \quad j = 0,1. \tag{7.47}$$

The basic idea is to set the kernel variance σ_j in such a way that the median distance of a point randomly drawn from a kernel to the origin matches the the estimated mean distance $\hat{d}_j$, for $j = 0,1$. This implies that half of the probability mass (i.e., half of the test points) of the bolstering kernel will be farther from the center than the estimated mean distance and the other half will be nearer. Assume for the moment a unit-variance kernel covariance matrix $D_0 = I_d$ (to fix ideas, we consider class label 0). Let R_0 be the random variable corresponding to the distance of a point randomly selected from the kernel density to the origin with cumulative distribution function $F_{R_0}(x)$. The median distance of a point randomly drawn from a kernel to the origin is given by $\alpha_{d,0} = F_{R_0}^{-1}(1/2)$, where the subscript d indicates explicitly that $\alpha_{d,0}$ depends on the dimensionality. For a kernel covariance matrix $D_0 = \sigma_0^2 I_d$, all distances get multiplied by σ_0. Hence, σ_0 is the solution of the equation $\sigma_0 F_{R_0}^{-1}(1/2) = \sigma \alpha_{d,0} = \hat{d}_0$. The argument for class label 1 is obtained by replacing 0 by 1 throughout; hence, the kernel standard deviations are set to

$$\sigma_j = \frac{\hat{d}_j}{\alpha_{d,j}}, \quad j = 0, 1.$$ (7.48)

As sample size increases, it is easy to see that $\hat{d}_0$ and $\hat{d}_1$ decrease, so that the kernel variances σ_0 and σ_1 also decrease, and there is less bias correction introduced by bolstered resubstitution. This is in accordance with the fact that resubstitution tends to be less optimistically biased as the sample size increases. In the limit, as sample size grows to infinity, the kernel variances tend to zero, and the bolstered resubstitution converges to the plain resubstitution.

For the specific case where all bolstering kernels are multivariate spherical Gaussian densities, the distance variables R_0 and R_1 are both distributed as *chi random variables* with d degrees of freedom, with density

$$f_{R_0}(r) = f_{R_1}(r) = \frac{2^{1-d/2} r^{d-1} e^{-r^2/2}}{\Gamma(\frac{d}{2})},$$ (7.49)

where Γ denotes the gamma function. For $d = 2$, this becomes the well-known Rayleigh density. The cumulative distribution function F_{R_0} or F_{R_1} can be computed by numerical integration of (7.49) and the inverse at point $1/2$ can be found by a simple binary search procedure (using the fact that cumulative distribution functions are monotonically increasing), yielding the dimensional constant α_d. There is no subscript "j" in this case because the constant is the same for both class labels. The kernel standard deviations are then computed as in (7.48), with α_d in place of $\alpha_{d,j}$. The constant α_d depends only on the dimensionality, and in fact can be interpreted as a type of "dimensionality correction," which adjust the value of the estimated mean distance to account for the feature space dimensionality. In the spherical Gaussian case, the values of the dimensional constant up to five dimensions are $\alpha_1 = 0.674$, $\alpha_2 = 1.177$, $\alpha_3 = 1.538$, $\alpha_4 = 1.832$, $\alpha_5 = 2.086$.

To fit diagonal kernels, we need to determine the kernel variances $\sigma_{01}^2, \ldots, \sigma_{0d}^2$ and $\sigma_{11}^2, \ldots, \sigma_{1d}^2$. A simple approach is to estimate the kernel variances σ_{0k}^2 and σ_{1k}^2 separately for each k, using the univariate data $S_{n,k} = \{(X_{1k}, Y_1), \ldots, (X_{nk}, Y_n)\}$, for $k = 1, \ldots, d$, where X_{ik} is the $k$the feature (component) in vector $\mathbf{X}_i$. This is an application of the *Naive Bayes principle*. We define the mean minimum distance along feature k as

$$\hat{d}_{jk} = \frac{1}{n_j} \sum_{i=1}^{n} \left(\min_{\substack{i'=1,\ldots,n \\ i' \neq i, Y_{i'}=j}} \{ \|X_{ik} - X_{i'k}\| \| \} \right) I_{Y_i=j}, \quad j = 0, 1.$$ (7.50)

The kernel standard deviations are set to

$$\sigma_{jk} = \frac{\hat{d}_{jk}}{\alpha_{1,j}}, \quad j = 0, 1, \ k = 1, \ldots, d.$$ (7.51)

In the Gaussian case, one has $\sigma_{jk} = \hat{d}_{jk}/\alpha_1 = \hat{d}_{jk}/0.674$, for $j = 0, 1$, $k = 1, \ldots, d$.

When resubstitution is too optimistically biased due to overfitting classification rules, it is not a good idea to spread incorrectly classified data points because that increases the optimistic bias of the error estimator. Bias is reduced if one assigns no bolstering kernels to incorrectly classified points. This decreases bias but increases variance because there is less bolstering. This variant is called *semi-bolstered resubstitution*.

Finally, we remark that bolstering can be in principle applied to any error-counting error estimation method. For example, consider leave-one-out estimation. Let $A_j^i = \{\mathbf{x} \in R^d \mid \Psi_{n-1}(S_n - \{(\mathbf{X}_i, Y_i)\})(\mathbf{x}) = j\}$, for $j = 0, 1$, be the the decision regions for the classifier designed on the deleted sample $S_n - \{(\mathbf{X}_i, Y_i)\}$. The *bolstered leave-one-out* estimator can be computed via

$$\hat{\varepsilon}_n^{\text{bloo}} = \frac{1}{n} \sum_{i=1}^{n} \left(\int_{A_1^i} p_i^\diamond(\mathbf{x} - \mathbf{X}_i) \, d\mathbf{x} \, I_{Y_i=0} + \int_{A_0^i} p_i^\diamond(\mathbf{x} - \mathbf{X}_i) \, d\mathbf{x} \, I_{Y_i=1} \right). \tag{7.52}$$

When the integrals cannot be computed exactly, a Monte-Carlo expression similar to (7.41) can be employed.

7.8 Additional Topics

7.8.1 Convex Error Estimators

Given any two error estimation rules, a new one (in fact, an infinite number of new ones) can be obtained via convex combination. Usually, this is done in an attempt to reduce bias. Formally, given an optimistically biased estimator $\hat{\varepsilon}_n^{\text{opm}}$ and a pessimistically biased estimator $\hat{\varepsilon}_n^{\text{psm}}$, we define the *convex error estimator*

$$\hat{\varepsilon}_n^{\text{conv}} = w \hat{\varepsilon}_n^{\text{opm}} + (1 - w) \hat{\varepsilon}_n^{\text{psm}}, \tag{7.53}$$

where $0 \leq w \geq 1$, $\text{Bias}(\hat{\varepsilon}_n^{\text{opm}}) < 0$, and $\text{Bias}(\hat{\varepsilon}_n^{\text{psm}}) > 0$. The 0.632 bootstrap error estimator discussed in Section 7.6 is an example of a convex error estimator, with $\hat{\varepsilon}_n^{\text{opm}} = \hat{\varepsilon}_n^{\text{boot}}$, $\hat{\varepsilon}_n^{\text{psm}} = \hat{\varepsilon}_n^r$, and $w = 0.632$. However, according to the this definition, the 0.632+ bootstrap is not, because its weights are not constant.

A simple convex estimator employs the typically optimistically biased resubstitution estimator and the pessimistically biased cross-validation estimator with equal weights:

$$\hat{\varepsilon}_n^{\text{r, cv}(k)} = 0.5 \, \hat{\varepsilon}_n^{\text{r}} + 0.5 \, \hat{\varepsilon}_n^{\text{cv}(k)}. \tag{7.54}$$

The next theorem shows that, for unbiased convex error estimation, one should look for component estimators with low variances. The 0.632 bootstrap estimator addresses this by replacing the high-variance cross-validation in (7.54) with the lower-variance estimator $\hat{\varepsilon}_n^{\text{boot}}$.

Theorem 7.4. *For an unbiased convex error estimator,*

$$\mathrm{MSE}(\hat{\varepsilon}_n^{\mathrm{conv}}) \le \max\{\mathrm{Var}(\hat{\varepsilon}_n^{\mathrm{opm}}), \mathrm{Var}(\hat{\varepsilon}_n^{\mathrm{psm}})\}. \tag{7.55}$$

Proof.

$$\mathrm{MSE}(\hat{\varepsilon}_n^{\mathrm{conv}}) = \mathrm{Var}_{\mathrm{dev}}(\hat{\varepsilon}_n^{\mathrm{conv}}) = w^2 \mathrm{Var}_{\mathrm{dev}}(\hat{\varepsilon}_n^{\mathrm{opm}}) + (1-w)^2 \mathrm{Var}_{\mathrm{dev}}(\hat{\varepsilon}_n^{\mathrm{psm}}) +$$

$$2w(1-w)\,\rho(\hat{\varepsilon}_n^{\mathrm{opm}} - \varepsilon_n, \hat{\varepsilon}_n^{\mathrm{psm}} - \varepsilon_n)\sqrt{\mathrm{Var}_{\mathrm{d}}(\hat{\varepsilon}_n^{\mathrm{opm}})\mathrm{Var}_{\mathrm{d}}(\hat{\varepsilon}_n^{\mathrm{psm}})}$$

$$\le w^2 \mathrm{Var}_{\mathrm{dev}}(\hat{\varepsilon}_n^{\mathrm{opm}}) + (1-w)^2 \mathrm{Var}_{\mathrm{dev}}(\hat{\varepsilon}_n^{\mathrm{psm}} +$$

$$2w(1-w)\sqrt{\mathrm{Var}_{\mathrm{d}}(\hat{\varepsilon}_n^{\mathrm{opm}})\mathrm{Var}_{\mathrm{d}}(\hat{\varepsilon}_n^{\mathrm{psm}})} \tag{7.56}$$

$$= \left(w\sqrt{\mathrm{Var}_{\mathrm{d}}(\hat{\varepsilon}_n^{\mathrm{opm}})} + (1-w)\sqrt{\mathrm{Var}_{\mathrm{d}}(\hat{\varepsilon}_n^{\mathrm{psm}})}\right)^2$$

$$\le \max\{\mathrm{Var}(\hat{\varepsilon}_n^{\mathrm{opm}}), \mathrm{Var}(\hat{\varepsilon}_n^{\mathrm{psm}})\},$$

where the last inequality follows from the relation

$$(wx + (1-w)y)^2 \le \max\{x^2, y^2\}, \quad 0 \le w \le 1. \tag{7.57}$$

$\diamond$

The following theorem provides a sufficient condition for an unbiased convex estimator to be more accurate than either component estimator.

Theorem 7.5. *If $\hat{\varepsilon}_n^{\mathrm{conv}}$ is an unbiased convex estimator and*

$$|\mathrm{Var}(\hat{\varepsilon}_n^{\mathrm{psm}}) - \mathrm{Var}(\hat{\varepsilon}_n^{\mathrm{opm}})| \le \min\{\mathrm{Bias}(\hat{\varepsilon}_n^{\mathrm{opm}})^2, \mathrm{Bias}(\hat{\varepsilon}_n^{\mathrm{psm}})^2\}, \tag{7.58}$$

then

$$\mathrm{MSE}(\hat{\varepsilon}_n^{\mathrm{conv}}) \le \min\{\mathrm{MSE}(\hat{\varepsilon}_n^{\mathrm{opm}}), \mathrm{MSE}(\hat{\varepsilon}_n^{\mathrm{psm}})\}. \tag{7.59}$$

Proof. If $\mathrm{Var}(\hat{\varepsilon}_n^{\mathrm{opm}}) \le \mathrm{Var}(\hat{\varepsilon}_n^{\mathrm{psm}})$, then we use Theorem 7.4 to get

$$\mathrm{MSE}(\hat{\varepsilon}_n^{\mathrm{conv}}) \le \mathrm{Var}(\hat{\varepsilon}_n^{\mathrm{psm}}) \le \mathrm{MSE}(\hat{\varepsilon}_n^{\mathrm{psm}}). \tag{7.60}$$

In addition, if $\mathrm{Var}(\hat{\varepsilon}_n^{\mathrm{psm}}) - \mathrm{Var}(\hat{\varepsilon}_n^{\mathrm{opm}}) \le \mathrm{Bias}(\hat{\varepsilon}_n^{\mathrm{opm}})^2$, we have

$$\mathrm{MSE}(\hat{\varepsilon}_n^{\mathrm{conv}}) \le \mathrm{Var}(\hat{\varepsilon}_n^{\mathrm{psm}})$$

$$= \mathrm{MSE}(\hat{\varepsilon}_n^{\mathrm{opm}}) - \mathrm{Bias}(\hat{\varepsilon}_n^{\mathrm{opm}})^2 + \mathrm{Var}(\hat{\varepsilon}_n^{\mathrm{psm}}) - \mathrm{Var}(\hat{\varepsilon}_n^{\mathrm{opm}}) \tag{7.61}$$

$$\le \mathrm{MSE}(\hat{\varepsilon}_n^{\mathrm{opm}}).$$

The two previous equations result in (7.59). If $\mathrm{Var}(\hat{\varepsilon}_n^{\mathrm{psm}}) \le \mathrm{Var}(\hat{\varepsilon}_n^{\mathrm{opm}})$, an analogous derivation shows that (7.59) holds if $\mathrm{Var}(\hat{\varepsilon}_n^{\mathrm{opm}}) - \mathrm{Var}(\hat{\varepsilon}_n^{\mathrm{psm}}) \le \mathrm{Bias}^2(\hat{\varepsilon}_n^{\mathrm{psm}})$. Hence, (7.59) holds in general if (7.58) is satisfied. $\diamond$

While it may be possible to achieve a lower MSE by not requiring unbiasedness, these inequalities regarding unbiased convex estimators carry a good deal of insight. For instance, since the bootstrap variance is generally smaller than the cross-validation variance, (7.55) implies that convex combinations involving bootstrap error estimators are likely to be more accurate than those involving cross-validation error estimators when the weights are chosen in such a way that the resulting bias is very small.

7.8.2 Smoothed Error Estimators

Smoothed error estimators attempt to improve bias and variance of error estimators for linear classification rules. They are similar to bolstering in that they attempt to smooth the error count. First, note that the resubstitution estimator can be rewritten as

$$\hat{\varepsilon}_n^r = \frac{1}{n} \sum_{i=1}^{n} (\psi_n(\mathbf{X}_i) I_{Y_i=0} + (1 - \psi_n(\mathbf{X}_i)) I_{Y_i=1}) \,, \tag{7.62}$$

where $\psi_n = \Psi_n(S_n)$ is the designed classifier. The classifier ψ_n is a sharp 0-1 step function that can introduce variance by the fact that a point near the decision boundary can change its contribution from 0 to $\frac{1}{n}$ (and vice-versa) via a slight change in the training data, even if the corresponding change in the decision boundary is small, and hence so is the change in the true error. In small-sample settings, $\frac{1}{n}$ is relatively large.

The idea behind smoothed error estimation is to replace ψ_n in (7.62) by a suitably chosen smooth function taking values in the interval $[0, 1]$, thereby reducing the variance of the original estimator. Consider a linear discriminant $W_L(\mathbf{X}) = a^T \mathbf{X} + b$; the sign of W_L gives the decision region to which a point belongs, and its magnitude measures the robustness of that decision: it can be shown that $|W_L(x)|$ is related to the Euclidean distance from a point x to the separating hyperplane (see Exercise 6.1). To achieve smoothing of the error count, the idea is to use a monotone increasing function $r : R \to [0, 1]$ applied on W_L. The function r should be such that $r(-u) = 1 - r(u)$, $\lim_{u \to -\infty} r(u) = 0$, and $\lim_{u \to \infty} r(u) = 1$. The *smoothed resubstitution* estimator is given by

$$\hat{\epsilon}_n^{\mathrm{rs}} = \frac{1}{n} \sum_{i=1}^{n} ((1 - r(W_L(\mathbf{X}_i))) I_{Y_i=0} + r(W_L(\mathbf{X}_i)) I_{Y_i=1}) \,, \tag{7.63}$$

For instance, the Gaussian smoothing function is given by

$$r(u) = \Phi \left(\frac{u}{b \hat{\Delta}} \right) \tag{7.64}$$

where Φ is the cumulative distribution function of a standard Gaussian variable,

$$\hat{\Delta} = \sqrt{(\bar{\mathbf{X}}_1 - \bar{\mathbf{X}}_0)^T \hat{\Sigma}^{-1} (\bar{\mathbf{X}}_1 - \bar{\mathbf{X}}_0)} \tag{7.65}$$

is the estimated Mahalanobis distance between the classes (recalling that $\hat{\Sigma}$ is the pooled sample covariance matrix), and b is a free parameter that must be provided, which is typical of smoothing methods. Another example is the windowed linear function $r(u) = 0$ on $(-\infty, -b)$, $r(u) = 1$ on (b, ∞), and $r(u) = (u + b)/2b$ on $[-b, b]$. Generally, a choice of function r depends on tunable parameters, such as b in the previous examples. The choice of parameter is a major issue, which affects the bias and variance of the resulting estimator (see the Bibliographical Notes).

7.8.3   Bayesian Error Estimation

All error estimation rules surveyed in this chapter up to this point share the fact that they do not involve prior knowledge regarding the feature-label distribution in their computation. Following the approach for parametric Bayesian classification in Section 4.4.3, one can obtain a model-based Bayesian error estimation rule, as described briefly next.

As in Section 4.4.3, we assume that there is a family of probability density functions $\{p(\mathbf{x} \mid \boldsymbol{\theta}) \mid \boldsymbol{\theta} \in R^m\}$, where the true parameter values $\boldsymbol{\theta}_0$ and $\boldsymbol{\theta}_1$ are random variables. Here we also assume that $c = P(Y = 1)$ is a random variable. The problem is therefore parametrized by the vector $\Theta = (c, \boldsymbol{\theta}_0, \boldsymbol{\theta}_1)$, with a prior distribution $p(\Theta)$. For a given classifier ψ_n, consider the error rates

$$
\begin{aligned}
\varepsilon_n^0(\Theta) &= P_\Theta(\psi_n(\mathbf{X}) \neq Y \mid Y = 0, S_n) \\
\varepsilon_n^1(\Theta) &= P_\Theta(\psi(\mathbf{X}) \neq Y \mid Y = 1, S_n) \\
\varepsilon_n(\Theta) &= P_\Theta(\psi(\mathbf{X}) \neq Y \mid S_n) = (1 - c)\varepsilon_n^0(\Theta) + c\varepsilon_n^1(\Theta)
\end{aligned}
\tag{7.66}
$$

The *Bayesian error estimator* $\hat{\varepsilon}^{\text{bayes}}$ is defined as the function of the data S_n that minimizes the MSE with respect to the true error:

$$
\hat{\varepsilon}^{\text{bayes}} = \arg\min_{\xi(S_n)} E\left[(\varepsilon_n(\Theta) - \xi(S_n))^2 \mid S_n\right].
\tag{7.67}
$$

It is well-known that the solution is the conditional expectation:

$$
\hat{\varepsilon}^{\text{bayes}} = E\left[\varepsilon_n(\Theta) \mid S_n\right],
\tag{7.68}
$$

which is the definition of the Bayesian error estimator. The expectation is with respect to the posterior distribution of Θ given the data S_n:

$$
p(\Theta \mid S_n) = \frac{p(S_n \mid \Theta)p(\Theta)}{\int_\Theta p(S_n \mid \Theta)p(\Theta)d\Theta},
\tag{7.69}
$$

where the data likelihood is given by

$$
p(S_n \mid \Theta) = p(S_n \mid c, \boldsymbol{\theta}_0, \boldsymbol{\theta}_1) = \Pi_{i=1}^n (1 - c)^{1-y_i} p(\mathbf{x}_i \mid \boldsymbol{\theta}_0)^{1-y_i} c^{y_i} p(\mathbf{x}_i \mid \boldsymbol{\theta}_1)^{y_i}.
\tag{7.70}
$$

As in Section 4.4.3, if one assumes that the parameters are independent a priori, i.e., $p(c, \boldsymbol{\theta}_0, \boldsymbol{\theta}_1) = p(c)p(\boldsymbol{\theta}_0)p(\boldsymbol{\theta}_1)$, then it can be shown that the parameters remain independent after observing the data, i.e., $p(c, \boldsymbol{\theta}_0, \boldsymbol{\theta}_1 \mid S_n) = p(c \mid S_n)p(\boldsymbol{\theta}_0 \mid S_n)p(\boldsymbol{\theta}_1 \mid S_n)$ (see Exercise 7.12). In this case, the Bayesian error estimator can be written as

$$
\begin{aligned}
\hat{\varepsilon}^{\text{bayes}} &= E\left[\varepsilon_n(\Theta) \mid S_n\right] = E\left[(1-c)\varepsilon_n^0(\boldsymbol{\theta}_0) + c\varepsilon_n^1(\boldsymbol{\theta}_1) \mid S_n\right] \\
&= (1 - E[c \mid S_n])E[\varepsilon_n^0(\boldsymbol{\theta}_0) \mid S_n] + E[c \mid S_n]E[\varepsilon_n^1(\boldsymbol{\theta}_1) \mid S_n] \qquad (7.71) \\
&= (1 - \hat{c}^{\text{bayes}})\hat{\varepsilon}^{\text{bayes},0} + \hat{c}^{\text{bayes}}\hat{\varepsilon}^{\text{bayes},1},
\end{aligned}
$$

where

$$
\begin{aligned}
\hat{c}^{\text{bayes}} &= E[c \mid S_n] = \int_0^1 c\,p(c \mid S_n)dc, \\
\hat{\varepsilon}^{\text{bayes},0} &= E[\varepsilon_n^0(\boldsymbol{\theta}_0) \mid S_n] = \int_{R^m} \varepsilon_n^0(\boldsymbol{\theta}_0)p(\boldsymbol{\theta}_0 \mid S_n)d\boldsymbol{\theta}_0, \qquad (7.72) \\
\hat{\varepsilon}^{\text{bayes},1} &= E[\varepsilon_n^1(\boldsymbol{\theta}_1) \mid S_n] = \int_{R^m} \varepsilon_n^1(\boldsymbol{\theta}_1)p(\boldsymbol{\theta}_1 \mid S_n)d\boldsymbol{\theta}_1.
\end{aligned}
$$

Notice that these estimators depend on the separate posteriors $p(c \mid S_n)$, $p(\boldsymbol{\theta}_0 \mid S_n)$, and $p(\boldsymbol{\theta}_1 \mid S_n)$, respectively.

A common choice of prior for c is the beta prior, $p(c) = \text{Beta}(\alpha, \beta)$ (see Section A1.4), when it can be shown that the posterior is a so-called beta-binomial distribution, $p(c \mid S_n) \propto c^{n_0+\alpha-1}(1-c)^{n_1+\beta-1}$, in which case one obtains and

$$
\hat{c}^{\text{bayes}} = E[c \mid S_n] = \frac{n_1 + \beta}{n + \alpha + \beta}. \qquad (7.73)
$$

where n_1 is the observed number of sample points with label 1. The uninformative (uniform) prior corresponds to the special case $\alpha = \beta = 1$, so that

$$
\hat{c}^{\text{bayes}} = \frac{n_1 + 1}{n + 2}. \qquad (7.74)
$$

Finally, another case that often arises in practice is where c is known, in which case the prior and posterior are unit masses on the value of c, with $\hat{c}^{\text{bayes}} = c$.

Example 7.3. (Bayesian error estimator for the Discrete Histogram Rule.) As in Example 3.3, assume that $p(\mathbf{x})$ is concentrated over a finite number of points $\{\mathbf{x}^1, \ldots, \mathbf{x}^b\}$ in R^d. Provided b is not too large, we may use the raw parametrization $\boldsymbol{\theta}_0 = \{p(\mathbf{x}^1 \mid Y = 0), \ldots, p(\mathbf{x}^b \mid Y = 0)\}$ and $\boldsymbol{\theta}_1 = \{p(\mathbf{x}^1 \mid Y = 1), \ldots, p(\mathbf{x}^b \mid Y = 1)\}$ (one of the parameters in $\boldsymbol{\theta}_0$ and $\boldsymbol{\theta}_1$ is redundant, since the values in each case must add up to 1). An appropriate choice of prior is the *Dirichlet* distribution with *concentration* parameters $(\alpha_1, \ldots, \alpha_b) > 0$ and $(\beta_1, \ldots, \beta_b) > 0$ for $\boldsymbol{\theta}_0$ and $\boldsymbol{\theta}_1$, respectively:

$$
p(\boldsymbol{\theta}_0) \propto \prod_{j=1}^b \boldsymbol{\theta}_0(j)^{\alpha_j - 1} \quad \text{and} \quad p(\boldsymbol{\theta}_1) \propto \prod_{j=1}^b \boldsymbol{\theta}_1(j)^{\beta_j - 1}. \qquad (7.75)
$$

Using the same notation as in Example 3.3, it can be shown that the posteriors are again Dirichlet,

$$p(\boldsymbol{\theta}_0 \mid S_n) \propto \prod_{j=1}^{b} \boldsymbol{\theta}_0(j)^{U_j + \alpha_j - 1} \quad \text{and} \quad p(\boldsymbol{\theta}_1 \mid S_n) \propto \prod_{j=1}^{b} \boldsymbol{\theta}_1(j)^{V_j + \beta_j - 1}. \tag{7.76}$$

From this, one obtains

$$\hat{\varepsilon}^{\text{bayes},0} = E[\varepsilon_n^0(\boldsymbol{\theta}_0) \mid S_n] = \sum_{j=1}^{b} \frac{U_j + \alpha_j}{n_0 + \sum_{k=1}^{b} \alpha_k} I_{U_j < V_j}$$

$$\hat{\varepsilon}^{\text{bayes},1} = E[\varepsilon_n^1(\boldsymbol{\theta}_1) \mid S_n] = \sum_{j=1}^{b} \frac{V_j + \beta_j}{n_1 + \sum_{k=1}^{b} \beta_k} I_{U_j \geq V_j} \tag{7.77}$$

If a beta prior is assumed for c, the complete Bayesian error estimator can be written using (7.71) and (7.73):

$$\hat{\varepsilon}^{\text{bayes}} = \frac{n_0 + 1}{n + 2} \left(\sum_{j=1}^{b} \frac{U_j + 1}{n_0 + b} I_{U_j < V_j} \right) + \frac{n_1 + 1}{n + 2} \left(\sum_{j=1}^{b} \frac{V_j + 1}{n_1 + b} I_{U_j \leq V_j} \right). \tag{7.78}$$

$\diamond$

The next theorem, which is an application of Fubini's Theorem (change of order of integration) characterizes the Bayesian error estimator in terms of the predictive densities defined in (4.47), and repeated here for convenience

$$p_0(\mathbf{x} \mid S_n) = \int_{R^m} p(\mathbf{x} \mid \boldsymbol{\theta}_0) p(\boldsymbol{\theta}_0 \mid S_n) \, d\boldsymbol{\theta}_0 \quad \text{and} \quad p_1(\mathbf{x} \mid S_n) = \int_{R^m} p(\mathbf{x} \mid \boldsymbol{\theta}_1) p(\boldsymbol{\theta}_1 \mid S_n) \, d\boldsymbol{\theta}_1. \tag{7.79}$$

Theorem 7.6. *Let $A_j = \{\mathbf{x} \in R^d \mid \psi_n(\mathbf{x}) = j\}$, for $j = 0, 1$, be the decision regions for the designed classifier. Then the class-specific Bayesian error estimator in (7.72) can be written simply as*

$$\hat{\varepsilon}^{\text{bayes},0} = \int_{A_1} p_0(\mathbf{x} \mid S_n) \, d\mathbf{x} \quad \text{and} \quad \hat{\varepsilon}^{\text{bayes},0} = \int_{A_0} p_1(\mathbf{x} \mid S_n) \, d\mathbf{x}. \tag{7.80}$$

Proof. We derive the expression for $\hat{\varepsilon}^{\text{bayes},0}$, as the one for $\hat{\varepsilon}^{\text{bayes},1}$ is entirely analogous. We have

$$\varepsilon_n^0(\boldsymbol{\theta}_0) = \int_{A_1} p(x \mid \boldsymbol{\theta}_0) \, d\mathbf{x}. \tag{7.81}$$

Therefore

$$\begin{aligned} \hat{\varepsilon}^{\text{bayes},0} &= E[\varepsilon_n^0(\boldsymbol{\theta}_0) \mid S_n] \\ &= \int_{R^m} \int_{A_1} p(\mathbf{x} \mid \boldsymbol{\theta}_0) p(\boldsymbol{\theta}_0 \mid S_n) \, d\mathbf{x} \, d\boldsymbol{\theta}_0 \\ &= \int_{A_1} \int_{R^m} p(\mathbf{x} \mid \boldsymbol{\theta}_0) p(\boldsymbol{\theta}_0 \mid S_n) \, d\boldsymbol{\theta}_0 \, d\mathbf{x} \\ &= \int_{A_1} p_0(\mathbf{x} \mid S_n) \, d\mathbf{x} \end{aligned} \tag{7.82}$$

by change of order of integration. $\diamond$

The Bayesian error estimator can thus be written as

$$\hat{\varepsilon}^{\text{bayes}} = (1 - \hat{c}^{\text{bayes}}) \int_{A_1} p_0(\mathbf{x} \mid S_n) \, d\mathbf{x} + \hat{c}^{\text{bayes}} \int_{A_0} p_1(\mathbf{x} \mid S_n) \, d\mathbf{x} \,. \tag{7.83}$$

7.9   Bibliographical Notes

The subject of error estimation has a long history and has produced a large body of literature; four main review papers summarize major advances in the field up to 2000 [Toussaint, 1974; Hand, 1986; McLachlan, 1987; Schiavo and Hand, 2000]; recent advances in error estimation since 2000 include work on model selection [Bartlett et al., 2002], bolstering [Braga-Neto and Dougherty, 2004; Sima et al., 2005b], feature selection [Sima et al., 2005a; Zhou and Mao, 2006; Xiao et al., 2007; Hanczar et al., 2007], confidence intervals [Kaariainen and Langford, 2005; Kaariainen, 2005; Xu et al., 2006], model-based second-order properties [Zollanvari et al., 2011, 2012], and Bayesian error estimators [Dalton and Dougherty, 2011b,c]. A book entirely devoted to the subject was published in 2015 [Braga-Neto and Dougherty, 2015], which covers the classical studies as well as the developments after 2000. In that reference, a classification rule and error estimation rule pair is called a *pattern recognition rule*.

The resubstitution error estimation rule is usually attributed to Smith [1947]. Cross-validation on the other hand is variously attributed to Lachenbruch and Mickey [1968]; Cover [1969]; Toussaint and Donaldson [1970]; Stone [1974]. Complete cross-validation is named in Kohavi [1995]. For a proof of the near-unbiasedness property (7.25) of k-fold cross-validation, see Chapter 5 of Braga-Neto and Dougherty [2015]. The variance problem of cross-validation has been known for a long time; for further discussion, see Toussaint [1974]; Glick [1978]; Devroye et al. [1996]; Braga-Neto and Dougherty [2004]. Theorem 7.1 is credited by Devroye et al. [1996] to Devroye and Wagner [1976] and Rogers and Wagner [1978]. The form that appears here is a slight modification of Theorem 24.2 in Devroye et al. [1996].

The general bootstrap sampling methodology is due to Efron [1979]. The application of the bootstrap to classification error estimation, and the zero and 0.632 bootstrap error estimators, appear in Efron [1983]. The recommendation of a number of bootstrap sample B between 25 and 200 appears in the latter reference. The balanced bootstrap appears in Chernick [1999]. The 0.632+ bootstrap error estimator is due to Efron and Tibshirani [1997].

Bolstered error estimation was introduced in Braga-Neto and Dougherty [2004] and further studied in Sima et al. [2005b]; Vu et al. [2008]; Sima et al. [2014]; Jiang and Braga-Neto [2014]. The nonparametric estimator of the bolstering kernel variance in (7.48) follows a distance argument used in Efron [1983]. Theorem 7.2 and Equation (7.39) extends a similar expression proposed in Kim et al. [2002] in the context of LDA. The expression for the chi density in (7.49) is from Evans et al. [2000]. For an approach to estimating the kernel variances for the bolstered leave-one-out estimator, see Braga-Neto and Dougherty [2004]. The Naive-Bayes bolstered error estimator is due to Jiang and Braga-Neto [2014]. The Naive-Bayes principle is described in Dudoit et al. [2002].

Convex error estimation was studied in depth in Sima and Dougherty [2006]. Theorems 7.4 and 7.5 appear in that reference. See also Toussaint and Sharpe [1974], where the simple estimator (7.54) appears, and Raudys and Jain [1991]

Smoothed error estimation for LDA is credited to Glick [1978] being investigated inGlick [1978]; Snapinn and Knoke [1985, 1989]; Hirst [1996]. The same basic idea can be applied to any linear classification rule. A few approaches for choosing the smoothing parameter have been tried, namely, arbitrary choice [Glick, 1978; Tutz, 1985], arbitrary function of the separation between classes [Tutz, 1985], parametric estimation assuming normal populations [Snapinn and Knoke, 1985, 1989], and simulation-based methods [Hirst, 1996]. Extension of smoothing to nonlinear classification rules is not straightforward, since a suitable discriminant is not generally available. In Devroye et al. [1996], and under a different but equivalent guise in Tutz [1985], using a nonparametric estimator $\hat{\eta}(\mathbf{x})$ of the posterior-probability function $\eta(\mathbf{x})$, as in Chapter 5, and a monotone increasing function $r : [0, 1] \rightarrow [0, 1]$, such that $r(u) = 1 - r(1 - u)$, $r(0) = 0$, and $r(1) = 1$ (this may simply be the identity function $r(u) = u$), the smoothed resubstitution estimator is given as before by (7.63), with W_L replaced by $\hat{\eta}$. In the special case $r(u) = u$, one has the posterior-probability error estimator of Lugosi and Pawlak [1994]. However, this approach depends on the availability of an estimator $\hat{\eta}$ and lacks, in general, a geometric interpretation.

Bayesian error estimation was introduced in Dalton and Dougherty [2011b,c]. For a continuous feature-space counterpart of Example 7.3 using a Gaussian model, see Dalton and Dougherty [2011a, 2012a]. Relative to performance, a key difference between non-Bayesian and Bayesian error estimation rules is that the latter allow the definition of a RMS conditional on the training sample [Dalton and Dougherty, 2012a,b]. See also Chapter 8 of Braga-Neto and Dougherty [2015].

Finally, we comment on the distributional study of error estimation under Gaussianity (for a comprehensive treatment, see Braga-Neto and Dougherty [2015], as well as McLachlan [1992]). The first works in this area in English are Lachenbruch [1965]; Hills [1966], though results were published in Russian a few years earlier [Toussaint, 1974; Raudys and Young, 2004]. Hills [1966] provided an exact formula for the expected resubstitution error estimate in the univariate case that involves the bivariate Gaussian cumulative distribution. M. Moran extended this result to the multivariate case when$\Sigma$is known in the formulation of the discriminant [Moran, 1975]. Moran's result can also be seen as a generalization of a similar result given by John [1961] for the expectation of the true error. McLachlan provided an asymptotic expression for the expected resubstitution error in the multivariate case, for unknown covariance matrix [McLachlan, 1976], with a similar result having been provided by Raudys [1978]. Foley [1972] derived an asymptotic expression for the variance of the resubstitution error. Finally, Raudys applied a double asymptotic approach where both sample size and dimensionality increase to infinity, which we call the "Raudys-Kolmogorov method," to obtain a simple asymptotically exact expression for the expected resubstitution error [Raudys, 1978]. Recent contributions include Zollanvari et al. [2009a, 2010, 2011, 2012]; Vu et al. [2014]; Zollanvari and Dougherty [2014].

7.10 Exercises

7.1. Suppose that the classification error ε_n and an error estimator $\hat{\varepsilon}_n$ are jointly Gaussian, such that

$$\varepsilon_n \sim N(\varepsilon^* + 1/n, 1/n^2), \quad \hat{\varepsilon}_n \sim N(\varepsilon^* - 1/n, 1/n^2), \quad \mathrm{Cov}(\varepsilon_n, \hat{\varepsilon}_n) = 1/(2n^2),$$

where ε^* is the Bayes error. Find the bias, deviation variance, RMS, correlation coefficient and tail probabilities $P(\hat{\varepsilon}_n - \varepsilon_n < -\tau)$ and $P(\hat{\varepsilon}_n - \varepsilon_n > \tau)$ of $\hat{\varepsilon}_n$. Is this estimator optimistically or pessimistically biased? Does performance improve as sample size increases? Is the estimator consistent?

7.2. You are given that an error estimator $\hat{\varepsilon}_n$ is related to the classification error ε_n through the simple model

$$\hat{\varepsilon}_n = \varepsilon_n + Z, \tag{7.84}$$

where the conditional distribution of the random variable Z given the training data S_n is Gaussian, $Z \sim N(0, 1/n^2)$. Is $\hat{\varepsilon}_n$ randomized or nonrandomized? Find the internal variance and variance of $\hat{\varepsilon}_n$. What happens as the sample size grows without bound?

7.3. Obtain a sufficient condition for an error estimator to be consistent in terms of the RMS. Repeat for strong consistency.
Hint: Consider (7.10) and (7.11), and then use (7.7).

7.4. You are given that $\mathrm{Var}(\varepsilon_n) \leq 5 \times 10^{-5}$. Find the minimum number of test samples m that will guarantee that the standard deviation of the test-set error estimator $\hat{\varepsilon}_{n,m}$ will be at most 1%.

7.5. You are given that the error of a given classifier is $\varepsilon_n = 0.1$. Find the probability that the test-set error estimate $\hat{\varepsilon}_{n,m}$ will be exactly equal to ε_n, if $m = 20$ testing samples are available.

7.6. This problem concerns additional properties of test-set error estimation.

   (a) Show that
$$\text{Var}(\hat{\varepsilon}_{n,m}) = \frac{E[\varepsilon_n](1 - E[\varepsilon_n])}{m} + \frac{m-1}{m}\text{Var}(\varepsilon_n). \tag{7.85}$$
   From this, show that $\text{Var}(\hat{\varepsilon}_{n,m}) \to \text{Var}(\varepsilon_n)$, as the number of testing samples $m \to \infty$.

   (b) Using the result of part (a), show that
$$\text{Var}(\varepsilon_n) \leq \text{Var}(\hat{\varepsilon}_{n,m}) \leq E[\varepsilon_n](1 - E[\varepsilon_n]) \tag{7.86}$$

   In particular, this shows that when $E[\varepsilon_n]$ is small, so is $\text{Var}(\hat{\varepsilon}_{n,m})$.
   Hint: For any random variable X such that $0 \leq X \leq 1$ with probability 1, one has $\text{Var}(X) \leq E[X](1 - E[X])$.

   (c) Show that the tail probabilities given the training data S_n satisfy:
$$P(|\hat{\varepsilon}_{n,m} - \varepsilon_n| \geq \tau \mid S_n) \leq e^{-2m\tau^2}, \text{ for all } \tau > 0. \tag{7.87}$$

   Hint: Use Hoeffding's inequality (see Theorem A.14).

   (d) By using the Law of Large Numbers (see Theorem A.12), show that, given the training data S_n, $\hat{\varepsilon}_{n,m} \to \varepsilon_n$ with probability 1 (i.e., the test-set error estimator is universally strongly consistent with respect to the test sample size).

   (e) Repeat item (d), but this time using the result from item (c).

   (f) The bound on $\text{RMS}(\hat{\varepsilon}_{n,m})$ in (7.22) is distribution-free. Show the distribution-based bound
$$\text{RMS}(\hat{\varepsilon}_{n,m}) \leq \frac{1}{2\sqrt{m}}\min(1, 2\sqrt{E[\varepsilon_n]}). \tag{7.88}$$

7.7. Consider the discrete histogram rule of Example 3.3. Show that the resubstitution and leave-one-out error estimators for this rule can be written respectively as
$$\hat{\varepsilon}_n^r = \frac{1}{n}\sum_{j=1}^{b}\min\{U_j, V_j\} = \frac{1}{n}\sum_{i=1}^{b}\left[U_j I_{U_j < V_j} + V_j I_{U_j \geq V_j}\right], \tag{7.89}$$

and
$$\hat{\varepsilon}_n^l = \frac{1}{n}\sum_{j=1}^{b}\left[U_j I_{U_j \geq V_j} + V_j I_{U_j \geq V_j - 1}\right]. \tag{7.90}$$

where the random variables U_j and V_j, for $j = 1, \ldots, b$, were defined in Example 3.3.

7.8. Show that for the discrete histogram rule,
$$E[\hat{\varepsilon}_n^r] \leq \varepsilon^* \leq E[\varepsilon_n]. \tag{7.91}$$

Hence, the resubstitution estimator in this case is not only guaranteed to be optimistically biased, but its expected value provides a lower bound for the Bayes error.
Hint: Use (7.89) and apply Jensen's Inequality (A.66).

7.9. Despite the guaranteed optimistic bias of resubstitution in the case of discrete histogram classification, shown in the previous problem, it has very good large-sample properties.

(a) Show that $\hat{\varepsilon}_n^r \to \varepsilon_n$ with probability one as $n \to \infty$ regardless of the distribution, i.e., the resubstitution estimator is universally strongly consistent.
Hint: Use the Law of Large Numbers (see Thm. A.12).

(b) Show that this implies that $E[\hat{\varepsilon}_n^r] \to E[\varepsilon_n]$ as $n \to \infty$, i.e., the resubstitution estimator is asymptotically unbiased. In view of (7.91), this implies that $E[\varepsilon_n]$ converges to ε_{bay} from above, while $E[\hat{\varepsilon}_n^r]$ converges to ε_{bay} from below, as $n \to \infty$.

7.10. This problem illustrates the very poor (even paradoxical) performance of cross-validation with very small sample sizes. Consider the resubstitution and leave-one-out error estimators $\hat{\varepsilon}_n^r$ and $\hat{\varepsilon}_n^l$ for the 3NN classification rule, with a sample of size $n = 4$ from a mixture of two equally-likely Gaussian populations$\Pi_0 \sim N_d(\boldsymbol{\mu}_0, \Sigma)$ and$\Pi_1 \sim N_d(\boldsymbol{\mu}_1, \Sigma)$. Assume that $\boldsymbol{\mu}_0$ and $\boldsymbol{\mu}_1$ are far enough apart to make $\delta = \sqrt{(\boldsymbol{\mu}_1 - \boldsymbol{\mu}_0)^T \Sigma^{-1}(\boldsymbol{\mu}_1 - \boldsymbol{\mu}_0)} \gg 0$ (in which case the Bayes error is $\varepsilon_{\text{bay}} = \Phi(-\delta/2) \approx 0$).

(a) For a sample S_n with $N_0 = N_1 = 2$, which occurs $P(N_0 = 2) = \binom{4}{2}2^{-4} = 37.5\%$ of the time, show that $\varepsilon_n \approx 0$ but $\hat{\varepsilon}_n^l = 1$.

(b) Show that $E[\varepsilon_n] \approx 5/16 = 0.3125$, but $E[\hat{\varepsilon}_n^l] = 0.5$, so that $\text{Bias}(\hat{\varepsilon}_n^l) \approx 3/16 = 0.1875$, and the leave-one-out estimator is far from unbiased.

(c) Show that $\text{Var}_d(\hat{\varepsilon}_n^l) \approx 103/256 \approx 0.402$, which corresponds to a standard deviation of $\sqrt{0.402} = 0.634$. The leave-one-out estimator is therefore highly-biased and highly-variable in this case.

(d) Consider the correlation coefficient of an error estimator $\hat{\varepsilon}_n$ with the true error ε_n:

$$\rho(\varepsilon_n, \hat{\varepsilon}_n) = \frac{\text{Cov}(\varepsilon_n, \hat{\varepsilon}_n)}{\text{Std}(\varepsilon_n)\text{Std}(\hat{\varepsilon}_n)}. \tag{7.92}$$

Show that $\rho(\varepsilon_n, \hat{\varepsilon}_n^l) \approx -0.98$, i.e., the leave-one-out estimator is almost perfectly negatively correlated with the true error.

(e) For comparison, show that, although $E[\hat{\varepsilon}_n^r] = 1/8 = 0.125$, so that $\text{Bias}(\hat{\varepsilon}_n^r) \approx -3/16 = -0.1875$, which is exactly the negative of the bias of leave-one-out, we have $\text{Var}_d(\hat{\varepsilon}_n^r) \approx 7/256 \approx 0.027$, for a standard deviation of $\sqrt{7}/16 \approx 0.165$, which is several times smaller than the leave-one-out variance, and $\rho(\varepsilon_n, \hat{\varepsilon}_n^r) \approx \sqrt{3/5} \approx 0.775$, showing that the resubstitution estimator is highly positively correlated with the true error.

7.11. Show that bolstered resubstitution can be written as

$$\hat{\varepsilon}_n^{\text{br}} = \frac{n_0}{n} \int_{A_1} p_{n,0}^\diamond(\mathbf{x})\, d\mathbf{x} + \frac{n_1}{n} \int_{A_0} p_{n,1}^\diamond(\mathbf{x})\, d\mathbf{x}, \tag{7.93}$$

where

$$p^\diamond_{n,0}(\mathbf{x}) = \frac{1}{n_0} \sum_{i=1}^{n} p^\diamond_i(\mathbf{x} - \mathbf{X}_i) \, I_{Y_i=0} \quad \text{and} \quad p^\diamond_{n,1}(\mathbf{x}) = \frac{1}{n_1} \sum_{i=1}^{n} p^\diamond_i(\mathbf{x} - \mathbf{X}_i) \, I_{Y_i=1}. \tag{7.94}$$

Compare to the expression for the Bayes error estimator in (7.83).

7.12. Show that if the parameters in a Bayesian setting are independent a priori, i.e., $p(c, \boldsymbol{\theta}_0, \boldsymbol{\theta}_1) = p(c)p(\boldsymbol{\theta}_0)p(\boldsymbol{\theta}_1)$, then the parameters remain independent after observing the data, i.e., $p(c, \boldsymbol{\theta}_0, \boldsymbol{\theta}_1 \mid S_n) = p(c \mid S_n)p(\boldsymbol{\theta}_0 \mid S_n)p(\boldsymbol{\theta}_1 \mid S_n)$.
Hint: write $p(c, \boldsymbol{\theta}_0, \boldsymbol{\theta}_1 \mid S_n) = p(c \mid S_n, \boldsymbol{\theta}_0, \boldsymbol{\theta}_1)p(\boldsymbol{\theta}_0 \mid \boldsymbol{\theta}_1, S_n) = p(\boldsymbol{\theta}_1 \mid S_n)$, and write $S_n = (S_{n_0}, S_{n_1})$ where S_{n_0} and S_{n_1} are the class-specific subsamples.

7.13. Redo Example 7.3 by deriving the predictive densities in (7.79) and then applying Theorem 7.6.

7.14. For the discrete histogram rule, derive a relationship between the Bayes error estimator using Dirichlet priors in (7.78) and the resubstitution estimator in (7.89).

7.11 Python Assignments

7.15. This assignment concerns bolstered error estimation with uniform spherical kernels.

(a) Develop formulas equivalent to (7.44) and (7.49) for the uniform spherical kernel case. Show that the dimensionality correction factors are given by $\alpha_d = 2^{-1/d}$ in this case.
Hint: it is more convenient to work with the radius rather than the variance of the spherical kernel.

(b) Modify the code in c07_bolst.py to compute the bolstered and semi-bolstered resubstitution estimates for the classifier and data in Figure 7.4. Compare to the the plain resubstitution estimate and the true error of the LDA classifier (which can be found exactly, since the distribution is Gaussian).

(c) Repeat item (b) a total of $M = 1000$ times, using a different random seed each time, and compare the resulting average error rates.

(d) Repeat items (b) and (c) with the bolstered resubstitution estimate using Gaussian kernels and compare the results.

7.16. Consider the synthetic data model in Section A8.1 with homoskedasticity, equally-likely classes, $d = 6$, $\sigma = 1, 2$, $\rho = 0.2, 0.8$, $k = 2, 6$ (with $l_1 = \lambda_2 = 3$ if $k = 2$; the features are independent with $k = 6$). There are thus a total of six different models: independent, low-correlated and highly-correlated features, under low or high variance. Consider the LDA, QDA, 3NN, and linear SVM classification rules.

(a) Generate a large number (e.g., $N = 1000$) synthetic training data sets for each sample size $n = 20$ to $n = 100$, in steps of 10. Also generate a test set of size $M = 400$ to estimate the true classification error in each case. Plot the performance metrics (7.3)–(7.5) as a function of the sample size for the resubstitution, leave-one-out, 5-fold cross-validation, .632 bootstrap, and bolstered resubstitution estimators. Describe what you see.

(b) Obtain the deviation distributions (all error estimators on the same plot) for each case above, by fitting beta densities to 1000 values. What do you observe?

7.17. Parts (b)–(d) of Problems 5.8, 5.9, and 5.10 concern test-set error rates. Obtain the same results concerning the resubstitution error rates. The difference between the estimated expected resubstitution and test-set error rates gives an estimate of the bias of resubstitution, since the test-set error rate is unbiased. What is the behavior of the resubstitution bias as a function of sample size and of the various classification parameters? How can you explain this in terms of overfitting?

Chapter 8

Model Selection for Classification

"It can scarcely be denied that the supreme goal of all theory is to make the irreducible basic elements as simple and as few as possible without having to surrender the adequate representation of a single datum of experience."
–Albert Einstein, *On the Method of Theoretical Physics*, 1933.

In this chapter, we address an obvious question about classification: which classification rule should one choose for a given problem? A related question is how to pick the free parameters of a classification rule. For example, how to pick the number of neighbors k to use in kNN classification, or how to choose the kernel bandwidth in kernel classification, or even how to select the number of training epochs to train a neural network. These are all *model selection* questions. The answer to such questions depends, as we argue in this chapter, on the ratio of complexity to sample size, where complexity includes the degrees of freedom in the classification rule and the dimensionality of the problem. All model selection procedures look for classifiers that both fit the training data well and display a good complexity to sample size ratio. If sample size is small compared to the complexity of the classification rule, overfitting is bound to occur, and a good fit to the training data does not translate to good prediction of future data. Under small sample sizes, constrained classification rules, with smaller degrees of freedom, smaller dimensionality, and simpler decision boundaries, are therefore preferable. This chapter begins with an analysis of classification complexity and the Vapnik-Chervonenkis theory of classification, and then examines a few practical methods for model selection.

© Springer Nature Switzerland AG 2020
U. Braga-Neto, *Fundamentals of Pattern Recognition and Machine Learning*,
https://doi.org/10.1007/978-3-030-27656-0_8

8.1   Classification Complexity

Let $\mathcal{C}$ be a *classifier space*, i.e., an arbitrary set of classifiers. A classification rule Ψ_n picks a classifier ψ_n in $\mathcal{C}$ based on training data S_n. For example, the LDA, linear SVM and perceptron classification rules produce classifiers in the space $\mathcal{C}$ of linear classifiers in R^d, while the QDA classification rule produces classifiers in the space $\mathcal{C}'$ of quadratic classifiers in R^d. (In this example, $\mathcal{C} \subset \mathcal{C}'$.) Parameters such as the weights of a fixed-architecture NN or the direction of a LDA hyperplane are part of the design process, and are thus modeled as part of $\mathcal{C}$. However, *free* parameters, such as the number of hidden layers and neurons in an NN or the *dimensionality* of the LDA classifier are assumed fixed and thus lead to different classes $\mathcal{C}$.

The best classifier in $\mathcal{C}$ according to the classification error is

$$\psi_{\mathcal{C}} \;=\; \arg\min_{\psi \in \mathcal{C}} \varepsilon[\psi] = \arg\min_{\psi \in \mathcal{C}} P[\psi(X) \neq Y] \,, \tag{8.1}$$

with error $\varepsilon_{\mathcal{C}} = \varepsilon[\psi_{\mathcal{C}}]$. In addition, given S_n, the designed classifier is $\psi_{n,\mathcal{C}} = \Psi_n(S_n) \in \mathcal{C}$, with error $\varepsilon_{n,\mathcal{C}} = \varepsilon[\psi_{n,\mathcal{C}}]$.

The *approximation error* is the difference between the best error in the class and the Bayes error:

$$\Delta_{\mathcal{C}} \;=\; \varepsilon_{\mathcal{C}} - \varepsilon_d > 0 \tag{8.2}$$

This reflects how well the classification rule can approximate the Bayes error. The *design error* is the difference between the error of the designed classifier and the best error in the class:

$$\Delta_{n,\mathcal{C}} \;=\; \varepsilon_{n,\mathcal{C}} - \varepsilon_{\mathcal{C}} > 0 \,. \tag{8.3}$$

This reflects how good a job one can do with the available data to design the best possible classifier in $\mathcal{C}$. In practice, what concerns us is the error of the designed classifier from the available data $\varepsilon_{n,\mathcal{C}}$. Combining the previous two equations yields:

$$\varepsilon_{n,\mathcal{C}} \;=\; \varepsilon_d + \Delta_{\mathcal{C}} + \Delta_{n,\mathcal{C}} \tag{8.4}$$

See Figure 8.1 for a graphical representation of this decomposition.

As mentioned in Chapter 3, the generally accepted performance measure for a classification rule is the expected classification error, which does not depend on the particular particular data. Applying expectation on both sides of (8.4) yields:

$$E[\varepsilon_{n,\mathcal{C}}] \;=\; \varepsilon_d + \Delta_{\mathcal{C}} + E[\Delta_{n,\mathcal{C}}] \,. \tag{8.5}$$

In order to select the best $\mathcal{C}$, we want the the expected classification error $E[\varepsilon_{n,\mathcal{C}}]$ to be as small as possible. Therefore, we would like both the approximation error $\Delta_{\mathcal{C}}$ and the expected design error $E[\Delta_{n,\mathcal{C}}]$ to be small. Unfortunately, that is not in general possible, as we argue next.

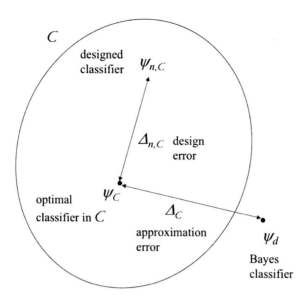

Figure 8.1: Designed classifier error decomposition into error of the Bayes classifier, approximation error, and design error.

As previously mentioned, the complexity of a classification rule has to do with the size of $\mathcal{C}$. A larger $\mathcal{C}$ (i.e., a more complex classification rule) is guaranteed to yield a smaller approximation error$\Delta_\mathcal{C}$ — if we make $\mathcal{C}$ big enough, eventually $\psi_d \in \mathcal{C}$ and we will obtain$\Delta_\mathcal{C} = 0$. However, a larger $\mathcal{C}$ generally means that the expected designed error will be larger. Making $\mathcal{C}$ smaller (i.e., choosing a simpler classification rule) will generally produce a smaller expected design error $E[\Delta_{n,\mathcal{C}}]$, but will increase the approximation error. This *complexity dilemma* is a manifestation in classification of the general bias-variance dilemma in statistics: one can generally control the bias or control the variance, but not both. Here, the "bias" is$\Delta_\mathcal{C}$ and the "variance" is $E[\Delta_{n,\mathcal{C}}]$.

The sample size dictates where the optimal operation point is in the complexity dilemma. For large samples, the design problem is minimized ($E[\Delta_{n,\mathcal{C}}]$ tends to be small), and the most important objective is to have a small approximation error$\Delta_\mathcal{C}$, so in this case the use of complex classification rules is warranted. However, in small-sample settings, which are prevalent in many scientific applications, the trade-offis tilted in the other direction. The design problem dominates ($E[\Delta_{n,\mathcal{C}}]$ tends to become large), and having a small approximation error$\Delta_\mathcal{C}$ becomes secondary. In this case the use of complex classification rules should be avoided. The situation is depicted graphically in Figure 8.2. Since $\mathcal{C}' \subset \mathcal{C}$, the latter is strictly more expressive than the former (e.g., $\mathcal{C}'$ and $\mathcal{C}$ could be the families of linear and quadratic classifiers, respectively). It is assumed that $n_2 \gg n_1$. We can see that under sample size n_2, $\mathcal{C}$ is the better classification rule. However, under sample size n_1,

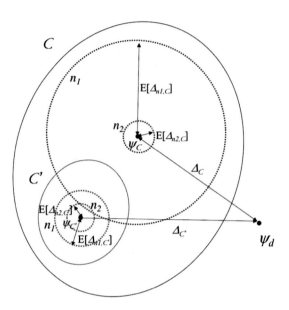

Figure 8.2: Complexity dilemma and sample size: $\mathcal{C}$ is better for n_2, but $\mathcal{C}'$ is better for n_1.

even though the approximation error for $\mathcal{C}'$ is larger, its expected design error is much smaller, making it overall better than $\mathcal{C}$.

The relationship between complexity and sample size can also be seen clearly in the *scissors plot*, already mentioned in Chapter 1. In Figure 8.3(a), we can see that, for $n > N_0$, the more complex class $\mathcal{C}$ is better, but for $n < N_0$, the simpler class $\mathcal{C}'$ is better instead. What the actual value of N_0 is depends on the distribution of the problem.

Increasing the dimensionality d has the effect of increasing the size of $\mathcal{C}$, that is, it increases classification complexity. For fixed n, the di ff erence

$$E[\varepsilon_{n,d}] - \varepsilon_d \,=\, \Delta_d + E[\Delta_{n,d}]\,. \tag{8.6}$$

will generally increase as d increases, as the expected design error $E[\Delta_{n,d}]$ cannot be controlled. The net result is that $E[\varepsilon_{n,d}]$ initially decreases, following the decrease in ε_d, but eventually increases, creating an optimal dimensionality d^*. See Figure 8.3(b) for an illustration. This is the phenomenon already referred to in Chapter 1, which has been variously called the "curse of dimensionality," the "peaking phenomenon," and the "Hughes Phenomenon." Notice that feature selection is a type of model selection, which ideally would reduce dimensionality to the optimal value d^*. This value is, however, usually unknown, as it depends on the sample size and distribution of the problem.

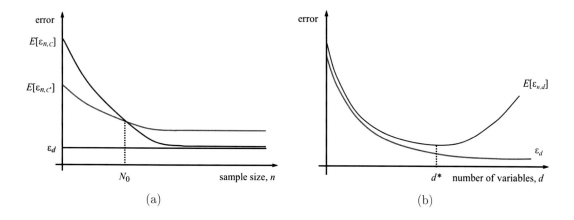

Figure 8.3: Complexity Dilemma. (a) Scissors plot: for $n > N_0$, the more complex class $\mathcal{C}$ is better, but for $n < N_0$, the simpler class $\mathcal{C}'$ is better. (b) Peaking phenomenon: as d increases, $E[\varepsilon_{n,d}]$ initially decreases, following the decrease in ε_d, but eventually increases.

8.2 Vapnik-Chervonenkis Theory

Vapnik-Chervonenkis (VC) theory introduces intuitively satisfying metrics for classification complexity, namely, the shatter coefficients and the VC dimension. The main result of VC theory is the VC theorem, which uniformly bounds the difference between empirical and true errors over an arbitrary family of classifiers, in a distribution-free manner, in terms of the shatter coefficients and VC dimension of the family. This leads to a practical model selection algorithm, known as Structural Risk Minimization, to be discussed in Section 8.3.

All results in VC theory are worst-case, as there are no distributional assumptions, and thus they can be very loose for a particular feature-label distribution and small sample sizes. Nevertheless, the VC theorem remains a powerful tool for the analysis of the large-sample deviation between the true and empirical (i.e., resubstitution) classification errors.

*8.2.1 Finite Model Selection

If space of classifiers $\mathcal{C}$ is finite, then model selection is simple, and the full generality of VC theory would not be needed, as we see in this section. Let $\psi \in \mathcal{C}$ and consider sample data $S_n = \{(\mathbf{X}_1, Y_1), \ldots, (\mathbf{X}_n, Y_n)\}$. The true error of ψ is $\varepsilon[\psi] = P(psi(\mathbf{X}) \neq Y)$, while its empirical error on S_n is

$$\hat{\varepsilon}[\psi] = \frac{1}{n} \sum_{i=1}^{n} I_{\psi(\mathbf{X}_i) \neq Y_i} \tag{8.7}$$

If ψ is not a function of S_n, then we can get a bound on the tail probability $P(|\hat{\varepsilon}[\psi] - \varepsilon[\psi]| \geq \tau)$ directly from Hoeffding's Inequality (Theorem A.14). To see this, consider the independent binary random variables $W_i = I_{\psi(\mathbf{X}_i) \neq Y_i}$, $i = 1, \ldots, n$, and let $Z_n = \sum_{i=1}^{n} W_i = n\hat{\varepsilon}[\psi]$. Since ψ is not a function of S_n, we have $E[\hat{\varepsilon}[\psi]] = E[\varepsilon[\psi]]$ (it is as if $\hat{\varepsilon}[\psi]$ is a test-set error estimator) so that $E[Z_n] = E[n\hat{\varepsilon}[\psi]] = nE[\hat{\varepsilon}[\psi]] = nE[\varepsilon[\psi]]$. Applying Hoeffding's Inequality to Z_n yields:

$$P(|\hat{\varepsilon}[\psi] - \varepsilon[\psi]| \geq \tau) \leq 2e^{-2n\tau^2}, \quad \text{for all } \tau > 0, \tag{8.8}$$

regardless of the distribution of $(\mathbf{X}, Y)$. This result is true whether $\mathcal{C}$ is finite or not. However, it fails if $\psi = \psi_n$ is a classifier selected in $\mathcal{C}$ by using S_n, as in this case $E[\hat{\varepsilon}[\psi_n]] \neq E[\varepsilon[\psi_n]]$ in general (here $\hat{\varepsilon}[\psi_n]$ is the resubstitution estimator). Since ψ_n is the classifier we are really interested in, the analysis needs to be modified. The solution is to bound $|\hat{\varepsilon}[\psi] - \varepsilon[\psi]|$ *uniformly* over $\mathcal{C}$, by a simple application of the Union Bound. The result is collected in the following theorem.

Theorem 8.1. *If the space of classifiers $\mathcal{C}$ is finite then, regardless of the distribution of $(\mathbf{X}, Y)$,*

$$P\left(\max_{\psi \in \mathcal{C}} |\hat{\varepsilon}[\psi] - \varepsilon[\psi]| > \tau\right) \leq 2|\mathcal{C}|e^{-2n\tau^2}, \quad \text{for all } \tau > 0. \tag{8.9}$$

Proof. First note that $\{\max_{\psi \in \mathcal{C}} |\varepsilon[\psi] - \hat{\varepsilon}[\psi]| > \tau\} = \bigcup_{\psi \in \mathcal{C}} \{|\varepsilon[\psi] - \hat{\varepsilon}[\psi]| > \tau\}$. We can then apply the Union Bound (A.10) to obtain: $P(\max_{\psi \in \mathcal{C}} |\varepsilon[\psi] - \hat{\varepsilon}[\psi]| > \tau) \leq \sum_{\psi \in \mathcal{C}} P(|\varepsilon[\psi] - \hat{\varepsilon}[\psi]| > \tau)$. Application of (8.8) then gives (8.9). $\diamond$

Now, the designed classifier ψ_n is a member of $\mathcal{C}$, therefore the bound in (8.9) applies:

$$P(|\hat{\varepsilon}[\psi_n] - \varepsilon[\psi_n]| > \tau) \leq 2|\mathcal{C}|e^{-2n\tau^2}, \quad \text{for all } \tau > 0. \tag{8.10}$$

This result states that the resubstitution estimator $\hat{\varepsilon}[\psi_n]$ gets arbitrarily close to the true error $\varepsilon[\psi_n]$, with high probability, as $n \to \infty$ (it also means that the resubstitution estimator is universally strongly consistent, by an application of Theorem A.8). Applying Lemma A.1 to (8.10) leads to

$$E\left[|\hat{\varepsilon}[\psi_n] - \varepsilon[\psi_n]|\right] \leq \sqrt{\frac{1 + \ln 2|\mathcal{C}|}{2n}}. \tag{8.11}$$

When the ratio between the complexity $\ln |\mathcal{C}|$ and the sample size n is small, the bound is tighter, and $E[\hat{\varepsilon}[\psi_n]]$ must be close to $E[\varepsilon[\psi_n]]$ (since $-E[|Z|] \leq E[Z] \leq E[|Z|]$). This implies that a smaller complexity relative to the sample size leads to less overfitting. As $n \to \infty$, with fixed $|C|$, $E[\hat{\varepsilon}[\psi_n]]$ tends to $E[\varepsilon[\psi_n]]$, and the bias of resubstitution disappears.

This analysis holds if $\mathcal{C}$ is finite. Although some classification rules correspond to finite $\mathcal{C}$ (e.g., the histogram rule with a finite number of zones), in most cases of practical interest, $|\mathcal{C}|$ is not finite. The full generality of VC theory is necessary to extend the previous analysis to arbitrary $\mathcal{C}$.

8.2.2 Shatter Coefficients and VC Dimension

We begin by defining the key concepts of shatter coefficients and VC dimension for families of sets. In the next section, the connection with classification is made.

Intuitively, the complexity of a family of sets should have to do with its ability to "pick out" subsets of a given set of points. We formalize this next. For a given n, consider a set of points $x_1, \ldots, x_n$ in R^d in *general position* (i.e., two points are not on top of each other, three points are not on a line, etc.) Given a set $A \subseteq R^d$, then

$$A \cap \{x_1, \ldots, x_n\} \subseteq \{x_1, \ldots, x_n\} \tag{8.12}$$

is the subset of $\{x_1, \ldots, x_n\}$ "picked out" by A. Now, consider a family $\mathcal{A}$ of subsets of R^d, and let

$$N_{\mathcal{A}}(x_1, \ldots, x_n) = |\{A \cap \{x_1, \ldots, x_n\} \mid A \in \mathcal{A}\}| , \tag{8.13}$$

i.e., the total number of subsets of $\{x_1, \ldots, x_n\}$ that can be picked out by sets in $\mathcal{A}$. The $n$the *shatter coefficient* of the family $\mathcal{A}$ is defined as

$$s(\mathcal{A}, n) = \max_{\{x_1, \ldots, x_n\}} N_{\mathcal{A}}(x_1, \ldots, x_n). \tag{8.14}$$

The shatter coefficients $\{s(\mathcal{A}, n); n = 1, 2, \ldots\}$ measure the size or complexity of $\mathcal{A}$. Note that $s(\mathcal{A}, n) \leq 2^n$ for all n. If $s(\mathcal{A}, n) = 2^n$, then there is a set of points $\{x_1, \ldots, x_n\}$ such that $N_{\mathcal{A}}(x_1, \ldots, x_n) = 2^n$, and we say that $\mathcal{A}$ *shatters* $\{x_1, \ldots, x_n\}$. This implies that $s(\mathcal{A}, m) = 2^m$, for all $m < n$, as well. On the other hand, if $s(\mathcal{A}, n) < 2^n$, then any set of points $\{x_1, \ldots, x_n\}$ contains at least one subset that cannot be picked out by any member of $\mathcal{A}$, in which case $s(\mathcal{A}, m) < 2^m$, for all $m > n$, as well. Therefore, there is a largest integer $k \geq 1$ such that $s(\mathcal{A}, k) = 2^k$. This integer is called the *VC dimension* $V_{\mathcal{A}}$ of $\mathcal{A}$ (assuming $|\mathcal{A}| \geq 2$). If $s(\mathcal{A}, n) = 2^n$ for all n, then $V_{\mathcal{A}} = \infty$. The VC dimension of $\mathcal{A}$ is thus the *maximal number of points in R^d that can be shattered by $\mathcal{A}$*. Like the shatter coefficients, $V_{\mathcal{A}}$ is a measure of the size of $\mathcal{A}$.

Example 8.1. For the class of half-lines $\mathcal{A}_1 = \{(-\infty, a] \mid a \in R\}$,

$$s(\mathcal{A}_1, n) = n + 1 \text{ and } V_{\mathcal{A}} = 1, \tag{8.15}$$

while for the class of intervals $\mathcal{A}_2 = \{[a, b] \mid a, b \in R\}$,

$$s(\mathcal{A}_2, n) = \frac{n(n+1)}{2} + 1 \text{ and } V_{\mathcal{A}} = 2. \tag{8.16}$$

These examples can be generalized as follows: for the class of "half-rectangles"

$$\mathcal{A}_d = \{(-\infty, a_1] \times \cdots \times (-\infty, a_d] \mid (a_1, \ldots, a_d) \in R^d\}, \tag{8.17}$$

it is easy to check that $V_{\mathcal{A}_d} = d$, while for the class of rectangles

$$\mathcal{A}_{2d} = \{[a_1, b_1] \times \cdots \times [a_d, b_d] \mid (a_1, \ldots, a_d, b_1, \ldots, b_d]) \in R^{2d}\}, \tag{8.18}$$

we have $V_{\mathcal{A}_{2d}} = 2d$. ◇

Note that in the example above, class $\mathcal{A}_m$ has m parameters and its VC dimension is m, i.e., the VC dimension is equal to the number of parameters. While this is intuitive, it is *not* true in general. This is demonstrated emphatically by the next well-known example.

Example 8.2. (One-parameter, infinite-VC family of sets.) Let

$$\mathcal{A} = \{A_\omega = \{x \in R \mid \sin(\omega x) > 0\} \mid \omega \in R\}. \tag{8.19}$$

Like $\mathcal{A}_1$ in Example 8.1, this is a family of one-dimensional sets A_ω indexed by a single parameter ω. However, we claim that $V_{\mathcal{A}} = \infty$. To see this, consider the set of points $\{x_1, \ldots, x_n\}$ such that $x_i = 10^{-i}$, for $i = 1, \ldots, n$, and consider any of its subsets of points $\{x_{i_1}, \ldots, x_{i_m}\}$. It is possible to select a ω such that $A_\omega \in \mathcal{A}$ "picks out" this subset, namely:

$$\omega = \pi \left(1 + \sum_{j=1}^n y_j 10^j\right), \tag{8.20}$$

where $y_{i_j} = 0$, for $j = 1, \ldots, m$, and $y_i = 1$, otherwise. This is so because $\sin(\omega x_i) > 0$ if $y_i = 0$ while $\sin(\omega x_i) < 0$ if $y_i = 1$ (Exercise 8.1). Hence, $\mathcal{A}$ shatters $\{x_1, \ldots, x_n\}$, for $n = 1, 2, \ldots$ ◇

The previous example shows that there is essentially no relationship between number of parameters and complexity.

A general bound for shatter coefficients is

$$s(\mathcal{A}, n) \leq \sum_{i=0}^{V_{\mathcal{A}}} \binom{n}{i}, \quad \text{for all } n. \tag{8.21}$$

Note that $\mathcal{A}_1$ and $\mathcal{A}_2$ in Example 8.1 achieve this bound, so it is tight. Assuming that $V_{\mathcal{A}} < \infty$, it follows directly from the Binomial Theorem that

$$s(\mathcal{A}, n) \leq (n+1)^{V_{\mathcal{A}}}. \tag{8.22}$$

8.2.3   VC Parameters of a Few Classification Rules

The preceding concepts can be apply to classification rules, through their associated space of classifiers $\mathcal{C}$. Given a classifier $\psi \in \mathcal{C}$, define the set

$$A_\psi = \{x \in R^d \mid \psi(x) = 1\}, \tag{8.23}$$

that is, the 1-decision region for ψ (this specifies the classifier completely, since the 0-decision region is simply A_ψ^c), and let $\mathcal{A}_\mathcal{C} = \{A_\psi \mid \psi \in \mathcal{C}\}$, that is, the family of all 1-decision regions produced by $\mathcal{C}$. Then the shatter coefficients $\mathcal{S}(\mathcal{C}, n)$ and VC dimension $V_\mathcal{C}$ for $\mathcal{C}$ are defined as

$$\mathcal{S}(\mathcal{C}, n) = s(\mathcal{A}_\mathcal{C}, n)$$
$$V_\mathcal{C} = V_{\mathcal{A}_\mathcal{C}}$$

(8.24)

All concepts defined in the previous section now have a classification interpretation. For example, a subset of $\{x_1, \ldots, x_n\}$ is "picked out" by ψ if ψ gives the label 1 to the points in the subset and label 0 to the points not in the subset. The set of points $\{x_1, \ldots, x_n\}$ is shattered by $\mathcal{C}$ if there are 2^n classifiers in $\mathcal{C}$ that produce all possible 2^n labeling assignments to $\{x_1, \ldots, x_n\}$. Furthermore, all the results discussed previously apply in the new setting; for example, following (8.22), if $V_\mathcal{C} < \infty$,

$$\mathcal{S}(\mathcal{C}, n) \le (n+1)^{V_\mathcal{C}} .$$

(8.25)

Hence, if $V_\mathcal{C}$ is finite, $s(\mathcal{C}, n)$ grows polynomially, not exponentially, with n, which will be important in Section 8.2.4.

Next, results on the VC dimension and shatter coefficients corresponding to commonly used classification rules are given.

Linear Classification Rules

Linear classification rules are those that produce classifiers with hyperplane decision boundaries. This includes NMC, LDA, Perceptrons, and Linear SVMs. Let $\mathcal{C}_d$ be the class of linear classifiers in R^d. Then it can be proved that

$$\mathcal{S}(\mathcal{C}_d, n) = 2 \sum_{i=0}^{d} \binom{n-1}{i}$$

(8.26)

$$V_{\mathcal{C}_d} = d + 1 .$$

The fact that $V_\mathcal{C} = d + 1$ means that there is a set of $d + 1$ points that can be shattered by linear classifiers in R^d, but no set of $d+2$ points can be shattered. Consider the case $d = 2$ as an example. From the formula above,

$$s(\mathcal{C}_2, 1) = 2 \binom{0}{0} = 2 = 2^1$$
$$s(\mathcal{C}_2, 2) = 2 \left[\binom{1}{0} + \binom{1}{1} \right] = 4 = 2^2$$
$$s(\mathcal{C}_2, 3) = 2 \left[\binom{2}{0} + \binom{2}{1} + \binom{2}{2} \right] = 8 = 2^3$$
$$s(\mathcal{C}_2, 4) = 2 \left[\binom{3}{0} + \binom{3}{1} + \binom{3}{2} \right] = 14 < 16 = 2^4$$

(8.27)

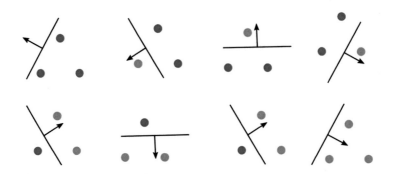

Figure 8.4: A set of $n = 3$ points in R^2 can be shattered, i.e., can be given all possible $2^3 = 8$ label assignments, by a linear classification rule. Adapted from Figure 1 in Burges [1998].

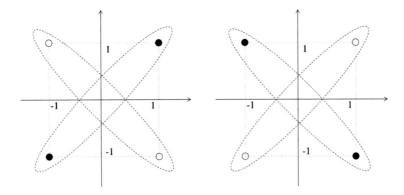

Figure 8.5: The two labeling assignments for a set of $n = 4$ points in R^2 that cannot be produced by linear classifiers.

Therefore, $V_{\mathcal{C}_2} = 3$. This means that there is a set of 3 points in R^2 that can be shattered by linear classifiers. In fact, any set of 3 points (in general position) in R^2 can be shattered by linear classifiers, as can be seen in Figure 8.4.

In addition, $V_{\mathcal{C}_2} = 3$ means that no set of 4 points (in general position) can be shattered— since there are at least $2^4 - s(\mathcal{C}_2, 4) = 16 - 14 = 2$ labeling assignments out of the possible $2^4 = 16$ that cannot be produced by linear classifiers. These correspond to the two variations of the XOR data set of Chapter 6. See Figure 8.5 for an illustration.

The VC dimension of linear classification rules increases linearly with the number of variables, which is an advantage of such rules. Note however that all linear classification rules having the same VC dimension does not mean they will always perform the same, particularly in small-sample cases.

kNN Classification Rule

For $k = 1$, clearly any set of n points can be shattered (giving label 1 to a subset of points and label 0 to its complement produces a 1NN classifier that "picks out" that subset), and thus the 1NN classification rule has infinite VC dimension. Some reflection shows that this is true for $k > 1$ as well, so the VC dimension of any kNN classification rule is infinite. Classes $\mathcal{C}$ with finite VC dimension are called *VC classes*. Thus, the class $\mathcal{C}_k$ of kNN classifiers is not a VC class, for each $k > 1$. Classification rules that have infinite VC dimension are not necessarily useless. For example, there is empirical evidence that 3NN and 5NN are good rules in small-sample cases. In addition, the Cover-Hart Theorem says that the asymptotic kNN error rate is around the Bayes error. However, the worst-case scenario if $V_{\mathcal{C}} = \infty$ is indeed quite bad, as discussed in Section 8.2.5.

Classification Trees

A binary tree with a depth of k levels of splitting nodes has at most $2^k - 1$ splitting nodes and at most 2^k leaves. Therefore, for a classification tree with data-independent splits (that is, fixed-partition tree classifiers), we have that

$$S(\mathcal{C}, n) = \begin{cases} 2^n, & n \leq 2^k \\ 2^{2^k}, & n > 2^k \end{cases} \tag{8.28}$$

and it follows that $V_{\mathcal{C}} = 2^k$. The shatter coefficients and VC dimension thus grow very fast (exponentially) with the number of levels. The case is different for data-dependent decision trees (e.g., CART and BSP). If stopping or pruning criteria are not strict enough, one may have $V_{\mathcal{C}} = \infty$ in those cases.

Non-Linear SVMs

It is easy to see that the shatter coefficients and VC dimension correspond to those of linear classification in the transformed high-dimensional space. More precisely, if the *minimal* space where the kernel can be written as a dot product is m, then $V_{\mathcal{C}} = m + 1$. For example, for the polynomial kernel

$$K(x, y) = (x^T y)^p = (x_1 y_1 + \cdots + x_d y_d)^p \tag{8.29}$$

we have $m = \binom{d+p-1}{p}$, i.e., the number of distinct powers of $x_i y_i$ in the expansion of $K(x, y)$, so $V_{\mathcal{C}} = \binom{d+p-1}{p} + 1$. For certain kernels, such as the Gaussian kernel,

$$K(x, y) = \exp(-|x - y|^2/\sigma^2) \tag{8.30}$$

the minimal space is infinitely dimensional, so $V_{\mathcal{C}} = \infty$.

Neural Networks

For the class $\mathcal{C}_k$ of neural networks with k neurons in one hidden layer, and arbitrary sigmoids, it can be shown that

$$V_{\mathcal{C}_k} \geq 2 \left\lfloor \frac{k}{2} \right\rfloor d \tag{8.31}$$

where $\lfloor x \rfloor$ is the largest integer $\leq x$. If k is even, this simplifies to $V_{\mathcal{C}} \geq kd$. For threshold sigmoids, $V_{\mathcal{C}_k} < \infty$. In fact,

$$\mathcal{S}(\mathcal{C}_k, n) \leq (ne)^\gamma \quad \text{and} \quad V_{\mathcal{C}_k} \leq 2\gamma \ln_2(e\gamma) \tag{8.32}$$

where $\gamma = kd + 2k + 1$ is the number of weights. The threshold sigmoid achieves the smallest possible VC dimension among all sigmoids. In fact, there are sigmoids for which $V_{\mathcal{C}_k} = \infty$ for $k \geq 2$.

Histogram Rules

For a histogram rule with a finite number of partitions b, it is easy to see that the the shatter coefficients are given by

$$\mathcal{S}(\mathcal{C}, n) = \begin{cases} 2^n, & n < b, \\ 2^b, & n \geq b. \end{cases} \tag{8.33}$$

Therefore, the VC dimension is $V_{\mathcal{C}} = b$.

8.2.4 Vapnik-Chervonenkis Theorem

The celebrated Vapnik-Chervonenkis theorem extends Theorem 8.1 to arbitrary $\mathcal{C}$ by using the shatter coefficients $\mathcal{S}(\mathcal{C}, n)$ instead of $|\mathcal{C}|$ as metrics of the size of $\mathcal{C}$. See Appendix A6 for a proof.

Theorem 8.2. *(Vapnik-Chervonenkis theorem.) Regardless of the distribution of (X, Y),*

$$P \left(\sup_{\psi \in \mathcal{C}} |\hat{\varepsilon}[\psi] - \varepsilon[\psi]| > \tau \right) \leq 8\,\mathcal{S}(\mathcal{C}, n) e^{-n\tau^2/32}, \quad \text{for all } \tau > 0. \tag{8.34}$$

If $V_{\mathcal{C}}$ is finite, we can use the inequality (8.25) to write the bound in terms of $V_{\mathcal{C}}$:

$$P \left(\sup_{\psi \in \mathcal{C}} |\hat{\varepsilon}[\psi] - \varepsilon[\psi]| > \tau \right) \leq 8(n+1)^{V_{\mathcal{C}}} e^{-n\tau^2/32}, \quad \text{for all } \tau > 0. \tag{8.35}$$

Therefore, if $V_{\mathcal{C}}$ is finite, the term $e^{-n\tau^2/32}$ dominates, and the bound decreases *exponentially* fast as $n \to \infty$. As in the previous section, the designed classifier ψ_n is a member of $\mathcal{C}$, therefore the bound in (8.35) applies:

$$P \left(|\hat{\varepsilon}[\psi_n] - \varepsilon[\psi_n]| > \tau \right) \leq 8(n+1)^{V_{\mathcal{C}}} e^{-n\tau^2/32}, \quad \text{for all } \tau > 0. \tag{8.36}$$

This is equation (7.12) in Chapter 7. In addition, applying Lemma A.1 to (8.36) leads to

$$E\left[|\hat{\varepsilon}[\psi_n] - \varepsilon[\psi_n]|\right] \leq 8\sqrt{\frac{V_{\mathcal{C}}\ln(n+1) + 4}{2n}}. \tag{8.37}$$

Hence, $E[|\hat{\varepsilon}[\psi] - \varepsilon[\psi]|]$ is $O(\sqrt{V_{\mathcal{C}}\ln n/n})$, from which we conclude that if $n \gg V_{\mathcal{C}}$, then $E[\hat{\varepsilon}[\psi]]$ is close to $E[\varepsilon[\psi]]$. It also means that, for $V_{\mathcal{C}} < \infty$, the resubstitution estimator is asymptotically unbiased as $n \to \infty$.

8.2.5   No-Free-Lunch Theorems

As was the case in Section 3.4, in a distribution-free scenario, one can pick a feature-label distribution that makes classification performance very poor. Here, we are concerned with the ratio $V_{\mathcal{C}}/n$ between complexity and sample size. Good classification performance demands that $n \gg V_{\mathcal{C}}$ (a well-known rule of thumb is $n > 20V_{\mathcal{C}}$), at least in the worst case, as is demonstrated by two well-known No Free-Lunch Theorems, stated here without proof.

Theorem 8.3. *Let $\mathcal{C}$ be a space of classifiers with $V_{\mathcal{C}} < \infty$ and let Ω be the set of all r.v.'s (X,Y) corresponding to $\varepsilon_{\mathcal{C}} = 0$. Then*

$$\sup_{(X,Y)\in\Omega} E[\varepsilon_{n,\mathcal{C}}] \geq \frac{V_{\mathcal{C}} - 1}{2en}\left(1 - \frac{1}{n}\right) \tag{8.38}$$

for $n \geq V_{\mathcal{C}} - 1$.

The previous theorem states that over all distributions that are separable by a classifier in $\mathcal{C}$ (e.g., all linearly-separable distributions in the of a linear classification rule), the worst-case performance can still be far from zero unless $n \gg V_{\mathcal{C}}$.

If, on the other hand, $V_{\mathcal{C}} = \infty$, then one cannot make $n \gg V_{\mathcal{C}}$, and a worst-case bound can be found that is independent of n (this means that there exists a situation where the classification error cannot be reduced no matter how large n may be). This is shown by the following theorem.

Theorem 8.4. *If $V_{\mathcal{C}} = \infty$, then for every $\delta > 0$, and every classification rule associated with $\mathcal{C}$, there is a feature-label distribution for (X,Y) with $\varepsilon_{\mathcal{C}} = 0$ but*

$$E[\varepsilon_{n,\mathcal{C}}] > \frac{1}{2e} - \delta, \text{ for all } n > 1. \tag{8.39}$$

The previous theorem shows that, even if infinite-VC classification rules, such as kNN, are not necessarily bad in practice, the worst-case performance of these rules can be indeed bad.

8.3   Model Selection Methods

The model selection problem can be formalized as follows. Assume that $\Psi_{n,k}$ uses training data S_n to pick a classifier $\psi_{n,k}$ in a corresponding space of classifiers $\mathcal{C}_k$, for $k = 1, \ldots, N$. Here we assume a finite number of choices N, for simplicity. This assumption is not too restrictive, since the choice is often between a small number of different classification rules, and any continuous free parameters can be discretized into a finite grid (the selection process is then usually called a *grid search*). The goal of model selection is to pick a classifier $\psi_n = \psi_{n,k_n} \in \mathcal{C}_{k_n}$ that has the smallest classification error among all the candidates. The selected classification rule is then Ψ_{n,k_n}. In the next few section we investigate model selection approaches that can be used in practice to solve this problem.

8.3.1   Validation Error Minimization

In this approach, the selected classifier ψ_n is the one that produces the smallest error count on a independent sample $S_m = \{(\mathbf{X}_1, Y_1), \ldots, (\mathbf{X}_m, Y_m)\}$, known as the *validation set*. In other words,

$$\psi_n = \arg\min_{\psi_{n,k}} \hat{\varepsilon}[\psi_{n,k}] = \arg\min_{\psi_{n,k}} \frac{1}{m} \sum_{i=1}^{m} |Y_i - \psi_{n,k}(\mathbf{X}_i)| \tag{8.40}$$

However, we really would like to select the classifier ψ_n^* that achieves the smallest true classification error:

$$\psi_n^* = \arg\min_{\psi_{n,k}} \varepsilon[\psi_{n,k}] = \arg\min_{\psi_{n,k}} E[|Y - \psi_{n,k}(\mathbf{X})] \tag{8.41}$$

We will see below that we can bound the difference between the true error $\varepsilon[\psi_n]$ of the sample-based classifier and the true error $\varepsilon[\psi_n^*]$ of the best classifier (which is all we care about). First, note that

$$\varepsilon[\psi_n] - \varepsilon[\psi_n^*] = \varepsilon[\psi_n] - \hat{\varepsilon}[\psi_n] + \hat{\varepsilon}[\psi_n] - \min_k \varepsilon[\psi_{n,k}]$$

$$= \varepsilon[\psi_n] - \hat{\varepsilon}[\psi_n] + \min_k \hat{\varepsilon}[\psi_{n,k}] - \min_k \varepsilon[\psi_{n,k}]$$

$$\leq \varepsilon[\psi_n] - \hat{\varepsilon}[\psi_n] + \max_k |\hat{\varepsilon}[\psi_{n,k}] - \varepsilon[\psi_{n,k}]| \leq 2 \max_k |\hat{\varepsilon}[\psi_{n,k}] - \varepsilon[\psi_{n,k}]| \tag{8.42}$$

where we used the facts that $\varepsilon[\psi_n^*] = \min_k \varepsilon[\psi_{n,k}]$ and $\hat{\varepsilon}[\psi_n] = \min_k \hat{\varepsilon}[\psi_{n,k}]$, as well the properties of minimum and maximum:

$$\min_k a_k - \min_k b_k \leq \max_k (a_k - b_k) \leq \max_k |a_k - b_k|. \tag{8.43}$$

Now, as in Theorem 8.1,

$$P\left(\max_k |\hat{\varepsilon}[\psi_{n,k}] - \varepsilon[\psi_{n,k}] > \tau\right) \leq 2Ne^{-2m\tau^2}, \quad \text{for all } \tau > 0. \tag{8.44}$$

Combining this with (8.42) proves the following result.

Theorem 8.5. *If a validation set with m points is used to select a classifier among N choices, the difference in true error between the selected classifier ψ_n and the best classifier ψ_n^* satisfies*

$$P\left(\varepsilon[\psi_n] - \varepsilon[\psi_n^*] > \tau\right) \leq 2Ne^{-m\tau^2/2}, \quad \text{for all } \tau > 0. \tag{8.45}$$

This result proves that as the size of the validation set increases to infinity, we are certain to make the right selection. The bound is tighter the larger m is compared to N. This agrees with the intuitive observation that it is easier to select among few choices than many choices. For example, an overly fine parameter grid search should be avoided and a rougher grid preferred if the classification error is not too sensitive to the choice of the parameter (a rougher grid also has obvious computational advantages).

Notice that $\hat{\varepsilon}[\psi_n]$ is *not* an unbiased estimator of $\varepsilon[\psi_n]$, and it does not have the other nice properties of the test-set error estimator either (but see Exercise 8.2), since the validation set is not independent of ψ_n (it is being used to select ψ_n). Indeed, $\hat{\varepsilon}[\psi_n]$ is likely to be optimistically-biased as an estimator of $\varepsilon[\psi]$. This situation is sometimes referred to as "training on the test data" and is one of the major sources of error and misunderstanding in supervised learning. As long as this is acknowledged, use of a validation set is a perfectly valid model selection method (and indeed very good one, if m is large compared to N).

In order to have an unbiased estimator of $\varepsilon[\psi_n]$, a test set that is independent of both the training and validation sets is required. This leads to a *training-validation-testing* three-way strategy to split the data. The classifiers are designed on the training data under different models, the validation data is used to select the best model, and the testing data is used at the end to assess the performance of the selected classifier. For example, this strategy can be used to select the number of epochs for training a Neural Network by backpropagation (see Chapter 6). For computational savings, training may be stopped (i.e., the number of training epochs selected) once the validation error reaches its first local minimum, rather than the global minimum. The test set must not be used to make this decision, or else the test-set error estimator is not unbiased.

8.3.2 Training Error Minimization

The approach described in the previous section is often unrealistic, for the same reasons given previously regarding test-set error estimation (Chapter 7): in scientific applications, it is often the case that data is scarce, and splitting the data into training and validation sets (let alone reserving another piece of the data for testing) becomes impractical because one needs to use all the available data for training. In this case, one may consider choosing a data-efficient error estimator $\hat{\varepsilon}_n$ (i.e., one that uses the training data to estimate the classification error) and pick a classifier that produces the smallest error estimate. In other words,

$$\psi_n = \arg\min_{\psi_{n,k}} \hat{\varepsilon}_n[\psi_{n,k}].$$ (8.46)

As before, let ψ_n^* be the classifier that achieves the smallest true classification error:

$$\psi_n^* = \arg\min_{\psi_{n,k}} \varepsilon[\psi_{n,k}].$$ (8.47)

The inequality (8.42) still holds:

$$\varepsilon[\psi_n] - \varepsilon[\psi_n^*] \leq 2\max_k |\hat{\varepsilon}_n[\psi_{n,k}] - \varepsilon[\psi_{n,k}]|.$$ (8.48)

The right-hand side bound is small, and good model selection performance is guaranteed, if the error estimator $\hat{\varepsilon}_n$ can estimate the error *uniformly* well over the set of classifiers $\{\psi_{n,k} \mid k = 1, \ldots, N\}$.

However, few theoretical performance guarantees exist regarding $\max_k |\hat{\varepsilon}_n[\psi_{n,k}] - \varepsilon[\psi_{n,k}]|$. It is common to use *cross validation* to pick the value of a free parameter in a grid search, due to its small bias. However, small bias does not guarantee that $\max_k |\hat{\varepsilon}_n[\psi_{n,k}] - \varepsilon[\psi_{n,k}]|$ is small. Nevertheless, the cross-validation grid search often works in practice, if the sample size is not too small.

8.3.3   Structural Risk Minimization

The *Structural Risk Minimization (SRM)* (SRM) principle is a model selection method that tries to balance a small empirical error against the complexity of the classification rule. Assuming $V_C < \infty$, let us start by rewriting the bound in (8.35), doing away with the supremum and the absolute value:

$$P\left(\varepsilon[\psi] - \hat{\varepsilon}[\psi] > \tau\right) \leq 8(n+1)^{V_C} e^{-n\tau^2/32}, \quad \text{for all } \tau > 0,$$ (8.49)

which holds for any $\psi \in \mathcal{C}$. Let ξ be the right-hand side, where $0 \leq \xi \leq 1$. Solving for τ gives:

$$\tau(\xi) = \sqrt{\frac{32}{n}\left[V_C \ln(n+1) - \ln\left(\frac{\xi}{8}\right)\right]}.$$ (8.50)

For a given ξ, we thus have

$$P\left(\varepsilon[\psi] - \hat{\varepsilon}[\psi] > \tau(\xi)\right) \leq \xi \implies P\left(\varepsilon[\psi] - \hat{\varepsilon}[\psi] \leq \tau(\xi)\right) \geq 1 - \xi.$$ (8.51)

In other word, the inequality

$$\varepsilon[\psi] \leq \hat{\varepsilon}[\psi] + \tau(\xi)$$ (8.52)

holds with probability at least $1 - \xi$. But since this holds for any $\psi \in \mathcal{C}$, it in particular holds for $\psi = \psi_{n,\mathcal{C}}$, a designed classifier in $\mathcal{C}$ trained on S_n. Therefore, we can say that the following inequality holds with probability at least $1 - \xi$:

$$\varepsilon[\psi_{n,\mathcal{C}}] \leq \hat{\varepsilon}[\psi_{n,\mathcal{C}}] + \sqrt{\frac{32}{n} \left[V_{\mathcal{C}} \ln(n+1) - \ln\left(\frac{\xi}{8}\right) \right]}. \tag{8.53}$$

The second term on the right-hand side of (8.53) is the so-called *VC confidence* and functions as a complexity penalty term. While the actual form of the VC confidence may vary according to the derivation, it always depends only on $V_{\mathcal{C}}$ and n, for a given ξ. In addition, it always becomes small if $n \gg V_{\mathcal{C}}$, and large otherwise.

Now, consider a nested sequence $\{\mathcal{C}_k\}$ of classifier spaces associated with classification rules $\Psi_{n,k}$, for $k = 1, \ldots, N$. Our goal is to pick a classification rule such that the classification error $\varepsilon[\psi_{n,\mathcal{C}_k}]$ is minimal. From (8.53), this can be done by picking a ξ sufficiently close to 1 (say, $\xi = 0.95$, for 95% confidence), computing the sum of empirical error and VC confidence

$$\hat{\varepsilon}[\psi_{n,\mathcal{C}_k}] + \sqrt{\frac{32}{n} \left[V_{\mathcal{C}_k} \ln(n+1) - \ln\left(\frac{\xi}{8}\right) \right]} \tag{8.54}$$

for each $k = 1, \ldots, N$, and then picking the k^* such that this is minimal. Therefore, we want to minimize $\hat{\varepsilon}_{n,\mathcal{C}^k}$ to achieve a good fit to the data, but will penalize large $V_{\mathcal{C}}$ compared to n. See Figure 8.6 for an illustration of the SRM method. We can see that, as the VC dimension of the classification rule increases with respect to the sample size, the empirical error decreases, but this is counter-balanced by an increasing complexity penalty term. The optimal model achieves a compromise between good fit to the data and small complexity.

8.4 Bibliographical Notes

The VC theorem is an extension of the Glivenko-Cantelli theorem and is part of the theory of empirical processes. An excellent reference on this topic is Pollard [1984]. See also Devroye et al. [1996], Vapnik [1998], and Castro [2020]. The proofs of results about shatter coefficients and VC dimension of linear, CART, and neural-network classification rules in Section 8.2.3 can be found in Devroye et al. [1996].

The no-free lunch Theorem 8.4 is Theorem 14.3 in Devroye et al. [1996], while the bound (8.21) is proved in Theorem 13.2 of Devroye et al. [1996]. The $n > 20V_{\mathcal{C}}$ rule of thumb appears in Vapnik [1998].

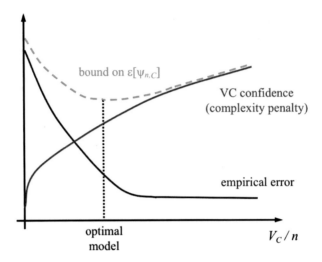

Figure 8.6: Structural Risk Minimization model selection method. As V_C/n increases, the empirical error decreases but the complexity penalty term increases. The optimal model achieves a compromise between good fit to the data and small complexity.

The minimum description length (MDL) principle is an information-theoretical model selection method proposed in Rissanen [1989], which closely resembles the SRM method. It is based on minimizing the sum of the bits necessary to encode the designed classifier and the number of bits necessary to encode the empirical error. Therefore, it attempts to maximize the fit to the data while penalizing complex classification rules. A plot similar to Figure 8.6 would apply to MDL as well.

8.5   Exercises

8.1. Provide the missing details in Example 8.2.

8.2. Using the same notation as in Section 8.3.1, prove that

$$E[\varepsilon[\psi_n] - \varepsilon[\psi_n^*]] \leq 2\sqrt{\frac{1 + \ln 2N}{2m}}\,. \tag{8.55}$$

In other words, the deviation in performance is $O(\sqrt{\ln N/m})$, and hence it vanishes as the ratio of the validation set size m over the number of choices N grows.

Hint: Use Theorem 8.5 and Lemma A.1.

8.3. Prove (8.44).

8.4. Using the same notation as in Section 8.3.1, prove that

$$P\left(|\hat{\varepsilon}[\psi_n] - \varepsilon[\psi_n]| > \tau\right) \leq 2Ne^{-2m\tau^2}, \quad \text{for all } \tau > 0. \tag{8.56}$$

Hence, even though $\hat{\varepsilon}[\psi_n]$ is generally optimistically biased, it is around $\varepsilon[\psi_n]$ with high probability if m is large (with respect to N). Now use Lemma A.1 to obtain an inequality for the expected absolute difference in errors.

Hint: To prove (8.56), obtain a suitable inequality and follow a derivation similar to the one to establish Theorem 8.5.

8.5. Suppose that a classification rule picks a classifier in a space $\mathcal{C}$ by minimizing the error on the training data set S_n. This is case, for example, of the the (discrete and continuous) histogram classifier and neural networks with weights picked to minimize the empirical error. Assuming that $V_{\mathcal{C}} < \infty$, prove that the design error (8.3) for this classification rule is controlled as

$$E\left[\Delta_{n,\mathcal{C}}\right] \leq 16\sqrt{\frac{V_{\mathcal{C}}\ln(n+1) + 4}{2n}}. \tag{8.57}$$

Therefore, the expected design error is guaranteed to be small if $n \gg V_{\mathcal{C}}$. This method of classifier design has been championed by Vapnik under the name *empirical risk minimization* (ERM).

Hint: First derive an equation similar to (8.48) for infinite $\mathcal{C}$ (replace min and max by inf and sup, respectively). Then apply the VC theorem and Lemma A.1.

8.6. Consider the class $\mathcal{C}$ of all quadratic classifiers in R^2, that is, classifiers of the form

$$\psi(x) = \begin{cases} 1, & x^T A x + b^T x + c > 0 \\ 0, & \text{otherwise.} \end{cases}$$

where A is a 2×2 matrix, $b \in R^2$, and $c \in R$.

(a) Show that this class has finite VC dimension, by finding a bound on $V_{\mathcal{C}}$.

Hint: Use the fact that the class of classifiers in R^d of the form

$$\psi(x) = \begin{cases} 1, & \sum_{i=1}^{r} a_i\phi_i(x) > 0 \\ 0, & \text{otherwise.} \end{cases}$$

where only the a_i are variable, and $\phi_i : R^d \to R$ are fixed functions of x, has VC dimension at most r. For example, this is clearly satisfied (with equality) for linear classifiers in R^2, in which case $r = 3$, with $\phi_1(x) = x_1$, $\phi_2(x) = x_2$, and $\phi_3(x) = 1$.

(b) Suppose that a quadratic classifier is designed by picking the parameters A, b and c that minimize the training error. Use the result from part (a) to show that, if the class-conditional densities are known to be 2-D Gaussians, with arbitrary means and covariance matrices, this classification rule is a strongly consistent rule, that is, $\varepsilon_n \to \varepsilon^*$ with probability 1 as $n \to \infty$.

8.7. In the figure below, assume that the classifier depicted on the left was designed by LDA, while the one on the right was designed by a nonlinear SVM with polynomial kernel $K(x, y) = (x^T y)^2$. Based on the principle of Structural Risk Minimization, explain which classifier you would pick. Assume a confidence $1 - \xi = 0.95$.

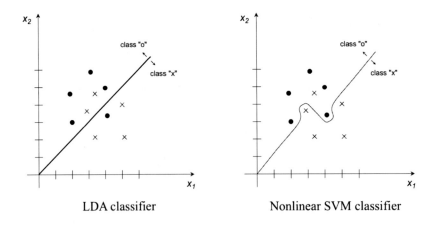

Figure 8.7: Diagram for Problem 8.7

8.8. Consider two one-dimensional classification rules Φ_1 and Φ_2. These rules produce classifiers with decision regions consisting of one and two intervals, respectively.

(a) Determine the VC dimension of Φ_1 and Φ_2.

(b) For given training data S_n, with $n = 10$, it is observed that the classifiers produced by Φ_1 and Φ_2 get 6 out of 10 and 9 out of 10 correct classifications, respectively. Based on the principle of Structural Risk Minimization, which classification rule you would pick? Assume a confidence level $1 - \xi = 0.95$, and use $\ln(11) = 2.4$ and $\ln(0.05/8) = -5$.

Chapter 9

Dimensionality Reduction

> "The feature which presents itself most forcibly to the
> untrained inquirer may not be that which is considered most
> fundamental by the experienced man of science; for the success
> of any physical investigation depends on the judicious selection
> of what is to be observed as of primary importance."
> –James Clerk Maxwell, *Scientific Papers*, 1831–1879.

It was seen in Chapter 2 that the Bayes error can only decrease, or stay the same, as more features are added. This seems to indicate that as many measurements as possible should be used in a classification problem. However, the expected error of a classification rule will often decrease at first and then increase after a certain point as more features are added; this is a famous counterintuitive fact, known as the *peaking phenomenon*, which was already mentioned in Chapter 1. Furthermore, peaking tends to occur earlier, i.e., the optimal dimensionality is smaller, under smaller sample sizes and more complex classification rules. Therefore, in order to improve classification accuracy, dimensionality reduction is necessary. This is especially true in applications where measurements are abundant, such as in high-resolution digital audio/image processing, high-throughput genomic/proteomic data, and long historical time-series data (e.g., weather, epidemiological, and stock pricing data). In addition to improving accuracy, other reasons to apply feature selection include the reduction of computational load (both in terms of execution time and data storage), improving the scientific interpretability of the designed classifiers, and visualizing the data in a two or three-dimensional space. In this chapter, we examine in detail well-known supervised and unsupervised dimensionality reduction techniques, including feature extraction, filter and wrapper feature selection, Principal Component Analysis (PCA), Multidimensional Scaling (MDS), and the Factor Analysis model.

© Springer Nature Switzerland AG 2020
U. Braga-Neto, *Fundamentals of Pattern Recognition and Machine Learning*,
https://doi.org/10.1007/978-3-030-27656-0_9

9.1   Feature Extraction for Classification

The dimensionality reduction problem can be formalized as follows. Given the original vector of measurements $\mathbf{X} \in R^p$, dimensionality reduction finds a transformation $T : R^p \to R^d$, where $d < p$, such that the new feature vector is $\mathbf{X}' = T(\mathbf{X}) \in R^d$. For example, in digital image processing, the vector $\mathbf{X}$ typically contains all pixel values in an image, while $\mathbf{X}'$ is a much smaller vector containing key descriptive image features (e.g., the orientation and shape of image regions). Following the terminology used in image processing, we call this process *feature extraction*.

The transformed feature vector $\mathbf{X}'$ should maximize a *class-separability criterion* $J(\mathbf{X}', Y)$ so as to minimize loss of discriminatory information. The following are examples of common class-separability criteria.

- The Bayes error:

$$J(\mathbf{X}', Y) = 1 - \varepsilon^*(\mathbf{X}', Y) = 1 - E[\min\{\eta(\mathbf{X}'), 1 - \eta(\mathbf{X}')\}] . \tag{9.1}$$

- F-errors:

$$J(\mathbf{X}', Y) = 1 - d_F(\mathbf{X}', Y) = 1 - E[F(\eta(\mathbf{X}'))] . \tag{9.2}$$

  This includes the asymptotic nearest-neighbor error ε_{NN}, the Matsushita error ρ, and the conditional entropy $H(Y \mid \mathbf{X}')$, all defined in Section 2.6.2.

- The Mahalanobis distance:

$$J(\mathbf{X}', Y) = \sqrt{(\boldsymbol{\mu}_1 - \boldsymbol{\mu}_0)^T \Sigma^{-1} (\boldsymbol{\mu}_1 - \boldsymbol{\mu}_0)} . \tag{9.3}$$

  where $\boldsymbol{\mu}_0$, $\boldsymbol{\mu}_1$, and Σ are the class means and the pooled sample covariance matrix, respectively.

- Given a classification rule Ψ_n, the designed classification error:

$$J(\mathbf{X}', Y) = 1 - \varepsilon_n(\mathbf{X}', Y) = 1 - E[|Y - \Psi_n(\mathbf{X}', S_n)|] . \tag{9.4}$$

- *Fisher's discriminant*: Here $\mathbf{X}' = T(\mathbf{X}) = \mathbf{w}^T \mathbf{X}$ and $\mathbf{w}$ is chosen to maximize

$$J(\mathbf{X}', Y) = \frac{\mathbf{w}^t S_B \mathbf{w}}{\mathbf{w}^t S_W \mathbf{w}} , \tag{9.5}$$

  where

$$S_B = (\hat{\boldsymbol{\mu}}_0 - \hat{\boldsymbol{\mu}}_1)(\hat{\boldsymbol{\mu}}_0 - \hat{\boldsymbol{\mu}}_1)^T \tag{9.6}$$

  is the *between-class scatter matrix* and

$$S_W = (n-2)\hat{\Sigma} = \sum_{i=1}^{n} [(\mathbf{X}_i - \hat{\boldsymbol{\mu}}_0)(\mathbf{X}_i - \hat{\boldsymbol{\mu}}_0)^T I_{Y_i=0} + (\mathbf{X}_i - \hat{\boldsymbol{\mu}}_1)(\mathbf{X}_i - \hat{\boldsymbol{\mu}}_1)^T I_{Y_i=1}] \tag{9.7}$$

is the *within-class scatter matrix*. It can be shown (see Exercise 9.3) that $\mathbf{w}^* = \Sigma_W^{-1}(\hat{\boldsymbol{\mu}}_1 - \hat{\boldsymbol{\mu}}_0)$ maximizes (9.5). Notice that $\mathbf{w}^*$ is colinear with the parameter $\mathbf{a}_n$ of the LDA classifier in in (4.15). Hence, with the appropriate choice of threshold, Fisher's discriminant leads to the LDA classification rule.

With the exception of Fisher's discriminant, all of the criteria listed above require knowledge of the feature-label distribution $p(\mathbf{X}', Y)$. In practice, they must be estimated from data. For example, the criterion in (9.4) can be approximated using any of the error estimation rules discussed in Chapter 7.

A feature extraction transformation $\mathbf{X}' = T(\mathbf{X})$ is said to be *lossless* for the assumed class-separability criterion if $J(\mathbf{X}', Y) = J(\mathbf{X}, Y)$. For example, if J is any F-error, which includes the Bayes error, it follows from Theorem 2.3 that T is lossless if it is invertible. Lossless feature extraction is considered further in Exercise 9.4.

In practice, obtaining an optimal feature extraction transformation is a difficult problem, given the extreme generality allowed in the transformation T. In image processing, the classical approach is to fashion T manually (e.g., compute various geometrical properties of segmented image regions). With the advent of convolutional neural networks (see Section 6.2.2), it has become possible to train the feature extraction process from data and select T automatically (see Python Assignment 6.12).

9.2 Feature Selection

The simplest way to reduce dimensionality is to discard features that are redundant or less informative about the label. This can be seen as a special case of feature extraction, where the transformation T is restricted to an *orthogonal projection*. Therefore, by definition, feature selection is sub-optimal with respect to feature extraction. However, in many scientific applications, interpretability of the resulting classifier is crucial, and complicated feature transformations are undesirable. For example, in genomics it is necessary to know the identity of genes in a small gene set that are predictive of a disease or condition, whereas a complicated black-box function of all genes in the starting high-dimensional space is undesirable, due to ethical and scientific considerations, even if it is more predictive. (We will see in Section 9.3 that PCA offers a compromise between these two extremes, by restricting the feature extraction transformation to be linear.) In the next few subsections, we will discuss various feature selection algorithms based on exhaustive and greedy searches. We will also examine the relationship between feature selection and classification complexity and error estimation.

9.2.1   Exhaustive Search

Let $A \subset \{1, \ldots, p\}$ be a subset of indices and define $\mathbf{X}^A$ be set of features indexed by A. For example, if $A = \{1, p\}$ then $\mathbf{X}^A = \{X_1, X_p\}$ is a 2-dimensional feature vector containing the first and last features in the original feature vector $\mathbf{X} \in R^p$. Let $J(A) = J(X^A, Y)$ be a class-separability criterion given A. The exhaustive feature selection problem is to find A^* such that

$$A^* = \arg\max_{|A|=d} J(A). \tag{9.8}$$

Since this is a finite optimization problem, the optimal solution is guaranteed to be reached in a finite time by exhaustive search: compute $J(A)$ for all possible subsets $A \subset \{1, \ldots, p\}$ of size d and pick the maximum.

If the classification error criterion (9.4) is used as the class-separability metric and the classification ruleΨ_n used in the criterion is the same classification rule used to train the final classifier on the selected feature set, then one is using the classification error (or, in practice, an estimate of it) directly to search the best feature set. This approach is called *wrapper feature selection*. All other cases are called *filter feature selection*. For example, using the Bayes error (or its estimate) as the class-separability criterion is a filter feature selection approach. Roughly speaking, filter feature selection is "independent" of the classification rule used to train the final classifier, while wrapper feature selection is not. Wrapper feature selection can fit the data better than filter feature selection due to "matching" the search to the final desired classifier. For the same reason, it can lead to selection bias and overfitting in small-sample cases, when filter feature selection would be more appropriate. (As elsewhere in the book, "small sample" means a small number of training points in comparison to the dimensionality or complexity of the problem)

Example 9.1. (Maximum Mutual-Information Feature Selection.) Suppose that the conditional entropy in (2.70) is used to define the class-separability criterion:

$$J(A) = 1 - H(Y \mid \mathbf{X}^A) \tag{9.9}$$

Then the optimization problem in (9.8) can be written as:

$$\begin{aligned} A^* &= \arg\max_{|A|=d} 1 - H(Y \mid \mathbf{X}^A) = \arg\max_{|A|=d} [H(Y) - H(Y \mid \mathbf{X}^A)] \\ &= \arg\max_{|A|=d} I(X^A; Y). \end{aligned} \tag{9.10}$$

where $H(Y) = -\sum_{y=0}^{1} P(Y = y) \ln_2 P(Y = y)$ is the entropy of the binary variable Y, which is a constant, and $I(X^A; Y) = H(Y) - H(Y \mid \mathbf{X}^A)$ is the *mutual information* between X^A and Y. This is a filter feature selection approach. In practice, the mutual information must be estimated from training data. $\diamond$

Clearly, the number of subsets to be evaluated in an exhaustive search is:

$$m = \binom{p}{d} = \frac{p!}{d!(p-d)!}.$$ (9.11)

The complexity of exhaustive feature selection is thus $O(p^d)$. The complexity of problem increases very quickly with increasing p and d. For example, with modest numbers $p = 100$ and $d = 10$, the total number of feature sets to be evaluated is greater than 10^{13}.

A famous result, known as the Cover-Van Campenhout Theorem, shows that any ordering, with respect to class separability, of the feature vectors of size d out of p features can occur. A proof of this result is based on constructing a simple discrete distribution with the required properties (see the Bibliographical Notes for more details).

Theorem 9.1. *(Cover-Van Campenhout Theorem.) Let $A_1, A_2, \dots A_{2^p}$ be any ordering of all possible subsets of $\{1, \dots, p\}$, satisfying only the constraint $A_i \subset A_j$ if $i < j$ (hence, $A_1 = \emptyset$ and $A_{2^p} = \{1, \dots, p\}$). Let $\varepsilon^*(A) = \varepsilon^*(X^A, Y)$ for short. There is a distribution of (X, Y) such that*

$$\varepsilon^*(A_1) > \varepsilon^*(A_2) > \cdots > \varepsilon^*(A_{2^p}).$$ (9.12)

The Cover-Van Campenhout theorem establishes that, in the absence of any knowledge about the feature-label distribution, the exponential complexity of exhaustive search cannot be avoided. The constraint $A_i \subset A_j$ if $i < j$ is necessary due to the monotonicity of the Bayes error. However, for feature sets of the same size, no restriction exists.

9.2.2 Univariate Greedy Search

In this and the next section, we consider several feature selection algorithms that are "greedy" in the sense that they attempt to find good feature sets without an exhaustive search. These methods are necessarily sub-optimal unless one has information about the feature-label distribution. Here, we will assume that no such information is available.

The simplest way to find feature sets quickly is to consider the features individually. For example, the following are heuristics used in practice to select features:

- Retain features that are highly variable across all classes; discard features that are nearly constant.

- Retain one feature from each cluster of mutually correlated features; avoid correlated features.

- Retain features that are strongly correlated with the label; discard "noisy features," i.e., features that are weakly correlated with label.

Notice that a feature may have high variance, be uncorrelated with all other features, but still be a noisy feature. Likewise, a feature may be highly correlated with the target but be redundant due to being correlated with other features as well.

In the *best individual features* approach, after perhaps "filtering" the set of all features using the previous heuristics, one applies a univariate class-separability criterion $J(X_i, Y)$ to each individual original feature X_i, and picks the d features with largest J. For example, a very popular method uses the absolute value of the test statistic of a two-sample t-test

$$J(X_i, Y) = \frac{|\hat{\mu}_{i,0} - \hat{\mu}_{i,1}|}{\sqrt{\frac{\hat{\sigma}^2_{i,0}}{n_0} + \frac{\hat{\sigma}^2_{i,1}}{n_1}}} \tag{9.13}$$

for ranking the features, where $\hat{\mu}_{i,j}$ and $\hat{\sigma}^2_{i,j}$ are the sample means and variances (equivalently, the p-value of the test can be used).

Univariate criteria are very simple to check and often work well, but they can also fail spectacularly, due to considering the features one at a time and ignoring *multivariate* effects among sets of features. A dramatic demonstration of this fact is provided by *Toussaint's Counter-Example*.

Theorem 9.2. *(Toussaint's Counter-Example.) Let $d = 3$. There is a distribution of $(\mathbf{X}, Y)$ such that X_1, X_2 and X_3 are conditionally-independent given Y and*

$$\varepsilon^*(X_1) < \varepsilon^*(X_2) < \varepsilon^*(X_3), \tag{9.14}$$

but such that

$$\varepsilon^*(X_1, X_2) > \varepsilon^*(X_1, X_3) > \varepsilon^*(X_2, X_3). \tag{9.15}$$

Therefore, the best 2 individual features form the worst 2-feature set, and the worst 2 individual features form the best 2-feature set. Furthermore, the best 2-feature set does not contain the best individual feature.

Proof. A distribution concentrated on the vertices of the unit cube $[0, 1]^3$ in R^3, with $P(Y = 0) = P(Y = 1) = 0.5$, and

$$\begin{aligned}
P(X_1 = 1 \mid Y = 0) &= 0.1 & P(X_1 = 1 \mid Y = 1) &= 0.9 \\
P(X_2 = 1 \mid Y = 0) &= 0.05 & P(X_2 = 1 \mid Y = 1) &= 0.8 \\
P(X_3 = 1 \mid Y = 0) &= 0.01 & P(X_3 = 1 \mid Y = 1) &= 0.71
\end{aligned} \tag{9.16}$$

and such that X_1, X_2, X_3 are independent given Y, has the required property. ◇

This result is striking because all the features are uncorrelated (in fact, independent) and still there are strong multivariate effects that swamp the univariate ones.

A different example in this class of phenomena is provided by the XOR problem (do not confuse this with the XOR data set in previous chapters).

Example 9.2. (XOR problem.) Consider two independent and identically distributed binary features $X_1, X_2 \in \{0, 1\}$ such that $P(X_1 = 0) = P(X_1 = 1) = 1/2$, and let $Y = X_1 \oplus X_2$, where "$\oplus$" denotes the logical XOR operation. Since Y is a function of (X_1, X_2), we have $\varepsilon^*(X_1, X_2) = 0$. However, $\eta(X_1) = P(Y = 1 \mid X_1) \equiv 1/2$, so that $\varepsilon^*(X_1) = 1/2$. Clearly, $\varepsilon^*(X_2) = 1/2$ also. Therefore, X_1 and X_2 are completely unpredictive of Y by themselves, but together they completely predict Y. $\diamond$

9.2.3 Multivariate Greedy Search

Multivariate greedy feature selection algorithm attempt to address the issues that plague univariate methods by evaluating feature vectors, rather than univariate features, while at the same time avoiding exhaustively searching the entire space of feature sets. Sequential methods generally outperform best individual feature selection (except in acute small-sample cases).

We consider here only *sequential* multivariate greedy search methods, where features are added or removed sequentially to the desired feature vector, until a stopping criterion is met. Sequential methods can be divided into *bottom-up* searches, where the length of the feature vector increases over time, and *top-down* searches, where it decreases.

Sequential Forward Search

Sequential Forward Search (SFS) is an algorithm that adds one feature at a time to the working feature set.

- Let $X_{(0)} = \emptyset$.

- Given the current feature set $X_{(k)}$, the criterion $J(X_{(k)} \cup X_i, Y)$ is evaluated for each $X_i \notin X_{(k)}$ and the X_i^* that maximizes this is added to the feature set: $X_{(k+1)} = X_{(k)} \cup X_i^*$.

- Stop if $k = d$ or if no improvement is possible.

Alternatively, the initial feature vector $\mathbf{X}_{(0)}$ might consist of a small vector ($d = 2$ or $d = 3$ are common) selected by exhaustive search among the initial p features. SFS is simple and fast but has a *finite horizon* problem: once a feature is added, it is "frozen" in place, i.e. it can never be removed from the working feature set.

Sequential Backward Search

Sequential Backward Search (SBS) is the top-down version of SFS.

- Let $X_{(0)} = X$.

- Given the current feature set $X_{(k)}$, the criterion $J(X_{(k)} \setminus X_i, Y)$ is evaluated for each $X_i \in X_{(k)}$ and the X_i^* that minimizes the drop

$$J(X_{(k)}, Y) - J(X_{(k)} \setminus X_i, Y)$$

  is removed from the feature set: $X_{(k+1)} = X_{(k)} \setminus X_i^*$.

- Stop at $k = d$.

As in the case of SFS, this method has a finite-horizon problem: once a feature is removed, it can never be added back. It has the additional disadvantage that feature sets of high dimensionality have to be considered in the initial steps of the algorithm — if the criterion J involves the classification error (e.g. wrapper feature selection), this would make SBS impractical for large p.

Generalized Sequential Search

This is a generalization of sequential search, where at each stage, all combinations Z_j of a small number r of features (values $r = 2$ and $r = 3$ are common) not in the current feature set $X_{(k)}$ are considered.

- In Generalized Sequential Forward Search (GSFS), the group Z_j^* that maximizes $J(X_{(k)} \cup Z_j, Y)$ is added to the current feature set: $X_{(k+1)} = X_{(k)} \cup Z_j^*$. This is a bottom-up approach.

- In Generalized Sequential Backward Search (GSBS), the group Z_j^* that minimizes the drop $J(X_{(k)}, Y) - J(X_{(k)} \setminus Z_j, Y)$ is removed from the current feature set: $X_{(k+1)} = X_{(k)} \setminus Z_j^*$. This is a top-down approach.

GSFS and GSBS are more accurate than regular SFS and SBS, at the expense of more computation. However, they still have the same disadvantages: a finite-horizon problem and high-dimensional feature vector evaluation in the case of GSBS.

Plus-l Take-r Search

This is an approach that solves the finite-horizon problem of the previous sequential search methods, at the expense of even more computation. Starting from an initial feature set, l features are added to the current feature set using SFS and then r features are removed using SBS, and the process is repeated until a stopping criterion is satisfied. This allows back-tracking, i.e., any feature can be added or removed from the working feature set as often as necessary. If $l > r$ this is a bottom-up search, while if $r > l$ this is a top-down search.

Floating Search

This method can be considered a development of the Plus-l Take-r Search method, where the values of l and r are allowed to vary, i.e., "float," at different stages of the feature selection process. The advantage of floating search is that one is allowed to backtrack in an "optimal" sense.

Sequential floating forward search (SFFS) is the bottom-up version, while sequential floating backward search (SFBS) is the top-down algorithm. The following procedure is a common instance the SFFS algorithm:

1. Let $X_{(0)} = \emptyset$ and find $X_{(1)}$ and $X_{(2)}$ by SFS.

2. At stage k, select the feature $X_i \notin X_{(k)}$ that maximizes $J(X_{(k)} \cup X_i, Y)$ and let $Z_{(k+1)} = X_{(k)} \cup X_i^*$

3. Find the feature $X_j \in Z_{(k+1)}$ that minimizes the drop $J(Z_{(k+1)}, Y) - J(Z_{(k+1)} \setminus X_j, Y)$. If $X_i^* = X_j^*$ then let $X_{(k+1)} = Z_{(k+1)}$, increment k and go back to step 2.

4. Otherwise keep decrementing k and removing samples from $Z_{(k)}$ to form samples $Z_{(k-1)}$ while $J(Z_{(k)}, Y) > J(X_{(k)}, Y)$ or until $k = 2$. Let $X_{(k)} = Z_{(k)}$, increment k and go back to step 2.

5. Stop if $k = d$ or if no improvement is possible.

9.2.4 Feature Selection and Classification Complexity

Feature selection can be understood as a form of regularization, or constraint, on a classification rule, with the purpose of decreasing classification complexity and improving the classification accuracy. First, notice that the composition of feature selection and classifier design defines a classification ruleΨ_n^D on the original high-dimensional feature space: if ψ_n^d denotes the classifier designed byΨ_n^d on the smaller space, then the classifier designed byΨ_n^D on the larger space is given by $\psi_n^D(\mathbf{X}) =$

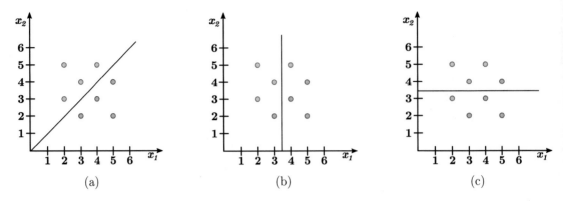

Figure 9.1: Constraint introduced by feature selection on a linear classification rule. (a) unconstrained classifier. (b) and (c) classifiers constrained by feature selection.

$\psi_n^d(\mathbf{X}') = \psi_n^d(T_n(\mathbf{X}))$, where T_n is a data-dependent. Since T is an orthogonal projection, the decision boundary produced by ψ_n^D is an extension at right angles of the one produced by ψ_n^d. If the decision boundary in the lower-dimensional space is a hyperplane, then it is still a hyperplane in the higher-dimensional space, but a hyperplane that is not allowed to be oriented along any direction, but must be orthogonal to the lower-dimensional space. See Figure 9.1 for an illustration. Therefore, feature selection corresponds to a constraint on the classification rule in the high-dimensional space, which can improve classification accuracy (in this case, due to the peaking phenomenon).

9.2.5   Feature Selection and Error Estimation

In this section, we consider when an error estimator can be computed in the reduced feature space selected by feature selection, and when it should be computed in the original space. This is a topic that has important practical consequences, as we discuss below.

Consider an error estimator $\hat{\varepsilon}_n^D = \Xi_n(\Psi_n^D, S_n^D, \xi)$, where Ψ_n^D is obtained from the combination of the feature selection transformation T_n and the classification rule Ψ_n^d in the smaller feature space. It is key to distinguish $\hat{\varepsilon}_n^D$ from the error estimator $\hat{\varepsilon}_n^d = \Xi_n(\Psi_n^d, S_n^d, \xi)$ obtained by replacing Ψ_n^D by Ψ_n^d and S_n^D by $S_n^d = \{(T_n(\mathbf{X}_1), Y_1), \ldots, (T_n(\mathbf{X}_n), Y_n)\}$. We say that the error estimation rule Ξ_n is *reducible* if $\hat{\varepsilon}_n^d = \hat{\varepsilon}_n^D$. For example, the resubstitution error estimation rule is reducible. This can be seen in Figure 9.1(b): the resubstitution error estimate is 0.25 whether one computes it in the original 2D feature space, or projects the data down to the selected variable and then computes it. The test-set error estimation rule is also reducible, if one projects the test sample to the same feature space selected in the feature selection step.

On the other hand, cross-validation is not a reducible error estimation rule. Consider leave-one-out: the estimator $\hat{\varepsilon}_n^{l,D}$ is computed by leaving out a point, applying the feature selection rule to the deleted sample, then applying Ψ_n^d in the reduced feature space, testing it on the deleted sample point (after projecting it into the reduced feature space), repeating the process n times, and computing the average error rate. This requires application of the feature selection process anew at each iteration, for a total of n times. This process, called by some authors *external cross-validation*, ensures that (7.26) is satisfied: $E[\hat{\varepsilon}_n^{l,D}] = E[\varepsilon_{n-1}^D]$. On the other hand, if feature selection is applied once at the beginning of the process and then leave-one-out proceeds in the reduced feature space while ignoring the original data, then one obtains an estimator $\hat{\varepsilon}_n^{l,d}$ that not only differs from $\hat{\varepsilon}_n^{l,D}$, but also does not satisfy any version of (7.26): neither $E[\hat{\varepsilon}_n^{l,d}] = E[\varepsilon_{n-1}^D]$ nor $E[\hat{\varepsilon}_n^{l,d}] = E[\varepsilon_{n-1}^d]$, in general. While the fact that the former identity does not hold is intuitively clear, the case of the latter identity is more subtle. This is also the identity that is mistakenly assumed to hold, usually. The reason it does not hold is that the feature selection process biases the reduced sample S_n^d, making it have a different sampling distribution than data independently generated in the reduced feature space. In fact, $\hat{\varepsilon}_n^{l,d}$ can be substantially optimistically biased. This phenomenon is called *selection bias*. As in the case of cross-validation, the bootstrap is not reducible. In the presence of feature selection, bootstrap resampling must be applied to the original data rather than the reduced data.

Bolstered error estimation may or may not be reducible. For simplicity, we focus here on bolstered resubstitution. First notice that the integrals necessary to find the bolstered error estimate in (7.39) in the original feature space R^D can be equivalently carried out in the reduced space R^d, if the kernel densities are comprised of independent components, in such a way that

$$f_i^{\Diamond,D}(\mathbf{x}) = f_i^{\Diamond,d}(\mathbf{x})f_i^{\Diamond,D-d}(\mathbf{x}), \quad \text{for } \mathbf{x} \in R^D, \, i=1,\ldots,n, \tag{9.17}$$

where $f_i^{\Diamond,D}(\mathbf{x})$, $f_i^{\Diamond,d}(\mathbf{x})$, and $f_i^{\Diamond,D-d}(\mathbf{x})$ denote the densities in the original, reduced, and difference feature spaces, respectively. One example would be Gaussian kernel densities with spherical or diagonal covariance matrices (see Section 7.7). For a given set of kernel densities satisfying (9.17),

$$
\begin{aligned}
\hat{\varepsilon}_n^{\mathrm{br},D} &= \frac{1}{n}\sum_{i=1}^n \left(I_{y_i=0} \int_{A_1} f_i^{\Diamond,d}(\mathbf{x}-\mathbf{X}_i)f_i^{\Diamond,D-d}(\mathbf{x}-\mathbf{X}_i)\,d\mathbf{x} \right.\\
&\qquad\qquad \left. + I_{y_i=1} \int_{A_0} f_i^{\Diamond,d}(\mathbf{x}-\mathbf{X}_i)f_i^{\Diamond,D-d}(\mathbf{x}-\mathbf{X}_i)\,d\mathbf{x} \right)\\
&= \frac{1}{n}\sum_{i=1}^n \left(I_{y_i=0} \int_{A_1^d} f_i^{\Diamond,d}(\mathbf{x}-\mathbf{X}_i)\,d\mathbf{x} \int_{R^{D-d}} f_i^{\Diamond,D-d}(\mathbf{x}-\mathbf{X}_i)\,d\mathbf{x} \right.\\
&\qquad\qquad \left. + I_{y_i=1} \int_{A_0^d} f_i^{\Diamond,d}(\mathbf{x}-\mathbf{X}_i)\,d\mathbf{x} \int_{R^{D-d}} f_i^{\Diamond,D-d}(\mathbf{x}-\mathbf{X}_i)\,d\mathbf{x} \right)\\
&= \frac{1}{n}\sum_{i=1}^n \left(I_{y_i=0} \int_{A_1^d} f_i^{\Diamond,d}(\mathbf{x}-\mathbf{X}_i)\,d\mathbf{x} + I_{y_i=1} \int_{A_0^d} f_i^{\Diamond,d}(\mathbf{x}-\mathbf{X}_i)\,d\mathbf{x} \right).
\end{aligned}
\tag{9.18}
$$

While being an important computation-saving device, this identity does not imply that bolstered resubstitution is always reducible whenever (9.17) is satisfied. Reducibility will also depend on the way that the kernel densities are adjusted to the sample data. The use of spheric kernel densities with variance determined by (7.48) results in a bolstered resubstitution estimation rule that is not reducible, even if the kernel densities are Gaussian. This is clear, since both the mean distance estimate and dimensional constant change between the original and reduced feature spaces, rendering $\hat{\varepsilon}_n^{\mathrm{br},d} \neq \hat{\varepsilon}_n^{\mathrm{br},D}$, in general. On the other hand, the "Naive Bayes" method of fitting diagonal kernel densities in (7.51) yields a reducible bolstered resubstitution error estimation rule, if the kernel densities are Gaussian. This is clear because (9.17) and (9.18) hold for diagonal kernels, and the "Naive Bayes" method produces the same kernel variances in both the original and reduced feature spaces, so that $\hat{\varepsilon}_n^{\mathrm{br},d} = \hat{\varepsilon}_n^{\mathrm{br},D}$.

9.3   Principal Component Analysis (PCA)

PCA is an extremely popular dimensionality reduction method, which is based on the previously-mentioned heuristic according to which low-variance features should be avoided. In PCA, a feature decorrelation step, known as the Karhunen-Loéve (KL) Transform, is applied first, after which the first d individual (transformed) features with the largest variance are retained.

PCA is classified as feature extraction, not feature selection, due to the extra step of feature decorrelation. However, since the KL Transform is linear, interpretability is not completely sacrificed in PCA: it is possible to express each selected feature as a linear combination of the original features. In this sense, PCA can be seen as a compromise between feature selection and the full generality of feature extraction. Additionally, the coefficients in the linear combination (which are arranged into a *loading matrix*) convey information about the relative importance of the original features.

The main issue with PCA for classification is that it is *unsupervised*, i.e., the dependence of Y on X is not considered. Even though supervised PCA-like algorithms exist (see the Bibliographical Notes), we will consider here only the traditional unsupervised version.

Consider a random vector $\mathbf{X} \in R^p$ with mean $\boldsymbol{\mu}_{\mathbf{X}}$ and covariance matrix $\Sigma_{\mathbf{X}}$. By virtue of being symmetric and positive semi-definite, $\Sigma_{\mathbf{X}}$ has a set of p orthonormal eigenvectors $\mathbf{u}_1, \ldots, \mathbf{u}_p$, with associated nonnegative eigenvalues $\lambda_1 \geq \lambda_2 \geq \cdots \geq \lambda_p \geq 0$ (see Sections A1.7 and A2). Consider the linear (affine) transformation given by

$$\mathbf{Z} = U^T(\mathbf{X} - \boldsymbol{\mu}_{\mathbf{X}}) \tag{9.19}$$

where $U = [\mathbf{u}_1 \ldots \mathbf{u}_p]$ is the matrix of eigenvectors. This is a rotation in the space R^p (preceded by

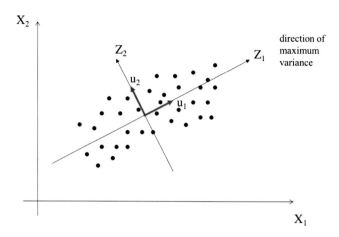

Figure 9.2: Karhunen-Loève Transform with $p = 2$. The data is uncorrelated in the rotated axis system $u_1 \times u_2$.

a translation to remove the mean), known as the (discrete) *Karhunen-Loève Transform*. Clearly,

$$E[\mathbf{Z}] = E\left[U^T(\mathbf{X} - \boldsymbol{\mu})\right] = U^T(E[\mathbf{X}] - \boldsymbol{\mu_X}) = 0 \tag{9.20}$$

and

$$\begin{aligned}
\Sigma_{\mathbf{Z}} &= E[\mathbf{ZZ}^T] = E\left[U^T(\mathbf{X} - \boldsymbol{\mu_X})(\mathbf{X} - \boldsymbol{\mu_X})^T U\right] \\
&= U^T E\left[(\mathbf{X} - \boldsymbol{\mu_X})(\mathbf{X} - \boldsymbol{\mu_X})^T\right] U = U^T \Sigma_{\mathbf{X}} U = \Lambda,
\end{aligned} \tag{9.21}$$

where Λ is the diagonal matrix of eigenvalues $\lambda_1, \ldots, \lambda_p$. With $\mathbf{Z} = (Z_1, \ldots, Z_p)$, the aforementioned derivation proves that

$$\begin{aligned}
E[Z_i] &= 0, \text{ for } i = 1, \ldots, p, \\
E[Z_i Z_j] &= 0, \text{ for } i, j = 1, \ldots, p, \; i \neq j, \\
E[Z_i^2] &= \mathrm{Var}(Z_i) = \lambda_i, \text{ for } i = 1, \ldots, p,
\end{aligned} \tag{9.22}$$

that is, the Z_i are zero-mean, uncorrelated random variables, with variance equal to the eigenvalues of $\Sigma_{\mathbf{X}}$. Since the eigenvalues were ordered in decreasing magnitude, Z_1 has the largest variance λ_1, and and the eigenvector u_1 points to the direction of maximal variation; Z_1 is called the first *principal component* (PC). The second PC Z_2 has the next largest variance λ_2, followed by the third PC Z_3, and so on. Though we do not prove it here, u_2 points in the direction of largest variance perpendicular to u_1, u_3 points in the direction of largest variance perpendicular to the space spanned by u_1 and u_2, and so on. The KL transformation is illustrated in Figure 9.2 in the case $p = 2$.

The PCA feature extraction transform $\mathbf{X}' = T(\mathbf{X})$ consists of applying the KL transform and then keeping the first d principal components: $\mathbf{X}' = (Z_1, \ldots, Z_d)$. In other words,

$$\mathbf{X}' = W^T(\mathbf{X} - \boldsymbol{\mu_X}) \tag{9.23}$$

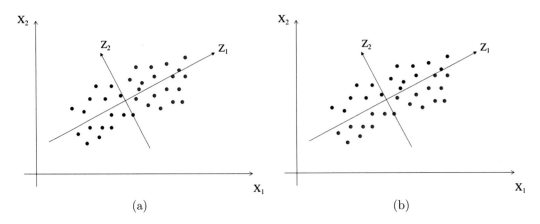

(a)   (b)

Figure 9.3: PCA transform for dimensionality reduction from $p = 2$ to $d = 1$ features. In (a), the first principal component Z_1 alone contains most of the discrimination information. But in (b), the discrimination information is contained in the second principal component Z_2, and PCA fails.

where $W = [\mathbf{u}_1 \cdots \mathbf{u}_d]$ is a rank-d matrix (therefore PCA is not in general invertible, and lossy with respect to the Bayes error criterion). The matrix W is called the loading matrix. Assuming $\mu_{\mathbf{X}}$ for simplicity (and without loss of generality), we can see that $Z_i = \mathbf{u}_i^T \mathbf{X}$, for $i = 1, \ldots, d$. Since $||\mathbf{u}_i|| = 1$, this is a weighted linear combination. The larger (in magnitude) a value in $\mathbf{u}_i$ is, the larger the relative importance of the corresponding feature is in the PC Z_i; this will be illustrated in Example 9.3 below.

In practice, the PCA transform is estimated using sample data $S_n = \{\mathbf{X}_1, \ldots, \mathbf{X}_n\}$, where the mean $\mu_{\mathbf{X}}$ and the covariance matrix $\Sigma_{\mathbf{X}}$ are replaced by their sample versions. When n is small relative to p, the sample covariance matrix is a poor estimator of the true covariance matrix. For example, it is known that the small (resp. large) eigenvalues of the sample covariance matrix are biased low (resp. high) with respect to the eigenvalues of the true covariance matrix; if n is small, this bias could be substantial.

As mentioned previously, the main issue with PCA for classification is that it is unsupervised, that is, it does not consider the label information (this could also turn into an advantage, in case unlabeled data is abundant). PCA is based on the heuristic that the discriminatory information is usually contained in the directions of largest variance. The issue is illustrated in Figure 9.3, where the same data as in Figure 9.2, with two different class labelings. Suppose that, in this artificial example, it is desired to perform dimensionality reduction from $p = 2$ to $d = 1$ features. The situation in Figure 9.3(a) is the most common in practice, and in this case the discriminatory information is preserved in Z_1. However, if the less common situation in Figure 9.3(b) occurs, then PCA fails: the discriminatory information is in the discarded feature Z_2, while Z_1 is a noise feature.

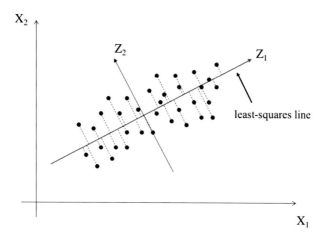

Figure 9.4: Least-squares interpretation of the PCA transform, for dimensionality reduction from $p = 2$ to $d = 1$ features. The average sum of squares is the sum of the lengths of all dashed lines divided by the number of points, which is minimum for the line corresponding to the first PC.

Finally, an alternative equivalent interpretation of (sample-based) PCA is that it produces the subspace that minimizes the least-squares distance with respect to the original data. More precisely, it can be shown that the PCA transform is the linear projection $T : R^d \to R^p$ that minimizes the average sum of squares

$$J = \frac{1}{n} \sum_{i=1}^{n} ||\mathbf{X}_i - T(\mathbf{X}_i)||^2 \tag{9.24}$$

In fact, it can be shown that the least square-error achieved is $J^* = \sum_{i=p+1}^{d} \lambda_i$ (the sum of discarded eigenvalues). Figure 9.4 shows an example in the case of dimensionality reduction from $p = 2$ to $d = 1$ features. We can see that the direction of the first PC is the line of best fit to the data and that the average sum of squares J, which is the sum of the lengths of all dashed lines divided by the number of points, is minimum in this case. Notice that with only two components, J is the sample variance in the direction of the second PC, that is λ_2.

Example 9.3. This example applies PCA to the soft magnetic alloy data set (see Section A8.5). In this data set, the features consist of atomic composition percentages and the annealing temperature for the material samples, while the response variable is their magnetic coercivity in A/m. After discarding all features (columns) that do not have at least 5% nonzero values and then discarding all entries (rows) that do not have a recorded coercivity value (i.e., discarding NaNs), one is left with a data matrix consisting of 12 features measured on 741 material samples. In order to attenuate systematic trends in the data, a small amount of zero-mean Gaussian noise is added to the measurements (with all resulting negative values clamped to zero). Then the data were normalized to have zero mean and unit variance in all features; this is recommended as PCA is sensitive to the scale of different features; here, the atomic composition and annealing temperatures are in different scales (0-100 scale for the former, and Kelvin scale for the latter). In addition, most of the alloys have large Fe and Si composition percentages. Without normalization, Fe, Si, and the annealing temperature would unduly dominate the analysis. Figures 9.5 (a–c) display the data set projected on all pairwise combinations of the first 3 PCs. In order to investigate the association between the PC features and the coercivity, the latter is categorized into three classes: "low" (coercivity ≤ 2 A/M), "medium" (2 A/M $<$ coercivity < 8 A/M), and "high" (coercivity ≥ 8 A/M), which are coded using red, green, and blue, respectively. We can see that most of the discriminatory information in the coercivity indeed seems to lie along the first PC. See Python Assignment 9.8. $\diamond$

9.4   Multidimensional Scaling (MDS)

Multidimensional Scaling (MDS), like PCA, is an unsupervised dimensionality reduction technique. The main idea behind MDS is to reduce dimensionality by finding points in the reduced space R^d that best approximate pairwise dissimilarities, e.g., the Euclidean distance or $1-$ the correlation, between points in the original space R^p. This is a nonlinear feature extraction so it can achieve greater compression than PCA. However, its complexity and lack of interpretability of the transformed features means that it is used almost exclusively as a data exploration tool: It allows the general structure of the high-dimensional data to be directly visualized in a space of dimensions $d = 2$ or $d = 3$.

Let d_{ij} be the *pairwise dissimilarities* between points $\mathbf{x}_i$ and $\mathbf{x}_j$ in the original data set. A dissimilarity may or may not be a true distance metric. In general, it is a nonnegative metric that should be equal to zero if $\mathbf{x}_i = \mathbf{x}_j$ and become larger as the points become dissimilar. Popular examples of pairwise dissimilarity metrics include the Euclidean distance and (one minus) the correlation coefficient between the vectors $\mathbf{x}_i$ and $\mathbf{x}_j$.

Now let δ_{ij} be the pairwise dissimilarities between the transformed points $\mathbf{x}_i' = T(\mathbf{x}_i)$ and $\mathbf{x}_j' = T(\mathbf{x}_j)$. MDS seeks a configuration of points $\mathbf{x}_i'$ in the reduced space that minimizes the *stress*:

$$S = \sqrt{\frac{\sum_{i,j}(\delta_{ij} - d_{ij})^2}{\sum_{i,j} d_{ij}^2}} . \tag{9.25}$$

Small values of stress indicates, in principle, a more faithful representation of the structure of the data in the original high-dimensional space. Values of stress less than 10% are considered excellent, and between 10%-15% are acceptable.

The MDS solution can be computed by starting with a randomly selected initial configuration of points and then minimizing the stress using gradient descent. However, the stress function is quite

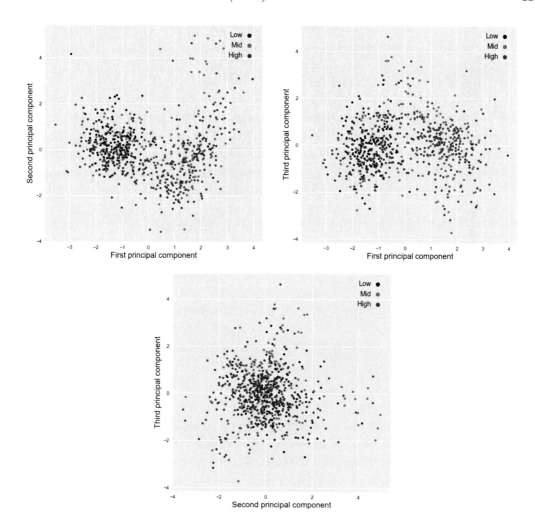

Figure 9.5: PCA example using the soft magnetic alloy data set. The plots of all pairwise combinations of the first 3 PCs show that most of the discriminatory information in the coercivity seems to lie along the first PC (plots generated by c09_PCA.py).

complex and the number of variables is $n \times d$, which makes gradient descent intractable if n and d are large. An alternative procedure, known as SMACOF (*Scaling by MAjorizing a COmplicated Function*), proceeds by finding an easy-to-minimize function that is equal to the stress at the current solution but larger everywhere else, and then letting the next solution be equal to the the minimum value of the majorating function (this is similar to the Expectation-Maximization algorithm to maximize the likelihood function; see Section 10.2.1). The process is stopped until there is no further improvement or a maximum number of iterations is achieved. The procedure needs to be repeated over a number of initial random configurations to avoid getting trapped in local minima.

Notice that the output of the MDS algorithm are the transformed points $\mathbf{x}'_i$ in the reduced space, *not* the feature transformation T. In fact, the input to the MDS algorithm could be simply the matrix of dissimilarities d_{ij} without reference to the original points; this problem is also known as *classical scaling*. For example, the input could be a matrix of pairwise distances between major U.S. cities and the output could be a two-dimensional map of the country (in this case, there would de no dimensionality reduction). Since the stress is invariant to rigid transformations, the map could be produced in unusual orientations (e.g. it could be upside-down).

The fact that the MDS transformation is not explicitly computed makes it hard to be used in feature extraction. For example, if a test point needs to be added, the process has to be repeated from scratch with the entire data (but the previous computed solution can be used as the new initial solution, speeding up execution somewhat).

Example 9.4. Here we use the dengue fever prognosis data set (see Section A8.2), already mentioned in Chapter 1. We apply MDS to the data matrix of 1981 gene expression measurements in peripheral blood mononuclear cells (PBMC) of dengue fever (DF), dengue hemorrhagic fever (DHF) and febrile nondengue (ND) patients, as classified by a clinician, and reduce the data to only two dimensions. Figure 9.6 displays the results using the correlation and Euclidean distance dissimilarity metrics. Both plots show the same general fact that the DHF patients are clustered tight, while the ND patients are more disparate, in agreement with the nonspecificity of this group. The DF patients appear to be divided into two groups, one that is similar to the DHF group and another that is similar to the ND group, illustrating the fact that there is a continuous spectrum from ND to DHF going through DF. These facts seem to be a bit more clear in the plot using correlation, which also achieves a smaller stress. If MDS to $d = 3$ instead of $d = 2$ was performed, we could expect the stresses to be smaller and the representations to be more accurate. See Python Assignment 9.9. ⋄

9.5   Factor Analysis

Suppose that one inverts the point of view of PCA and considers the following *generative model* for the data:

$$\mathbf{X} = W\mathbf{Z} + \boldsymbol{\mu}, \tag{9.26}$$

where $\mathbf{Z}$ is the vector of principal components, and $W = [\mathbf{u}_1 \cdots \mathbf{u}_p]$ is the loading matrix, as before. This inversion is only possible, of course, if no principal components are discarded and W is full rank. Now consider a general rank-d, $d \times p$ matrix C, called the *factor loading matrix*. Using C in

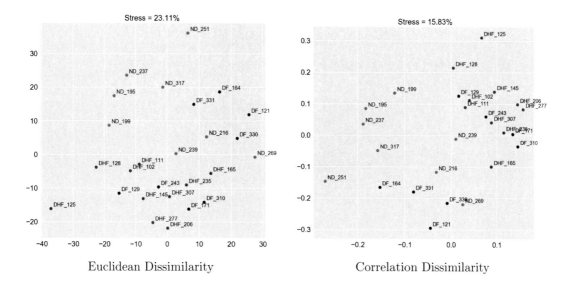

Figure 9.6: MDS example using the Dengue Fever gene-expression data set based on Euclidean and correlation dissimilarity metrics (plots generated by c09_MDS.py).

place of W incurs an error (e.g., see the dashed lines in Figure 9.4) and (9.26) needs to be modified by adding an error term:

$$\mathbf{X} = C\mathbf{Z} + \boldsymbol{\mu} + \boldsymbol{\varepsilon}. \qquad (9.27)$$

In the *Factor Analysis* model, the generative model in (9.27) is treated probabilistically: $\mathbf{Z}$ is assumed to be a zero-mean, uncorrelated Gaussian random vector, $\mathbf{Z} \sim \mathcal{N}(0, I_p)$, called the vector of factors, and $\boldsymbol{\varepsilon} \sim \mathcal{N}(0, \Psi)$ is a zero-mean Gaussian error term with arbitrary covariance structure. Clearly, the generative model is Gaussian, with

$$\mathbf{X} \sim \mathcal{N}(\boldsymbol{\mu}, CC^T + \Psi) \qquad (9.28)$$

This is a *latent-variable model*, since the observed data $\mathbf{X}$ are represented in terms of hidden variables $\mathbf{Z}$. See Figure 9.7 for an illustration.

In practice, the parameters C, $\boldsymbol{\mu}$, and Ψ need to be estimated from data, usually via maximum-likelihood estimation. The special case where $\Psi = \sigma^2 I_d$ is known as probabilistic PCA. In the limit case where $\sigma \to 0$, Probabilistic PCA reduces to classical PCA. The ML estimator of the parameters of the probabilistic PCA model can be obtained in closed form. A main interest in probabilistic PCA resides in applying the Expectation-Maximization (EM) algorithm (see Section 10.2.1) in the solution of the ML problem. Even though the solution obtained is approximate, it avoids the diagonalization the covariance matrix required by the closed-form solution, which is computational

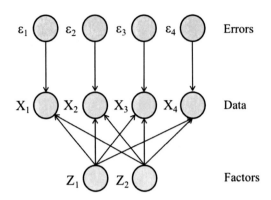

Figure 9.7: Factor Analysis model. The observed data is generated as a liner combination of latent factors plus noise. In general, the data mean must be added as well (Here the data is assumed to be zero mean for convenience).

intractable in high-dimensional spaces. In the limit when $\sigma \to 0$, this EM algorithm computes the solution of classical PCA. (See the Bibliographical Notes for more details.)

9.6   Bibliographical Notes

The papers in Jain and Zongker [1997]; Kohavi and John [1997]; Kudo and Sklansky [2000] are well-known references on feature selection.

Fisher's discriminant can be generalized to $c > 2$ classes, in which case it extracts a feature vector of dimensionality $c - 1$; see Section 3.8.3 of Duda et al. [2001] for the details.

Theorem 9.1 is due to Cover and van Campenhout [1977]. This is Theorem 32.1 in Devroye et al. [1996], where a proof is given based on a simple discrete distribution with the required properties. Interestingly, the proof given by Cover and van Campenhout [1977] is based on multivariate Gaussians instead, which shows that Gaussianity is not a sufficient constraint to avoid exponential complexity in feature selection.

Branch-and-bound algorithms [Narendra and Fukunaga, 1977; Hamamoto et al., 1990] attempt to circumvent the exponential complexity of exhaustive feature selection by using monotonicity or other properties of the class-separability criterion $J(A)$. For example, the Bayes error is monotone in the sense that $A \subseteq B \Rightarrow J(B) \leq J(A)$. Branch-and-bound algorithms attempt to exploit this to cut the searches along monotone chains in the space of feature vectors. In some cases, this can find the optimal feature vector in polynomial time. However, this does not violate the Cover-Van Campenhout Theorem, since the worst-case performance of branch-and-bound algorithms is still exponential. With additional constraints on the distributions, Nilsson et al. [2007] claim to find the optimal feature set, with the Bayes error as class-separability criterion, in polynomial time

Theorem 9.2 (Toussaint's Counter-Example) and its proof appear in Toussaint [1971]. It improves on an earlier example due to Elashoff et al. [1967], which has the same set-up, but shows only $\varepsilon^*(X_1, X_2) > \varepsilon^*(X_1, X_3)$. Cover [1974] goes a step further in and gives an example where X_1 and X_2 are conditionally independent given Y, $\varepsilon^*(X_1) < \varepsilon^*(X_2)$, but $\varepsilon(X_1) > \varepsilon(X_2, X_2')$, where X_2' is an independent realization of X_2; i.e., the best individual feature is worse than repeated measurements of the worst feature.

It seems that the terminology "wrapper" and "filter" feature selection was introduced by John et al. [1994]. This reference also discusses strongly and weakly relevant features. The XOR problem example is an example of intrinsically multivariate prediction, discussed in Martins et al. [2008].

Sima et al. [2005a] showed, in an empirical study, that in small-sample cases, simple SFS feature selection can be as accurate as the much more complex SFFS algorithm, depending on the properties of the error estimator used in the wrapper search; i.e., superior error estimation compensates for less computational effort in the search.

The selection bias of cross-validation was demonstrated in Ambroise and McLachlan [2002]. A useful model for generating synthetic data to investigate feature selection performance was published in Hua et al. [2009].

The mathematical properties of the PCA transform and several of its variants are covered in detail in Webb [2002]. Maximum-likelihood estimation of the parameters in Factor Analysis and Probabilistic PCA, including its EM implementation, is covered in detail in Bishop [2006]. As mentioned in the text, this generates an EM algorithm for computing the solution of classical PCA as well, which avoids diagonalization of the covariance matrix; this algorithm can be viewed as iteratively adjusting the reduced PCA space to the data until the least-squares fit solution is obtained (as can be seen in Figure 9.4).

The general Karhunen-Loève Transform refers to the expansion of a continuous-parameter stochastic process into a sum of orthogonal functions multiplied by an uncorrelated sequence of random variables [Stark and Woods, 1986]. The discrete HL transform used in the PCA transform is the specialization of that for finite-dimensional random vectors.

The MDS discussed in the text is also called *metric MDS*. In *non-metric MDS*, instead of trying to match the dissimilarities in the original and transformed space, one tries to simply to match their rankings. This would be appropriate in case the magnitudes of the dissimilarities are not important but only their relative rankings. The SMACOF algorithm for computation of the MDS transform can be applied to both metric and nonmetric MDS. See an excellent review of the SMACOF algorithm in Groenen et al. [2016]. See also De Leeuw and Mair [2009].

9.7   Exercises

9.1. Let $\mathbf{X} \in R^d$ be a feature set of size d. An additional feature $X_0 \in R$ is a redundant or "noisy" feature if there is no improvement in discrimination upon joining X_0 to $\mathbf{X}$, i.e., if $\varepsilon^*(\mathbf{X}, Y) = \varepsilon^*(\mathbf{X}', Y)$, where $\mathbf{X}' = (\mathbf{X}, X_0) \in R^{d+1}$. Show that a sufficient condition for this undesirable situation is that X_0 be independent of $(\mathbf{X}, Y)$.

9.2. Consider the standard Gaussian model in R^p, where the classes are equally likely and the class-conditional densities are spherical unit-variance Gaussians (i.e., $\Sigma_i = I$, for $i = 0, 1$). The model is specified by the class means $\mu_1 = \delta a$ and $\mu_0 = -\delta a$, where $\delta > 0$ is a separation parameter and $a = (a_1, \ldots, a_p)$ is a parameter vector. Without loss of generality, assume that $||a|| = 1$.

 (a) Find the optimal classifier and the optimal error in the original feature space R^p.

 (b) Find the Bayes error for a subset $\mathbf{X}' = (X_{i_1}, \ldots, X_{i_d})$ of the original p features in terms of the corresponding coefficients $a' = (a_{i_1}, \ldots, a_{i_d})$.

 (c) If the criterion for feature selection is the Bayes error, how would you select the vector $\mathbf{X}'$ to obtain the optimal feature set of size d?

9.3. Given data $S_n = \{(\mathbf{x}_1, y_1), \ldots, (\mathbf{x}_n, y_n)\}$, Fisher's discriminant seeks the direction vector $\mathbf{w} \in R^d$ such that the projected data $\tilde{S}_n = \{(\mathbf{w}^T \mathbf{x}_1, y_1), \ldots, (\mathbf{w}^T \mathbf{x}_n, y_n)\}$ is maximally separated, in the sense that the criterion

$$J(\mathbf{w}) = \frac{|m_1 - m_0|^2}{s_0^2 + s_1^2} \tag{9.29}$$

is maximized, where m_0 and m_1 are the class-specific sample means of the projected data and

$$s_0 = \sum_{i=1}^{n} (\mathbf{w}^T \mathbf{x}_i - m_0)^2 I_{y_i=0} \quad \text{and} \quad s_1 = \sum_{i=1}^{n} (\mathbf{w}^T \mathbf{x}_i - m_1)^2 I_{y_i=1} \tag{9.30}$$

measure the scatter of the data around m_0 and m_1, respectively. This is a linear dimensionality reduction transformation like PCA. But unlike PCA, it takes the labels into account, so it is supervised. Hence, when reducing dimensionality to $p = 1$ features, Fisher's discriminant is preferable to PCA.

(a) Show that $J(\mathbf{w})$ in (9.29) can be written as in (9.5).

(b) By using direct differentiation, show that the solution $\mathbf{w}^*$ must satisfy the so-called generalized eigenvalue problem:

$$S_B \mathbf{w} = \lambda S_W \mathbf{w} \tag{9.31}$$

for some $\lambda > 0$, where S_B and S_W are defined in (9.6) and (9.7).

Hint: use the vectorial differentiation formula $(\mathbf{w}^T A \mathbf{w})' = 2A\mathbf{w}$.

(c) Assuming that S_W is nonsingular, then $S_W^{-1} S_B \mathbf{w} = \lambda \mathbf{w}$, i.e., $\mathbf{w}^*$ is an eigenvector of matrix $S_W^{-1} S_B$. Show that $\mathbf{w}^* = \hat{\Sigma}_W^{-1}(\hat{\boldsymbol{\mu}}_1 - \hat{\boldsymbol{\mu}}_0)$ is the eigenvector being sought and the solution. Furthermore, $\lambda = J(\mathbf{w}^*)$ is the largest eigenvalue of $S_W^{-1} S_B$ (indeed, it is its only nonzero eigenvalue).

Hint: Use the expansion $S_B = \mathbf{v}\mathbf{v}^T$, where $\mathbf{v} = \hat{\boldsymbol{\mu}}_0 - \hat{\boldsymbol{\mu}}_1$. Matrix S_B is a *rank-one matrix*, and thus so is $S_W^{-1} S_B$. (See Section A2 for a review of basic matrix theory.)

9.4. A feature extraction transformation is lossless if the class-separability criterion is unchanged: $J(\mathbf{X}', Y) = J(\mathbf{X}, Y)$. Assume that J is the Bayes error.

(a) Show that $T(\mathbf{X}) = \eta(\mathbf{X})$ is a lossless transformation $T : R^p \to [0, 1]$.

(b) Show that a transformation $T : R^p \to R^d$ is lossless if there is a (Borel-measurable) function $G : R^d \to R$ such that

$$\eta(\mathbf{X}) = G(T(\mathbf{X})) \quad \text{with probability 1}, \tag{9.32}$$

i.e., the posterior-probability function depends on $\mathbf{X}$ only through $T(\mathbf{X})$. The transformed feature $\mathbf{X}' = T(\mathbf{X})$ is called a *sufficient statistic* in this case. An example of sufficient statistic was seen in Example 2.2.

(c) Use the result in item (b) to find a lossless univariate feature if $\eta(X) = e^{-c\|X\|}$, for some *unknown* $c > 0$. This shows that lossless feature extraction can achieved with only *partial* knowledge about the distribution of the problem.

(d) Find a lossless feature vector of size $d = 2$ if $\eta(X_1, X_2, X_3) = H(X_1 X_2, X_2 X_3)$, for a fixed but unknown function $H : R^3 \to R^2$.

(e) Find a lossless feature vector of size $d = 2$ if

$$\begin{aligned} p(\mathbf{X} \mid Y = 0) &= k_0 \ln(1 + \|X + b_0\|) \\ p(\mathbf{X} \mid Y = 1) &= k_1 \ln(1 + \|X - b_1\|) \end{aligned} \tag{9.33}$$

are equally-likely class-conditional densities, where $k_0, k_1 > 0$ are unknown.

9.5. Verify that the distribution specified in the proof of Toussaint's Counter-Example (see Theorem 9.2) indeed has the required property.

9.6. Obtain, by inspection, the first and second PCs Z_1 and Z_2 as a function of $\mathbf{X} = (X_1, X_2, X_3, X_4)$ and the percentage of variance explained by Z_1 and Z_2 in the following cases.

(a)

$$\mu_{\mathbf{X}} = \begin{bmatrix} 1 \\ 2 \\ -1 \\ 3 \end{bmatrix} \quad \text{and} \Sigma \ \mathbf{X} = \begin{bmatrix} 1 & 0 & 0 & 0 \\ 0 & 3 & 0 & 0 \\ 0 & 0 & 5 & 0 \\ 0 & 0 & 0 & 2 \end{bmatrix}.$$

(b)

$$\mu_{\mathbf{X}} = \begin{bmatrix} 1 \\ -1 \\ 2 \\ 3 \end{bmatrix} \quad \text{and} \Sigma \ \mathbf{X} = \begin{bmatrix} 2 & 0 & 0 & 0 \\ 0 & 4 & 0 & 0 \\ 0 & 0 & 3 & 0 \\ 0 & 0 & 0 & 1 \end{bmatrix}.$$

9.8   Python Assignments

9.7. This computer project applies wrapper feature selection to the Breast Cancer Prognosis gene-expression data (see Section A8.3) in order to find gene-expression signatures for good prognosis, and estimate their accuracy using testing data. The criterion for the search will be simply the resubstitution error estimate of the classifier designed on each feature set. Divide the available data into 60% training data and 40% testing data. Using the training data, find the

- top 2 genes using exhaustive search,
- top 3–5 genes using sequential forward search (starting from the feature set in the previous item),

corresponding to the following classification rules:

- LDA, $p = 0.75$,
- Linear SVM, $C = 10$,
- Nonlinear SVM with RBF kernel, $C = 10$ and gamma set to 'auto',
- NN with 5 neurons in one hidden layer with logistic nonlinearities and the lbfgs solver.

Also find these classifiers using all genes (no feature selection). If at any step of feature selection, two candidate feature sets have the same minimum apparent error, pick the one with the smallest indices (in "dictionary order"). Generate a 20×3 table containing in each row one of the 20 classifiers, and in the columns the genes found in each case and the resubstitution and test-set errors of the selected feature set. How do you compare the different classification rules? How do you compare the resubstitution and test-set error estimates obtained?

9.8. This assignment concerns the application of PCA to the soft magnetic alloy data set (see Section A8.5).

(a) Reproduce the plots in Figure 9.5 by running `c09_PCA.py`.

(b) Plot the percentage of variance explained by each PC as a function of PC number. This is called the *scree plot*. Now plot the cumulative percentage of variance explained by the PCs as a function of PC number. How many PCs are needed to explain 95% of the variance?

Coding hint: use the attribute `explained_variance_ratio_` and the `cusum()` method.

(c) Print the loading matrix W (this is the matrix of eigenvectors, ordered by PC number from left to right). The absolute value of the coefficients indicate the relative importance of each original variable (row of W) in the corresponding PC (column of W).

(d) Identify which two features contribute the most to the discriminating first PC and plot the data using these top two features. What can you conclude about the effect of these two features on the coercivity? This is an application of PCA to feature selection.

9.9. This assignment applies MDS to the dengue fever prognosis data set (see Section A8.2).

(a) Reproduce the plots in Figure 9.6 by running `c09_PCA.py`.

Coding hint: As of version 0.21.3 of `sklearn`, the `MDS` class returns an unnormalized stress value, which is not very useful. In order to compute the normalized stress in (9.25), the following "hack" is needed: add the line

`stress = np.sqrt(stress / ((disparities.ravel() ** 2).sum() / 2))`

before the `return` statement of function `_smacof_single()` in the `mds.py` file (in a local installation of the anaconda distribution, this file is in a directory similar to `$HOME/opt/anaconda3/lib/python3.7/site-packages/sklearn/manifold/`).

(b) What happens if the data are normalized to have zero mean and unit variance in all features, as was done in Example 9.3, prior to computation of the MDS? Based on these results, is it recommended to apply normalization in the MDS case? Contrast this with the PCA case.

(c) Obtain and plot the 3D MDS plots for both the correlation and the Euclidean dissimilarities. What do you observe about the stress values, as compared to the values for the 2D MDS.

(d) Obtain plots of the data set projected on the first and second PCs (applying normalization). Are the classes as well separated as in the MDS plots? If not, how do you explain this?

Chapter 10

Clustering

> "The first step in wisdom is to know the things themselves; this notion consists in having a true idea of the objects; objects are distinguished and known by classifying them methodically and giving them appropriate names."
> –Carl Linnaeus, *Systema Naturae*, 1735.

In some situations, training data are available without labeling. This could be because of the expense of labeling the data, or the unavailability of reliable labels, or because the data are perceived to come from a single group. This is the domain of *unsupervised learning*. In Chapter 9, we reviewed unsupervised dimensionality reduction techniques (PCA and MDS). Our interest in the current chapter is on unsupervised learning to identify the structure of the underlying data distribution in the original feature space. These techniques can be used to find subgroups (*clusters*) in the data and build hierarchical data representations. If label information is available, clustering can be used to detect previously unknown classes. In this chapter we review the basic non-hierarchical clustering algorithm, namely, the K-Means Algorithm, followed by Gaussian-Mixture Modeling (GMM), which can be seen as a probabilistic version of K-Means. Then we consider hierarchical clustering algorithms and, finally, we describe the Self-Organizing Map (SOM) clustering algorithm.

10.1 K-Means Algorithm

Given data $S_n = \{\mathbf{X}_1, \ldots, \mathbf{X}_n\}$, the objective of the K-Means algorithm is to find K cluster centers $\boldsymbol{\mu}_1, \ldots, \boldsymbol{\mu}_K$ (K is given) and, for each point $\mathbf{X}_i$, find an assignment to one of the K clusters.

© Springer Nature Switzerland AG 2020
U. Braga-Neto, *Fundamentals of Pattern Recognition and Machine Learning*,
https://doi.org/10.1007/978-3-030-27656-0_10

Cluster assignment is made by means of vectors $\mathbf{r}_1, \ldots, \mathbf{r}_n$ where each $\mathbf{r}_i$ is a vector of size K using a "one-hot" encoding scheme:

$$\mathbf{r}_i = (0, \ 0, \ \cdots, 1, \cdots, \ 0, \ 0)^T, \quad \text{for } i = 1, \ldots, n, \tag{10.1}$$

where $\mathbf{r}_i(k) = 1$ if and only if $\mathbf{X}_i$ belongs to cluster k, for $k = 1, \ldots, K$. (Each point can belong to only one cluster.) For example, with $K = 3$ and $n = 4$, we might have $\mathbf{r}_1 = (1, 0, 0)$, $\mathbf{r}_2 = (0, 0, 1)$, $\mathbf{r}_3 = (1, 0, 0)$, $\mathbf{r}_4 = (0, 1, 0)$, in which case $\mathbf{X}_1, \mathbf{X}_3$ are assigned to cluster 1, $\mathbf{X}_2$ is assigned to cluster 3, while $\mathbf{X}_4$ is assigned to cluster 2.

The K-means algorithm seeks the vectors $\{\boldsymbol{\mu}_i\}_{i=1}^n$ and $\{\mathbf{r}_i\}_{k=1}^K$ that minimize a score based on the normalized sum of distances of all points to their corresponding centers:

$$J = \frac{1}{n} \sum_{i=1}^n \sum_{k=1}^K \mathbf{r}_i(k) ||\mathbf{X}_i - \boldsymbol{\mu}_k||^2 \,. \tag{10.2}$$

The solution can be obtained iteratively with two optimizations at each step:

1. Holding the current values $\{\boldsymbol{\mu}_k\}_{k=1}^K$ fixed, find the values $\{\mathbf{r}_i\}_{i=1}^n$ that minimize the score J ("E-Step").

2. Holding the current values $\{\mathbf{r}_i\}_{i=1}^n$ fixed, find the values $\{\boldsymbol{\mu}_k\}_{k=1}^K$ that minimize the score J ("M-Step").

The nomenclature "E-step" and "M-step" is due to an analogy with the EM ("Expectation-Maximization" algorithm for Gaussian mixtures, to be discussed in the next section.

In the "E-step," with the current values $\{\boldsymbol{\mu}_k\}_{k=1}^K$ fixed, the values $\{\mathbf{r}_i\}_{i=1}^n$ that minimize J can be found by inspection:

$$\mathbf{r}_i(k) = \begin{cases} 1, & \text{if } k = \arg\min_{j=1,\ldots,K} ||\mathbf{X}_i - \boldsymbol{\mu}_j||^2, \\ 0 & \text{otherwise.} \end{cases} \tag{10.3}$$

for $i = 1, \ldots, n$. In other words, we simply assign each point to the closest cluster mean.

In the "M-step," with the current values $\{\mathbf{r}_i\}_{i=1}^n$ fixed, the values $\{\boldsymbol{\mu}_k\}_{k=1}^K$ that minimize J can be found by simple differentiation:

$$\frac{\partial J}{\partial \boldsymbol{\mu}_k} = 2 \sum_{i=1}^n \mathbf{r}_i(k) ||\mathbf{X}_i - \boldsymbol{\mu}_k|| = 0, \tag{10.4}$$

which gives

$$\boldsymbol{\mu}_k = \frac{\sum_{i=1}^n \mathbf{r}_i(k) \mathbf{X}_i}{\sum_{i=1}^n \mathbf{r}_i(k)}, \tag{10.5}$$

for $k = 1, \ldots, K$. In other words, we simply assign to $\boldsymbol{\mu}_k$ the mean value of all training points assigned to cluster k in the previous "E-Step." The E and M steps should be repeated until there is no significant change in the score J. The detailed procedure is summarized below.

Algorithm 1 K-Means.

1: Initialize K, $\tau > 0$ and $\{\boldsymbol{\mu}_k^{(0)}\}_{k=1}^K$.

2: **repeat**

3:     E-Step: Update cluster assignments:

$$\mathbf{r}_i^{(m+1)}(k) = \begin{cases} 1, & \text{if } k = \arg\min_{j=1,\ldots,K} ||\mathbf{X}_i - \boldsymbol{\mu}_j^{(m)}||^2, \\ 0 & \text{otherwise.} \end{cases} \quad , \quad \text{for } i = 1, \ldots, n.$$

4:     M-Step: Update cluster centers:

$$\boldsymbol{\mu}_k^{(m+1)} = \frac{\sum_{i=1}^n \mathbf{r}_i^{(m+1)}(k)\mathbf{X}_i}{\sum_{i=1}^n \mathbf{r}_i^{(m+1)}(k)}, \quad \text{for } k = 1, \ldots, K.$$

5:     Calculate score:

$$J^{(m+1)} = \frac{1}{n} \sum_{i=1}^n \sum_{k=1}^K \mathbf{r}_i^{(m+1)}(k)||\mathbf{X}_i - \boldsymbol{\mu}_k^{(m+1)}||^2.$$

6: **until** $|J^{(m+1)} - J^{(m)}| < \tau$.

Example 10.1. We apply the K-means algorithm to the soft magnetic alloy data set (see Example 9.3 and Section A8.5). To enable visualization of the results, we consider only the iron (Fe) and Boron (B) atomic features, which were determined to be important in the analysis in Example 9.3, and take only the first 250 of the 741 points in the data set. The algorithm stops when the absolute difference between consecutive values of J is smaller than 0.005. In this case, the algorithm stopped after six iterations. We can see the presence of two clusters, which mostly correspond to high and low magnetic coercivity. Figure 10.1 displays the initial configuration and the results after the E-step of each iteration. ⋄

One of the issues with the K-means algorithm is the presence of multiple minima in the score function J. In order to guarantee that a good solution is found, the simplest method is to run the algorithm multiple times with random initialization of the centers and select the solution with the minimum score. This is sometimes called the *random restart method*. As an illustration, Figure 10.2 displays several solutions achieved with $K = 3$ on the same data set in Example 10.1, after a fixed number of 10 iterations and multiple random restarts (In Example 10.1, the solution obtained for the case $K = 2$ is already the best one). We can see that the score J can vary considerably, showing the need for multiple random restarts. Notice also that the same solution can appear multiple times with different labelings.

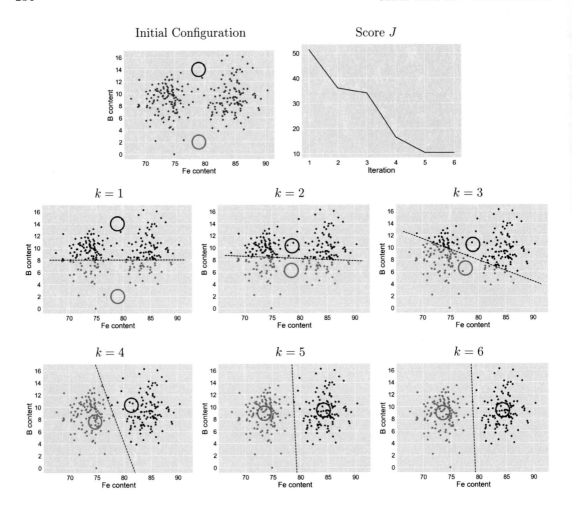

Figure 10.1: K-Means example using the Iron (Fe) and Boron (B) features of the soft magnetic alloy data set. The top left plot displays the original data and initial means (located at the center of the circles). The top right plot displays the value of the score J after the E-step of each iteration. The plots below display the results after the E-step of each iteration; the algorithm converges after 6 iterations (plots generated by c10_Kmeans.py).

We have avoided the issue of the choice of the number of clusters K until now. In practice, this is a difficult problem, not unlike choosing the number of features to keep in feature selection or the number of principal components to use in PCA. One cannot simply vary K and choose the one that gives the least score J because, just as in feature selection, large K compared to sample size n creates overfitting and an artificially small J, so that the plot of J against the K is in general monotonically decreasing. Hence, one needs to look for a small score J while at the same time penalizing for

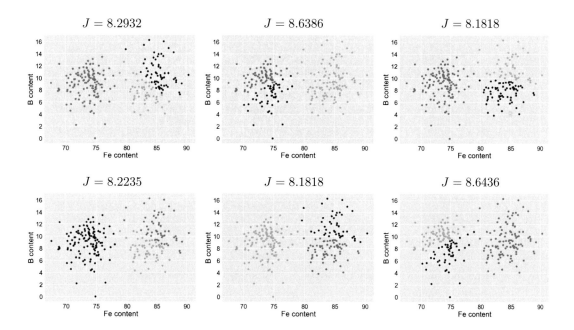

Figure 10.2: K-Means solutions for $K = 3$ clusters for different random initializations of the centers with J score achieved after 10 iterations, using the same data set as in Example 10.1. Notice that two of the solutions are the same, but with different labelings (plots generated by c10_Kmeans_rndstart.py).

large K. In practice, a criterion that is often used is to select a value K^* corresponding to the first occurrent of a sharp "elbow" (a large drop followed by stable values) in the plot of J against K. For other criteria, see the Bibliographical Notes.

Before ending our discussion of the K-means algorithm, we mention a popular variant of it, known as *fuzzy c-means*. This is a fuzzy version of the K-means algorithm, where each point is not assigned to a single cluster, but has instead a fuzzy *degree of membership* to each cluster. More precisely, each vector $\mathbf{r}_i$ can assume nonnegative values such that $\sum_{k=1}^{K} \mathbf{r}_i(k) = 1$, where the value $0 \le \mathbf{r}_i(k) \le 1$ gives the degree of membership of point $\mathbf{X}_i$ to cluster k. The algorithm seeks vectors $\{\boldsymbol{\mu}_i\}_{i=1}^{n}$ and $\{\mathbf{r}_i\}_{k=1}^{K}$ that minimize the score

$$J = \sum_{i=1}^{n} \sum_{k=1}^{K} \mathbf{r}_i(k)^s ||\mathbf{X}_i - \boldsymbol{\mu}_k||^2 \tag{10.6}$$

where $s \ge 1$ is a parameter that controls the "fuzziness" of the resulting clusters. Solutions can be obtained by a similar process as the usual K-means algorithm, with E and M steps.

10.2   Gaussian Mixture Modeling

The K-means algorithm makes no assumption about the distribution of the data. A different *model-based* approach to clustering is to assume a particular parametric shape for the distribution and estimate the parameters from the data (this is similar to the Parametric Classification Rules in Chapter 4). The appropriate assumption for clustering is a *mixture* of probability densities. We will consider in this section the important particular case of the Gaussian Mixture Model (GMM), using the Expectation-Maximization algorithm to obtain maximum-likelihood estimates of the parameters.

The GMM for the overall data distribution is:

$$p(\mathbf{x}) = \sum_{k=1}^{K} \pi_k \mathcal{N}(\mathbf{x} \mid \boldsymbol{\mu}_k, \Sigma_k), \tag{10.7}$$

where K is the desired number of clusters and $\pi_1, \ldots, \pi_K$ are nonnegative numbers, with $\sum_{i=1}^{K} \pi_k = 1$, called the *mixing parameters*. The mixing parameter π_k is simply the *a-priori* probability that a given random point $\mathbf{X}$ belongs to cluster C_k:

$$\pi_k = P(\mathbf{X} \in C_k), \tag{10.8}$$

for $k = 1, \ldots, K$. Bayes' theorem allows one to compute the *a-posteriori* probabilities of cluster membership given the data:

$$\begin{aligned} \gamma_k(\mathbf{x}) &= P(\mathbf{X} \in C_k \mid \mathbf{X} = \mathbf{x}) \\ &= \frac{p(\mathbf{X} = \mathbf{x} \mid \mathbf{X} \in C_k) P(\mathbf{X} \in C_k)}{\sum_{k=1}^{K} p(\mathbf{X} = \mathbf{x} \mid \mathbf{X} \in C_k) P(\mathbf{X} \in C_k)} = \frac{\pi_k \mathcal{N}(\mathbf{x} \mid \boldsymbol{\mu}_k, \Sigma_k)}{\sum_{k=1}^{K} \pi_k \mathcal{N}(\mathbf{x} \mid \boldsymbol{\mu}_k, \Sigma_k)}, \end{aligned} \tag{10.9}$$

for $k = 1, \ldots, K$. The key quantity $\gamma_k(\mathbf{x}) > 0$ gives the *cluster membership* of point $\mathbf{x}$ in cluster k (also known in the literature as the "cluster responsibility"). Note that $\sum_{k=1}^{K} \gamma_k(\mathbf{x}) = 1$, for all $\mathbf{x} \in R^d$. Hence, the cluster memberships are nonnegative and add up to one, i.e., they are in fact probabilities. Indeed, notice the similarity with the posterior probabilities $\eta_k(\mathbf{x}) = P(Y = k \mid \mathbf{X} = \mathbf{x})$ in classification. Just as in classification, "hard" cluster membership can be obtained by assigning a point $\mathbf{x}$ to the cluster k with the largest "soft" cluster membership $\gamma_k(\mathbf{x})$. Thus, traditional clustering can be performed by estimating the cluster memberships $\{\gamma_k(\mathbf{X}_n)\}_{i=1}^{n}$, for $k = 1, \ldots, K$.

To obtain the cluster memberships for the observed data, we need to find estimates of the parameters $\{\pi_k, \boldsymbol{\mu}_k, \Sigma_k\}_{i=1}^{K}$. Given independence of the data points, the likelihood function can be written as

$$p(S_n \mid \{\pi_k, \boldsymbol{\mu}_k, \Sigma_k\}_{i=1}^{K}) = \prod_{i=1}^{n} p(\mathbf{X}_i \mid \{\pi_k, \boldsymbol{\mu}_k, \Sigma_k\}_{i=1}^{K}) = \prod_{i=1}^{n} \left(\sum_{k=1}^{K} \pi_k \mathcal{N}(\mathbf{X}_i \mid \boldsymbol{\mu}_k, \Sigma_k) \right). \tag{10.10}$$

Therefore, the log-likelihood function is given by:

$$\ln p(S_n \,|\, \{\pi_k, \boldsymbol{\mu}_k, \Sigma_k\}_{i=1}^K) \;=\; \sum_{i=1}^n \ln\left(\sum_{k=1}^K \pi_k \,\mathcal{N}(\mathbf{X}_i \,|\, \boldsymbol{\mu}_k, \Sigma_k)\right), \tag{10.11}$$

and the maximum-likelihood parameter estimates are determined by:

$$\{\hat{\pi}_k, \hat{\boldsymbol{\mu}}_k, \hat{\Sigma}_k\}_{i=1}^K \;=\; \arg\max_{\{\pi_k, \boldsymbol{\mu}_k, \Sigma_k\}_{i=1}^K} \sum_{i=1}^n \ln\left(\sum_{k=1}^K \pi_k \,\mathcal{N}(\mathbf{X}_i \,|\, \boldsymbol{\mu}_k, \Sigma_k)\right). \tag{10.12}$$

In the case of a single Gaussian ($K = 1$), this maximization can be accomplished in closed form, resulting in the usual sample means and sample covariance matrix estimators ($\pi_1 = 1$ and there are no mixing parameters to estimate). However, for $K \geq 2$, no analytical expressions for the maximizing parameters are known, and maximization must proceed numerically. The reason is the presence of the inner summation in (10.11), which prevents us from applying the log to the Gaussian densities directly. In the next section we examine a well-known numerical approach to solve this hard maximum-likelihood problem.

10.2.1 Expectation-Maximization Approach

One possibility to maximize the log-likelihood (10.11) is to apply a gradient-descent algorithm. Here we describe a different hill-climbing approach, known as the *Expectation-Maximization Algorithm*, which implements maximum-likelihood estimation in models with "hidden" variables. The EM algorithm is an iterative algorithm, which is guaranteed to converge to a local maximum of the likelihood function (a proof of convergence of the EM algorithm is given in Section A7). As we will see below, the M-step of the EM algorithm allows us one to "interchange" log and the inner sum in (10.11), rendering optimization in closed-form possible (for the M step).

First we state the EM procedure in the general case and then specialize it to GMM. Let $\mathbf{X}$ denote the observed training data, and let $\mathbf{Z}$ denote hidden variables that are not directly observable, but on which $\mathbf{X}$ depends. Also let $\boldsymbol{\theta} \in \Theta$ be a vector of model parameters. Maximum likelihood estimation attempts to find the value of $\boldsymbol{\theta}$ that maximizes the log-likelihood function $L(\boldsymbol{\theta}) = \ln p_{\boldsymbol{\theta}}(\mathbf{X})$. The justification behind the EM algorithm is that the maximization of this *incomplete* log-likelihood $L(\boldsymbol{\theta})$ is difficult; however, the maximization of the *complete* log-likelihood $\ln p_{\boldsymbol{\theta}}(\mathbf{Z}, \mathbf{X})$ would be easy (perhaps even yielding a closed-form solution), if only we knew the value of $\mathbf{Z}$. For simplicity, we will assume that $\mathbf{Z}$ is discrete (this will be the case in GMM).

Since the hidden variable $\mathbf{Z}$, and thus the complete likelihood $p_{\boldsymbol{\theta}}(\mathbf{Z}, \mathbf{X})$, are not directly available, the EM algorithm prescribes considering instead the function

$$Q(\boldsymbol{\theta}, \boldsymbol{\theta}^{(m)}) \;=\; E_{\boldsymbol{\theta}^{(m)}}[\ln p_{\boldsymbol{\theta}}(\mathbf{Z}, \mathbf{X}) \,|\, \mathbf{X}] \;=\; \sum_{\mathbf{Z}} \ln p_{\boldsymbol{\theta}}(\mathbf{Z}, \mathbf{X})\, p_{\boldsymbol{\theta}^{(m)}}(\mathbf{Z} \,|\, \mathbf{X}) \tag{10.13}$$

where $\boldsymbol{\theta}^{(m)}$ is the current estimate of $\boldsymbol{\theta}$. The score in (10.13) is the expected value of $p_{\boldsymbol{\theta}}(\mathbf{Z}, \mathbf{X})$ with respect to the conditional distribution of $\mathbf{Z}$ given $\mathbf{X}$ at the current estimated value $\boldsymbol{\theta}^{(m)}$. Hence, the unknown hidden variable $\mathbf{Z}$ is "averaged out" by the expectation. It is shown in Section A7 that maximization of (10.13) with respect to $\boldsymbol{\theta}$,

$$\boldsymbol{\theta}^{(m+1)} = \arg\max_{\boldsymbol{\theta}} Q(\boldsymbol{\theta}, \boldsymbol{\theta}^{(m)}), \tag{10.14}$$

necessarily improves the log-likelihood, i.e., $L(\boldsymbol{\theta}^{(m+1)}) > L(\boldsymbol{\theta}^{(m)})$, unless one is already at a local maximum of $L(\theta)$, in which case $L(\boldsymbol{\theta}^{(m+1)}) = L(\boldsymbol{\theta}^{(m)})$. The EM algorithm corresponds to make an initial guess $\boldsymbol{\theta} = \boldsymbol{\theta}^{(0)}$ and then iterating between the steps of computing (10.13) at the current value of the estimate $\boldsymbol{\theta}^{(m)}$ (known as the "E-step") and the maximization (10.14) to obtain the next estimate $\boldsymbol{\theta}^{(m+1)}$. The procedure is summarized below.

Algorithm 2 Expectation-Maximization (EM).

1: Initialize $\boldsymbol{\theta}^{(0)}$ and $\tau > 0$.

2: **repeat**

3:    E-Step: Compute the Q score:

$$Q(\boldsymbol{\theta}, \boldsymbol{\theta}^{(m)}) = E_{\boldsymbol{\theta}^{(m)}}[\ln p_{\boldsymbol{\theta}}(\mathbf{Z}, \mathbf{X}) \mid \mathbf{X}] = \sum_{\mathbf{Z}} \ln p_{\boldsymbol{\theta}}(\mathbf{Z}, \mathbf{X}) \, p_{\boldsymbol{\theta}^{(m)}}(\mathbf{Z} \mid \mathbf{X}).$$

4:    M-Step: Update the parameters:

$$\boldsymbol{\theta}^{(m+1)} = \arg\max_{\boldsymbol{\theta}} Q(\boldsymbol{\theta}, \boldsymbol{\theta}^{(m)}).$$

5:    Calculate the log-likelihood:

$$L(\boldsymbol{\theta}^{(m+1)}) = \ln p_{\boldsymbol{\theta}^{(m+1)}}(\mathbf{X}).$$

6: **until** $|L(\boldsymbol{\theta}^{(m+1)}) - L(\boldsymbol{\theta}^{(m)})| < \tau$.

We describe next how to carry out the estimation of parameters of the Gaussian-Mixture model using the EM methodology. Let $\boldsymbol{\theta} = \{\pi_k, \boldsymbol{\mu}_k, \Sigma_k\}_{i=1}^{K}$ be the parameter vector, and let $\mathbf{X} = \{\mathbf{X}_1, \ldots, \mathbf{X}_n\}$ be the observed data, which is assumed to be independent and identically distributed as in (10.7). The hidden variables $\mathbf{Z} = \{\mathbf{Z}_1, \ldots, \mathbf{Z}_n\}$ here indicate the "true" cluster memberships:

$$\mathbf{Z}_i = (0, \ 0, \ \cdots, \ 1, \ \cdots, \ 0, \ 0), \quad \text{for } i = 1, \ldots, n, \tag{10.15}$$

such that $\mathbf{Z}_i(k) = 1$ if $\mathbf{X}_i$ belongs to cluster k, for $k = 1, \ldots, K$ (each point can belong to only one cluster). Comparing this to (10.1), we observe that this is the same one-hot encoding scheme used in the K-means algorithm.

The incomplete log-likelihood function is given by (10.11),

$$L(\boldsymbol{\theta}) = \ln p_{\boldsymbol{\theta}}(\mathbf{X}) = \sum_{i=1}^{n} \ln \left(\sum_{k=1}^{K} \pi_k \mathcal{N}(\mathbf{X}_i \mid \boldsymbol{\mu}_k, \Sigma_k) \right). \tag{10.16}$$

As mentioned previously, due to the presence of the summation, the log does not apply directly to the Gaussian densities, making the maximization of $L(\boldsymbol{\theta})$ nontrivial. However, the complete log-likelihood is given by

$$\ln p_{\boldsymbol{\theta}}(\mathbf{Z}, \mathbf{X}) = \ln \left(\prod_{i=1}^{n} \prod_{k=1}^{K} \prod_k^{\mathbf{Z}_i(k)} \mathcal{N}(\mathbf{X}_i \mid \boldsymbol{\mu}_k, \Sigma_k)^{\mathbf{Z}_i(k)} \right) = \sum_{i=1}^{n} \sum_{k=1}^{K} \mathbf{Z}_i(k) \ln \left(\pi_k \mathcal{N}(\mathbf{X}_i \mid \boldsymbol{\mu}_k, \Sigma_k) \right). \tag{10.17}$$

Part of the "magic" of EM in this case is to allow the interchange of log and summation in (10.17) in comparison with (10.16). The Q function in (10.13) is computed as follows (E-step):

$$Q(\boldsymbol{\theta}, \boldsymbol{\theta}^{(m)}) = E_{\boldsymbol{\theta}^{(m)}} \left[\sum_{i=1}^{n} \sum_{k=1}^{K} \mathbf{Z}_i(k) \ln \left(\pi_k \mathcal{N}(\mathbf{X}_i \mid \boldsymbol{\mu}_k, \Sigma_k) \right) \,\middle|\, \mathbf{X} \right]$$

$$= \sum_{i=1}^{n} \sum_{k=1}^{K} E_{\boldsymbol{\theta}^{(m)}}[\mathbf{Z}_i(k) \mid \mathbf{X}] \ln \left(\pi_k \mathcal{N}(\mathbf{X}_i \mid \boldsymbol{\mu}_k, \Sigma_k) \right), \tag{10.18}$$

where

$$E_{\boldsymbol{\theta}^{(m)}}[\mathbf{Z}_i(k) \mid \mathbf{X}] = E_{\boldsymbol{\theta}^{(m)}}[\mathbf{Z}_i(k) \mid \mathbf{X}_i] = P_{\boldsymbol{\theta}^{(m)}}[\mathbf{Z}_i(k) = 1 \mid \mathbf{X}_i]$$

$$= \gamma_k^{(m)}(\mathbf{X}_i) = \frac{\pi_k^{(m)} \mathcal{N}(\mathbf{X}_i \mid \boldsymbol{\mu}_k^{(m)}, \Sigma_k^{(m)})}{\sum_{k=1}^{K} \pi_k^{(m)} \mathcal{N}(\mathbf{X}_i \mid \boldsymbol{\mu}_k^{(m)}, \Sigma_k^{(m)})}, \tag{10.19}$$

i.e., the cluster membership of point $\mathbf{X}_i$ in cluster k under the current value of the parameter vector $\boldsymbol{\theta}^{(m)}$, for $k = 1, \ldots, K$ and $i = 1, \ldots, n$. Substituting this into (10.18) and recalling that $\boldsymbol{\theta} = \{\pi_k, \boldsymbol{\mu}_k, \Sigma_k\}_{i=1}^{K}$ leads to:

$$Q(\{\pi_k, \boldsymbol{\mu}_k, \Sigma_k\}_{i=1}^{K}, \{\pi_k^{(m)}, \boldsymbol{\mu}_k^{(m)}, \Sigma_k^{(m)}\}_{i=1}^{K}) = \sum_{i=1}^{n} \sum_{k=1}^{K} \gamma_k^{(m)}(\mathbf{X}_i) \ln \left(\pi_k \mathcal{N}(\mathbf{X}_i \mid \boldsymbol{\mu}_k, \Sigma_k) \right). \tag{10.20}$$

The M-step prescribes that

$$\{\pi_k^{(m+1)}, \boldsymbol{\mu}_k^{(m+1)}, \Sigma_k^{(m+1)}\}_{i=1}^{K} = \underset{\{\pi_k, \boldsymbol{\mu}_k, \Sigma_k\}_{i=1}^{K}}{\arg\max} \; Q(\{\pi_k, \boldsymbol{\mu}_k, \Sigma_k\}_{i=1}^{K}, \{\pi_k^{(m)}, \boldsymbol{\mu}_k^{(m)}, \Sigma_k^{(m)}\}_{i=1}^{K}). \tag{10.21}$$

This maximization is straightforward. By differentiation with respect to $\boldsymbol{\mu}_k$ and Σ_k^{-1}, we have

$$\frac{\partial Q(\boldsymbol{\theta}, \boldsymbol{\theta}^{(m)})}{\partial \boldsymbol{\mu}_k} = \sum_{i=1}^{n} \gamma_k^{(m)}(\mathbf{X}_i) \Sigma_k^{-1}(\mathbf{X}_i - \boldsymbol{\mu}_k) = 0,$$

$$\frac{\partial Q(\boldsymbol{\theta}, \boldsymbol{\theta}^{(m)})}{\partial \Sigma_k^{-1}} = -\frac{1}{2} \sum_{i=1}^{n} \gamma_k^{(m)}(\mathbf{X}_i)(\Sigma_k - (\mathbf{X}_i - \boldsymbol{\mu}_k)(\mathbf{X}_i - \boldsymbol{\mu}_k)^T) = 0, \tag{10.22}$$

for $k = 1, \ldots, K$. Solving this system of equations we obtain

$$
\boldsymbol{\mu}_k^{(m+1)} = \frac{\sum_{i=1}^n \gamma_k^{(m)}(\mathbf{X}_i)\,\mathbf{X}_i}{\sum_{i=1}^n \gamma_k^{(m)}(\mathbf{X}_i)},
$$

$$
\Sigma_k^{(m+1)} = \frac{\sum_{i=1}^n \gamma_k^{(m)}(\mathbf{X}_i)(\mathbf{X}_i - \hat{\boldsymbol{\mu}}_k^{(m+1)})(\mathbf{X}_i - \hat{\boldsymbol{\mu}}_k^{(m+1)})^T}{\sum_{i=1}^n \gamma_k^{(m)}(\mathbf{X}_i)},
$$

(10.23)

for $k = 1, \ldots, K$. It can be verified that this stationary point is indeed a maximum point. As for maximization with respect to π_k, we introduce a Lagrange multiplier λ for the constraint $\sum_{k=1}^K \pi_k = 1$ and therefore look for a stationary point of $Q(\boldsymbol{\theta}, \boldsymbol{\theta}^{(m)}) + \lambda(\sum_{k=1}^K \pi_k - 1)$. Differentiating with respect to π_k gives:

$$
\frac{\partial Q(\boldsymbol{\theta}, \boldsymbol{\theta}^{(m)})}{\partial \pi_k} = \sum_{i=1}^n \frac{\gamma_k^{(m)}(\mathbf{X}_i)}{\pi_k} + \lambda = 0
$$

(10.24)

for $k = 1, \ldots, K$. Multiplying each of these K equations on both sides by π_k (we are assuming throughout that none of the π_k are zero) and adding them together leads to

$$
\sum_{i=1}^n \sum_{k=1}^K \gamma_k^{(m)}(\mathbf{X}_i) + \lambda \sum_{k=1}^K \pi_k = 0 \ \Rightarrow \ \lambda = -n,
$$

(10.25)

where we used that fact that $\sum_{k=1}^K \gamma_k^{(m)}(\mathbf{X}_i)$ and $\sum_{k=1}^K \pi_k$ are both equal to 1. Substituting $\lambda = -n$ back into (10.24) and solving that equation leads to

$$
\pi_k^{(m+1)} = \frac{1}{n} \sum_{i=1}^n \gamma_k^{(m)}(\mathbf{X}_i).
$$

(10.26)

for $k = 1, \ldots, K$. This process is repeated until the the log-likelihood $L(\boldsymbol{\theta})$ does not change significantly. At this point, hard clustering assignments can be made, if so desired by assigning point $\mathbf{X}_i$ to the cluster k where it has the largest cluster membership. Notice that alternative hard clustering assignments can be made. For example, the algorithm may refuse to assign a point to any cluster unless the largest cluster membership exceeds a prespecified threshold. The entire EM procedure is summarized below.

Algorithm 3 Clustering using EM and the Gaussian Mixture Model.

1: Initialize K, $\tau > 0$, $\{\pi_k^{(0)}, \boldsymbol{\mu}_k^{(0)}, \text{and} \Sigma_k^{(0)}\}_{i=1}^K$.

2: **repeat**

3:    E-Step: Update cluster memberships:

$$\gamma_k^{(m)}(\mathbf{X}_i) = \frac{\pi_k^{(m)} \mathcal{N}(\mathbf{X}_i \mid \boldsymbol{\mu}_k^{(m)}, \Sigma_k^{(m)})}{\sum_{k=1}^K \pi_k^{(m)} \mathcal{N}(\mathbf{X}_i \mid \boldsymbol{\mu}_k^{(m)}, \Sigma_k^{(m)})}.$$

4:    M-Step: Re-estimate model parameters:

$$\boldsymbol{\mu}_k^{(m+1)} = \frac{\sum_{i=1}^n \gamma_k^{(m)}(\mathbf{X}_i)\, \mathbf{X}_i}{\sum_{i=1}^n \gamma_k^{(m)}(\mathbf{X}_i)},$$

$$\Sigma_k^{(m+1)} = \frac{\sum_{i=1}^n \gamma_k^{(m)}(\mathbf{X}_i)(\mathbf{X}_i - \hat{\boldsymbol{\mu}}_k^{(m+1)})(\mathbf{X}_i - \hat{\boldsymbol{\mu}}_k^{(m+1)})^T}{\sum_{i=1}^n \gamma_k^{(m)}(\mathbf{X}_i)},$$

$$\pi_k^{(m+1)} = \frac{1}{n} \sum_{i=1}^n \gamma_k^{(m)}(\mathbf{X}_i).$$

5:    Calculate log-likelihood:

$$L(\{\pi_k^{(m+1)}, \boldsymbol{\mu}_k^{(m+1)}, \Sigma_k^{(m+1)}\}_{i=1}^K) = \sum_{i=1}^n \ln\left(\sum_{k=1}^K \pi_k^{(m+1)} \mathcal{N}(\mathbf{X}_i \mid \boldsymbol{\mu}_k^{(m+1)}, \Sigma_k^{(m+1)})\right).$$

6: **until** $|L(\{\pi_k^{(m+1)}, \boldsymbol{\mu}_k^{(m+1)}, \Sigma_k^{(m+1)}\}_{i=1}^K) - L(\{\pi_k^{(m)}, \boldsymbol{\mu}_k^{(m)}, \Sigma_k^{(m)}\}_{i=1}^K)| < \tau$.

Example 10.2. We apply the EM algorithm for GMM to the soft magnetic alloy data set, used in Example 10.1, with $K = 2$. Figure 10.3 displays the initial configuration, a plot of the log-likelihood after the E-step of each iteration against iteration number, and a uniform sampling of the results after the E-step of each iteration. The data points are colored by a quadratic Beziér curve interpolation of green, red, and blue, where red indicated intermediate values of cluster membership. Each Gaussian is represented by its 0.5, 1, and 1.5-standard-deviation contours of constant density. We can see that uncertainty in the clustering, represented by the presence of red points, decreases at later stages of the fitting procedure. In this example, the absolute difference between consecutive values of L is required to be less than 5×10^{-8} to terminate the algorithm. This very small value is needed in order to avoid premature termination in the flat stretch of the log-likelihood function. $\diamond$

An issue with maximum-likelihood estimation, in general, is overfitting. In the case of GMM, it is possible to generate artificially large values of the log-likelihood if one of the Gaussian densities "collapses" to one of the data points. To see this, assume that one of the Gaussian means $\boldsymbol{\mu}_j$ coincides with training point $\mathbf{X}_1$. Then this point contributes a term to the log-likelihood of the

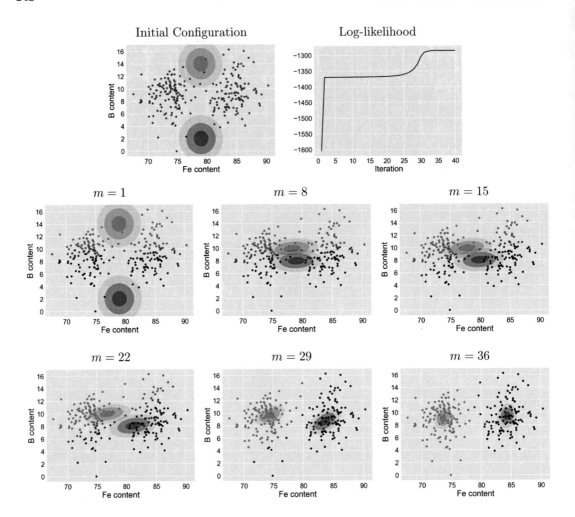

Figure 10.3: Gaussian mixture modeling example using the Iron (Fe) and Boron (B) features of the soft magnetic alloy data set. The top left plot displays the original data and initial Gaussians. The top right plot displays the value of the log-likelihood after the E-step of each iteration. The plots below display the results after the E-step of each iteration. The data points are colored by a quadratic Beziér curve interpolation of green, red, and blue, where red indicated intermediate values of cluster membership. The Gaussian are represented by their 0.5, 1, and 1.5-standard-deviation contours of constant density (plots generated by `c10_GMM.py`).

form

$$\mathcal{N}(\mathbf{X}_1 \,|\, \boldsymbol{\mu}_j = \mathbf{X}_1, \Sigma_j) \;=\; \frac{1}{(2\pi)^{d/2}|\Sigma_j|^{1/2}} \,. \tag{10.27}$$

By letting $|\hat{\Sigma}_j| \to 0$, one can increase L without bound. In order to avoid that, a sophisticated implementation of the GMM algorithm checks whether any of the cluster covariances is collapsing to one of the data points and if so, reinitializes the mean and covariance means of that cluster.

10.2.2 Relationship to K-Means

We saw in Example 10.2 that convergence is more complicated and slower than in the case of the K-means algorithm. This is partly because the K-means algorithm needs to estimate only the cluster centers, whereas there are many more parameters to be estimated in GMM fitting.

In fact, K-means can be seen as a limiting case of GMM clustering. The relationship is revealed by considering the case where $\Sigma_k = \sigma^2 I$ (spherical covariance matrices). In this case, the cluster memberships for $\mathbf{X}_i$ become:

$$\gamma_k(\mathbf{X}_i) = \frac{\pi_k \exp(-||\mathbf{x} - \boldsymbol{\mu}_k||^2/2\sigma^2)}{\sum_{k=1}^{K} \pi_k \exp(-||\mathbf{x} - \boldsymbol{\mu}_k||^2/2\sigma^2)}, \tag{10.28}$$

for $k = 1, \ldots, K$. Let $\boldsymbol{\mu}_j$ be the mean vector closest to $\mathbf{X}_i$. Then it is easy to see that, if one lets $\sigma \to 0$, then $\gamma_j(\mathbf{X}_i) \to 1$, while all other memberships go to zero. In other words, $\gamma_k(\mathbf{X}_i) \to \mathbf{r}_i(k)$, and cluster assignment is done as in the K-means algorithm. This shows that K-means tends to look for spherical clusters, while GMM has the added flexibility of being able to detect elongated elliptical clusters.

Finally, just as in the case of the K-means algorithm, the choice of the number K of Gaussians is a difficult model selection question. Increasing K introduces overfitting and artificially large values of the log-likelihood. Looking for an "elbow" in the plot of the log-likelihood against K is often a simple and effective solution.

10.3 Hierarchical Clustering

The previous clustering methods produce a fixed assignment of data points to clusters. In addition, as we discussed at the end of the previous section, they tend to look for spherical or elliptical clusters. In this section, we discuss *hierarchical clustering*, which removes both restrictions. Here, different clustering results are obtained by adopting an iterative process of cluster creation. In addition, the shape of the clusters obtained is in principle arbitrary. The process could be

- **Agglomerative** (Bottom-up): start with each point in a separate cluster and iteratively merge clusters.

- **Divisive** (Top-down): start with all the data in a single cluster and iteratively split clusters.

Here we will focus on agglomerative hierarchical clustering, which is the most common form of hierarchical clustering. Given two clusters C_i and C_j (these are just disjoint, nonempty sets of data points), agglomerative hierarchical clustering is based on a *pairwise dissimilarity* metric $d(C_i, C_j)$. The algorithm starts with n singleton clusters corresponding to each of the data points, and merges into a new cluster the two clusters (in this case, points) that are the most similar, i.e., the two clusters that minimize the given pairwise dissimilarity metric. Next, among the current clusters, the two most similar ones are merged into a new cluster. The process is repeated until there is a single cluster containing all data, after $n - 1$ merging steps. The result of this process is usually presented as a *dendrogram*, which is an acyclic tree where each node represents a merge, the leaf nodes are the individual data points, and the root node is the cluster containing all data.

Typically, the dendrogram is plotted in such a way that the height of each node is equal to the dissimilarity between the children clusters. The *cophenetic distance* between two data points is defined as the height of their lowest common parent node. Cutting the dendrogram at a given height produces a traditional assignment of points to clusters. Cutting a different heights produces different, nested clusterings. See Figure 10.4 for an illustration.

Naturally, different dendrograms are produced by different pairwise cluster dissimilarity metrics. Let $d(\mathbf{x}, \mathbf{x}')$ be a distance metric, e.g., the Euclidean distance or $1-$ the correlation, between points $\mathbf{x}, \mathbf{x}' \in R^d$. The most common pairwise cluster dissimilarity metrics used in hierarchical clustering are:

- Single-Linkage Dissimilarity:

$$d_s(C_i, C_j) \;=\; \min\{d(\mathbf{X}, \mathbf{X}') \mid \mathbf{X} \in C_i, \mathbf{X}' \in C_j\}. \tag{10.29}$$

- Complete-Linkage Dissimilarity:

$$d_s(C_i, C_j) \;=\; \max\{d(\mathbf{X}, \mathbf{X}') \mid \mathbf{X} \in C_i, \mathbf{X}' \in C_j\}. \tag{10.30}$$

- Average-Linkage Dissimilarity:

$$d_c(C_i, C_j) \;=\; \frac{1}{|C_i||C_j|} \sum_{\mathbf{X} \in C_i} \sum_{\mathbf{X}' \in C_j} d(\mathbf{X}, \mathbf{X}'). \tag{10.31}$$

Comparing the dissimilarities above, we observe that single-linkage is a form of nearest-neighbor metric, which therefore tends to produce long elongated clusters, as we will see below. This phenomenon is known as *chaining*. Complete linkage on the other hand is a maximum-distance criterion, which tends to produce round, compact clusters, as in the case of K-means and GMM clustering. Average linkage has an intermediate behavior between these two extremes.

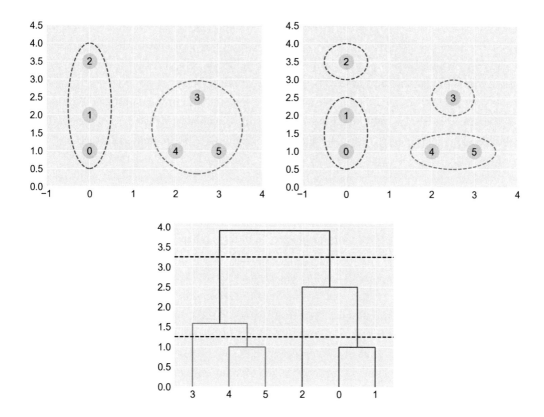

Figure 10.4: Agglomerative hierarchical clustering. Top: Original data set and two nested clusterings. Bottom: Dendrogram; the nested clusterings displayed at the top are obtained by cutting the dendrogram at the heights indicated by the horizontal dashed lines (plots generated by c10_hclust.py).

Given the sequence of clusters produced in the hierarchical clustering process, and the pairwise cluster dissimilarities between them, a dendrogram can be produced. Notice that the algorithm does not really need the original data to construct a dendrogram, but only the matrix of pairwise distances $d(\mathbf{X}, \mathbf{X}')$ between all pairs of data points.

Example 10.3. Here we continue Example 9.4 by applying hierarchical clustering to the dengue fever prognosis data set. Figure 10.5 displays the dendrograms produced by the single-linkage, complete-linkage, and average-linkage pairwise dissimilarity metrics based on the correlation distance between data points. The dissimilarities read on the vertical axis are therefore between 0 and 1, with largest number for complete linkage and smallest numbers for single linkage. A hard clustering assignment is obtained by cutting the dendrograms at 85% of the maximum height, which is displayed as a horizontal dashed line. The obtained clusters are painted with distinct colors. Chaining is clearly seen in the single-linkage result, which obtains a single large cluster (and an outlier singleton cluster). The complete-linkage result, by comparison produces two clearly defined clusters. The average-linkage result is intermediary between the two. If we compare this to the 2D MDS plot using the correlation distance in Figure 9.6, we can see that they are essentially in agreement. The left cluster contains all the DHF cases, plus 4 of the DF cases, and only two of the ND cases. One of the ND cases, case 199, is only loosely associated with the cluster (as are two of the DHF cases, 125 and 128). The right cluster contains exclusively DF and ND cases, and no DHF cases. This is in agreement with the MDS plot and the observation made in Example 9.4 that the DF cases are divided into 2 groups, one of which is very similar to the DHF group in terms of gene expression, reflecting the difficulty of labeling these diseases reliably, as they are on a spectrum. ◇

As a final remark, notice that the dendrograms produces a visualization of the structure of the data in the original high-dimensional space, so in this sense they are similar to dimensionality reduction methods used to visualize data in 2D or 3D space, such as PCA and MDS. A common mistake however is to consider that two data points are close in the original space if they are next to each other in the dendrogram. For instance, the DF_331 and ND_251 cases appear next to each other in the average-linkage dendrogram in Figure 10.5, but they are in fact quite dissimilar, as their cophenetic distance is around 0.25, which is nearly 80% of the maximum in the data set (this is confirmed by the location of these data points in the correlation-metric MDS plot of Figure 9.6).

10.4 Self-Organizing Maps (SOM)

Self-organizing maps (SOM) is a popular alternative to the clustering algorithms discussed previously. SOM iteratively adapts grid of nodes to the data. The grid is typically two-dimensional and rectangular, but this is not necessary. At first the nodes of the grid are arbitrarily positioned in the feature space. Then a data point is selected at random, and the grid node closest to that point is made to move a certain distance towards it, followed by smaller moves towards it by the neighboring nodes in the grid. This process is repeated for a large number of iterations (20,000-50,000) until the grid adapts itself to the data. The data points closest to each node then define the clusters. The achieved clustering maintains the general structure of the initial grid. See Figure 10.6 for an illustration. SOM becomes useful as a dimensionality reduction tool when the dimensionality of the data is larger than that of the map.

We give next a more precise description of the SOM algorithm. Let $\mathbf{u}_1, \ldots, \mathbf{u}_N$ be the SOM grid nodes, and let $D(\mathbf{u}_i, \mathbf{u}_j)$ denote the distance between any two nodes $\mathbf{u}_i$ and $\mathbf{u}_j$. Let $f_k(\mathbf{u}_i) \in R^d$ be the position to which node $\mathbf{u}_i$ is mapped in step k. (The initial map f_0 is chosen arbitrarily.) Let $\mathbf{X}^{(k)}$ be a randomly-selected data point in step k, and let $\mathbf{u}^{(k)}$ be the grid node that maps closest to

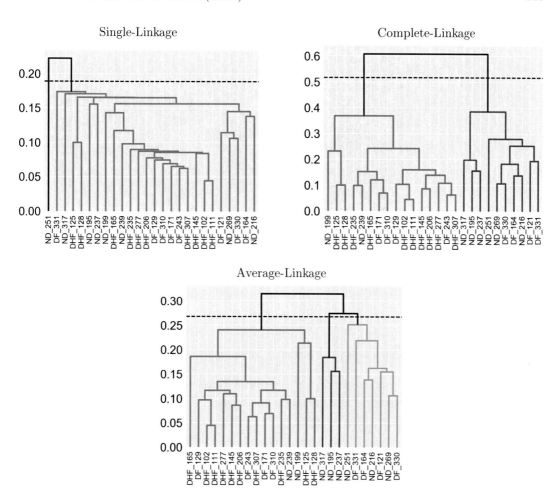

Figure 10.5: Hierarchical clustering applied to the Dengue Fever prognosis data set. Dendrograms produced by the single-linkage, complete-linkage, and average-linkage pairwise dissimilarity metrics and the correlation distance between data points. A hard clustering assignment is obtained by cutting the dendrograms at 85% of the maximum height, which is displayed as a horizontal dashed line. The obtained clusters are indicated by distinct colors. Chaining is clearly seen in the single-linkage result, while complete linkage produces two clearly defined clusters. The average-linkage result is intermediary between the two (plots generated by `c10_DF_hclust.py`).

$\mathbf{X}^{(k)}$, that is, such that $||f_k(\mathbf{u}^{(k)}) - \mathbf{X}^{(k)}||$ is minimum. Then the mapped grid is adjusted by moving points as follows:

$$f_{k+1}(\mathbf{u}_i) = f_k(\mathbf{u}_i) + \lambda(D(\mathbf{u}_i, \mathbf{u}^{(k)}), k)||f_k(\mathbf{u}_i) - \mathbf{X}^{(k)}||, \tag{10.32}$$

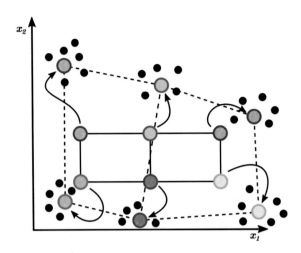

Figure 10.6: Self-organizing map with a 2×3 grid. The grid iteratively adjust itself to the data, and the clusters are defined as the set of points closest to each node.

for $i = 1, \ldots, N$, where the learning rate λ is a decreasing function of both parameters. In other words, the point $\mathbf{u}_i$ is moved towards $\mathbf{X}^{(k)}$ less when $D(\mathbf{u}_i, \mathbf{u}^{(k)})$ is large or at later iterations.

A SOM can be seen as a neural network, where the grid nodes $\mathbf{u}_1, \ldots, \mathbf{u}_N$ are the neurons, with weights equal to the coordinates of the mapped points $f_k(\mathbf{u}_1), \ldots, f_k(\mathbf{u}_N)$, where each neuron is constrained to respond similarly to its neighbors. The process of adjusting the mapped points in (10.32) is akin to network training. This neural network is called a *Kohonen network*.

10.5   Bibliographical Notes

Well-known book-length treatments of clustering are provided by Jain et al. [1988] and Kaufman and Rousseeuw [1990]. See also Chapter 10 of Webb [2002] and Section 10.3 of James et al. [2013]. Clustering is also known as *vector quantization* in the information theory field, where it is used to achieve data compression (each cluster center providing a prototype for the entire cluster). See Section 10.5.3 of Webb [2002] for more details. The K-means algorithm was published in Lloyd [1982], although it had been in use, by Lloyd and others, long before that. The EM algorithm was originally introduced in Dempster et al. [1977]. See Chapter 3 of McLachlan and Krishnan [1997] for an excellent review of its theoretical properties.

The idea of hierarchical clustering can be said to go back, in modern times, to the work of Linneaus on the taxonomy of nature [Linnaeus, 1758]. The idea of agglomerative hierarchical clustering using pairwise dissimilarities was proposed in Ward Jr [1963]. Modern *phylogeny trees* used to classify living species are examples of hierarchical clustering dendrograms; in the *metagenomics* field of Bioinformatics, phylogeny trees are used to organize unknown microbial species, with the dendrogram leaves being called into *operational taxonomic units* (OTUs) [Tanaseichuk et al., 2013].

Picking the number of clusters is a difficult model selection problem in clustering. Two methods used to do that are the *silhouette* [Kaufman and Rousseeuw, 1990] and *CLEST* [Dudoit and Fridlyand, 2002] algorithms. In Zollanvari et al. [2009b], it was found that when the data is bimodal (two true clusters) then both the silhouette and CLEST give the correct answer, but for larger numbers of true clusters, CLEST performs better.

For more on the Self-Organizing Map (SOM) algorithm, see Tamayo et al. [1999]. Another application of the SOM algorithm to Bioinformatics appears in Zollanvari et al. [2009b].

Unlike in classification and regression, defining clustering error quantitatively is an unsettled problem. This topic has been generally called the *cluster validity* in the literature, and it is typically treated in ad-hoc fashion. In Dougherty and Brun [2004], the problem of cluster validity was addressed rigorously using the theory of random labeled point processes, which allows the definition of optimal clustering operators, as well as training and testing of clustering algorithms.

10.6 Exercises

10.1. For the data set below, find all possible solutions of the K-means algorithm (using different starting points) for $K = 2$ and $K = 4$. What if $K = 3$?

Hint: look for center positions that are invariant to the E and M steps.

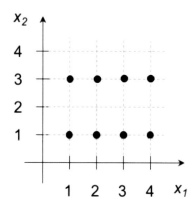

Figure 10.7: Diagram for Problem 10.1

10.2. Show that the M-step equations for estimating the parameters of the Gaussian-Mixture model (see Algorithm 3) can be obtained via the following informal optimization process. Assuming that the cluster memberships $\{\gamma_k(\mathbf{X}_i)\}_{i=1}^{n}$ are known and fixed:

(a) Find the values $\{\boldsymbol{\mu}_k\}_{k=1}^{K}$ that maximize the log-likelihood L, and plug in current estimates of other quantities.

Hint: use differentiation.

(b) Using the mean estimates in the previous step, find the values $\{\Sigma_k\}_{k=1}^{K}$ that maximize the log-likelihood L, and plug in current estimates of π_k.

Hint: use differentiation.

(c) Find the values $\{\pi_k\}_{k=1}^{K}$ that maximize the log-likelihood L.

Hint: This necessitates a slightly more complex optimization process than in the previous two steps, involving Lagrange multipliers, due to the constraints $\pi_k \geq 0$ and $\sum_k \pi_k = 1$.

The E-step corresponds simply to updating the cluster membership estimates given the estimates $\{\pi_k, \boldsymbol{\mu}_k, \Sigma_k\}_{i=1}^{K}$ obtained in the M-step.

10.3. Derive the E and M steps of the GMM algorithm under the the constraint that the covariance matrices must be

(a) Equal to each other.

(b) Spherical but not equal to each other.

(c) Spherical and equal to each other.

10.4. Compute manually the dendrograms corresponding to single-linkage, complete-linkage, and average-linkage pairwise cluster dissimilarity metrics for the data below. What do you observe in the comparison among the dendrograms?

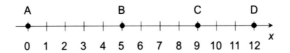

Figure 10.8: Diagram for Problem 10.4

10.5. *Ward's Method.* Consider a hierarchical clustering method based on a pairwise dissimilarity metric $d(C_i, C_j)$ defined recursively as follows. If both clusters are singletons, then

$$d(\{\mathbf{X}_i\}, \{\mathbf{X}_j\}) = ||\mathbf{X}_i - \mathbf{X}_j||^2, \tag{10.33}$$

that is, the squared Euclidean distance. Otherwise, if C_i was previously obtained by merging two clusters C_{i1} and C_{i2}, define

$$d(C_i, C_j) = \frac{|C_{i1}| + |C_j|}{|C_i| + |C_j|} d(C_{i1}, C_j) + \frac{|C_{i2}| + |C_j|}{|C_i| + |C_j|} d(C_{i2}, C_j) - \frac{|C_j|}{|C_i| + |C_j|} d(C_{i1}, C_{i2}), \quad (10.34)$$

where $|C|$ denotes the number of points in cluster C. Show that this is a minimum variance criterion, which, at each iteration of the agglomerative process, increases the within-variance of the clusters by the least amount possible.

10.6. *Lance-Williams Algorithms* are a broad class of hierarchical clustering algorithms, which are based on a pairwise dissimilarity metric $d(C_i, C_j)$ that can be defined recursively as follows. The distances $d(\{\mathbf{X}_i\}, \{\mathbf{X}_j\})$ between all pairs of points are given as input. At a later stage of the computation, assume that C_i was previously obtained by merging two clusters C_{i1} and C_{i2} and define

$$d(C_i, C_j) = \alpha_{i1} d(C_{i1}, C_j) + \alpha_{i2} d(C_{i2}, C_j) + \beta d(C_{i1}, C_{i2}) + \gamma |d(C_{i1}, C_j) - d(C_{i2}, C_j)|, \quad (10.35)$$

where α_{i1}, α_{i2}, β, and γ are parameters that may depend on the cluster sizes. Show that single-linkage, complete-linkage, average-linkage and Ward's hierarchical clustering methods are in the family of Lance-Williams algorithms, by giving the values of $d(\{\mathbf{X}_i\}, \{\mathbf{X}_j\})$ and the parameters α_{i1}, α_{i2}, β, and γ.

10.7 Python Assignments

10.7. This assignment concerns the application of K-means clustering to the soft magnetic alloy data set.

(a) Reproduce all the plots in Figure 10.1 by running `c10_Kmeans.py`. What happens if random initialization of centers is used? Can it affect the results and how?
Note: scikit-learn has a class `cluster.Kmeans`. However, it is not straightforward to expose the solutions after the E-step of each iteration, as in Figure 10.1. Hence, the K-means algorithm is coded from scratch in `c10 Kmeans.py`.

(b) Color code each point according to "low" coercivity (≤ 2 A/M), "medium" coercivity (between 2 A/M and 8 A/M), and "high" coercivity (≥ 8 A/M), using red, green, and blue, respectively. Are the two clusters associated with the coercivity values, and how?

(c) Reproduce the plots in Figure 10.2. Now try it with $K = 4, 5, 6$. What do you observe?

10.8. This assignment concerns the application of Gaussian mixture modeling clustering to the soft magnetic alloy data set.

    (a) Reproduce all the plots in Figure 10.3 by running `c10_GMM.py`. What happens if random initialization of the Gaussians is used? Can it affect the results and how?

    Note: Once again, scikit-learn has a class `mixture.GMM`, but coding the GMM algorithm directly from scratch is not difficult, and affords full control of the results.

    (b) Extend the result to $K = 3, 4, 5, 6$. What do you observe?

    (c) Modify the code to use spherical and equal covariance matrices. Can you obtain results that are close to the K-means results if the variances are small?

10.9. Apply hierarchical clustering to to the dengue fever prognosis data set.

    (a) Reproduce the plots in Figure 10.5 by running `c10_DF_hclust.py`. What happens if Ward linkage is used? (See Exercise 10.5.)

    (b) Repeat part (a) using the Euclidean distance instead of the correlation. Compare with the previous results, as well as the 2D MDS plots in Example 9.4.

10.10. Apply hierarchical clustering to the soft magnetic alloy data set.

    (a) Deterministically sample the data set by taking rows $0, 12, 24, \ldots, 132$. Construct dendrograms for single, average, and complete linkage for the Fe and Si features using the Euclidean distance. Label the leaves of the dendrograms with "low" (coercivity < 2 A/M), "medium" (2 A/M ≤ coercivity ≤ 8 A/M), and "high" (coercivity > 8 A/M).

    (b) Cut the dendrograms at 85%, 70% and 60% of the total height and obtain hard clusterings. What can you infer from this?

Chapter 11

Regression

"The most probable value of the unknown quantities will be
that in which the sum of the squares of the differences between
the actually observed and the computed values multiplied by
numbers that measure the degree of precision is a minimum."
–Carl Friedrich Gauss, *Theoria Motus Corporum Coelestium in
Sectionibus Conicis Solem Ambientium*, 1809.

In regression, the objective is to predict the value of a target $Y \in R$ given a feature vector $\mathbf{X} \in R^d$, using sample data and/or information about the distribution of the random vector $(\mathbf{X}, Y)$. The major difference with respect to classification is that in regression, the target Y is a numerical quantity, rather than a discrete label coding for different categories. This apparently small difference, however, makes regression quite different than classification, both in theory and in practice. For example, the classification error no longer applies, and there is no single gold standard performance criterion. Nevertheless, there are enough similarities that much of the material on classification covered in Chapters 2-9 could be retraced for regression. This is done in the present chapter, although we have room here to cover only the most relevant topics. The chapter begins with a discussion of optimal regression, and then examines general properties of sample-based regression. Next, we discuss regression algorithms: parametric, nonparametric, and function-approximation regression. We cover in some detail parametric linear least-squares estimation, which is a very classical and useful tool in regression analysis. The section on nonparametric regression focuses mainly on Gaussian-process regression, a topic that recently has become quite popular and useful in many different areas. The chapter also discusses regression error estimation and variable selection.

© Springer Nature Switzerland AG 2020
U. Braga-Neto, *Fundamentals of Pattern Recognition and Machine Learning*,
https://doi.org/10.1007/978-3-030-27656-0_11

11.1   Optimal Regression

In what follows, we assume that $(\mathbf{X}, Y)$ is a jointly distributed continuous random vector in R^{d+1}, so that the feature-target distribution $P_{\mathbf{X},Y}$ is determined by a density function $p(\mathbf{x}, y)$ on R^{d+1}. In some regression contexts, especially in classical statistics, $\mathbf{X}$ is not random (e.g., a univariate X could be the months of the year). We focus instead on the case where $\mathbf{X}$ is random, which is the most common case in supervised learning. The following simple result proves to be important in what follows.

Theorem 11.1. *If $E[|Y|] < \infty$, there is a function $f : R^d \to R$ and a random variable ε such that*

$$Y = f(\mathbf{X}) + \varepsilon, \tag{11.1}$$

where $E[\varepsilon \mid \mathbf{X}] = 0$.

Proof. Let $f(\mathbf{x}) = E[Y \mid \mathbf{X} = \mathbf{x}]$, for $\mathbf{x} \in R^d$, and let $\varepsilon = Y - E[Y \mid \mathbf{X}]$. Then $Y = f(\mathbf{X}) + \varepsilon$, and

$$E[\varepsilon \mid \mathbf{X}] = E[Y - E[Y \mid \mathbf{X}] \mid \mathbf{X}] = E[Y \mid \mathbf{X}] - E[Y \mid \mathbf{X}] = 0. \tag{11.2}$$

as required. The integrability condition $E[|Y|] < \infty$ guarantees that $E[Y \mid \mathbf{X}]$ is well defined. ◇

The previous result states that Y can be decomposed into a deterministic function f of $\mathbf{X}$ and a *zero-mean additive noise* term ε. If ε is independent of $(\mathbf{X}, Y)$, the model is called *homoskedastic*, otherwise, it is said to be *heteroskedastic*.

Example 11.1. Let X be uniformly distributed in the interval $[0, 2\pi]$, $f(X) = \sin(X)$, and $\varepsilon \mid X = x \sim \mathcal{N}(0, \sigma^2(x))$, where

$$\sigma(x) = A(2\pi x - x^2), \quad x \in [0, 2\pi], \tag{11.3}$$

for $A > 0$. This is a heteroskedastic model, as the noise variance σ^2 is a function of x. Indeed, it is easy to check that σ^2 is zero at the extremes of the interval $x = 0$ and $x = 2\pi$, and maximal at the midpoint $x = \pi$. Samples from the model (11.1) for $A = 0.02$ (low noise) and $A = 0.1$ (high noise), with $n = 120$ points each, are displayed in Figure 11.1. Also displayed are f and 1-standard-deviation bands. We can see that f is a very good predictor of Y if the noise amplitude A is small, but not quite if A is large. ◇

We saw in the previous example that f was a good predictor of Y as long as the noise variance was not too large. We now examine this issue formally. The *conditional regression error* of a predictor f at a point $\mathbf{x}$ is given by:

$$L[f](\mathbf{x}) = E[\ell(Y, f(\mathbf{X})) \mid \mathbf{X} = \mathbf{x}] = \int \ell(y, f(\mathbf{x}))p(y \mid \mathbf{x})dy, \quad \mathbf{x} \in R^d, \tag{11.4}$$

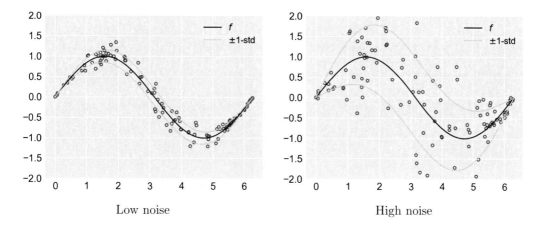

Figure 11.1: Regression example with noisy sinusoidal data. We observe that the function f is a very good predictor of Y if the noise intensity is small, but not if it is large (plots generated by c11_sine.py).

where $\ell : R \times R \to R$ is an appropriate *loss function*, examples of which are the *quadratic loss* $\ell(y, f(\mathbf{x})) = (y - f(\mathbf{x}))^2$, the *absolute loss* $\ell(y, f(\mathbf{x})) = |y - f(\mathbf{x})|$, and the *Minkowski loss* $\ell(y, f(\mathbf{x})) = |y - f(\mathbf{x})|^q$, for $q > 0$.

In order to obtain a criterion that is independent of a specific value of $\mathbf{X}$, the loss function must be averaged over both $\mathbf{X}$ and Y. Accordingly, the (unconditional) *regression error* of f is defined as:

$$L[f] = \iint \ell(y, f(\mathbf{x}))\, p(\mathbf{x}, y)\, d\mathbf{x}\, dy = E[L[f](\mathbf{X})]. \tag{11.5}$$

An *optimal regression function* f^* for a given loss function satisfies

$$f^* = \arg\min_{f \in F} L[f], \tag{11.6}$$

where F is the set of all (Borel-measurable) functions on R^d. The *optimal regression error* is $L^* = L[f^*]$.

The most common loss function in regression is the quadratic loss. However, proponents of the absolute loss point to the fact that it is more immune to outliers than the quadratic loss. The Minkowski loss is actually a family of functions, which includes the previous two as special cases. In this chapter, we will focus on the quadratic loss. But it is clear that, unlike in the case of classification where the classification error is the gold standard, there *is* no universally accepted optimality criterion in regression.

The next result, which can be seen as a counterpart of Theorem 2.1, is a fundamental theorem not only in the theory of regression, but also in signal processing generally.

Theorem 11.2. *If $E[|Y|] < \infty$,*

$$f^*(\mathbf{x}) = E[Y \mid \mathbf{X} = \mathbf{x}], \quad \mathbf{x} \in R^d. \tag{11.7}$$

is an optimal regression function for the quadratic loss.

Proof. We have to show that $L[f^*] \leq L[f]$, with the quadratic loss, for any $f \in F$. For any $\mathbf{x} \in R^d$,

$$\begin{aligned}
L[f](\mathbf{x}) - L[f^*](\mathbf{x}) &= \int \left((y - f(\mathbf{x}))^2 - (y - f^*(\mathbf{x}))^2\right) p(y \mid \mathbf{x}) \, dy \\
&= \int \left((y - f^*(\mathbf{x}) + f^*(\mathbf{x}) - f(\mathbf{x}))^2 - (y - f^*(\mathbf{x}))^2\right) p(y \mid \mathbf{x}) \, dy \\
&= (f^*(\mathbf{x}) - f(\mathbf{x}))^2 - 2(f^*(\mathbf{x}) - f(\mathbf{x})) \underbrace{\int (y - f^*(\mathbf{x})) \, p(y \mid \mathbf{x}) \, dy}_{=0} \\
&= (f^*(\mathbf{x}) - f(\mathbf{x}))^2 \geq 0,
\end{aligned} \tag{11.8}$$

with equality only if $f(\mathbf{x}) = f^*(\mathbf{x})$. Integrating (11.8) over $\mathbf{X}$ proves the claim. The integrability condition $E[|Y|] < \infty$ guarantees that $E[Y \mid \mathbf{X}]$ is well defined. ◇

With the quadratic loss, $L[f]$ is also known as the *mean square error* (MSE) of f, and the *conditional mean* f^* is called a *minimum mean-square error* (MMSE) regression function. There is not a unique MMSE regression function, since the value of f^* can be changed over a set of probability zero without changing the value of $L[f^*]$. It can similarly be shown that the *conditional median* is an optimal regression function for the absolute loss, called a *minimum absolute difference (MAD)* regression function, whereas the *conditional mode* is an optimal regression function for the Minkowski loss with $q \to 0$, also known as a *maximum-a-posteriori (MAP)* regression function.

The proof of Theorem 11.2 also shows that, just as a Bayesian classifier, an MMSE regression function minimizes $L[f](\mathbf{x})$ at each point $\mathbf{x} \in R^d$, in addition to minimizing $L[f]$. From (11.4), (11.9), and Thm 11.1, the optimal value at each value of $\mathbf{x} \in R^d$ is

$$\begin{aligned}
L^*(\mathbf{x}) = L[f^*](\mathbf{x}) &= \int |y - E[Y \mid \mathbf{X} = \mathbf{x}]|^2 p(y \mid \mathbf{x}) dy = \mathrm{Var}(Y \mid \mathbf{X} = \mathbf{x}) \\
&= \mathrm{Var}(f(\mathbf{X}) + \varepsilon \mid \mathbf{X} = \mathbf{x}) = \mathrm{Var}(\varepsilon \mid \mathbf{X} = \mathbf{x}).
\end{aligned} \tag{11.9}$$

This is a lower bound on the performance of all regression functions at each given value $\mathbf{x} \in R^d$. The optimal regression error is then

$$L^* = L[f^*] = E[L^*(\mathbf{X})] = E[\mathrm{Var}(Y \mid \mathbf{X})] = E[\mathrm{Var}(\varepsilon \mid \mathbf{X})], \tag{11.10}$$

which gives a lower bound on the overall performance of all regression functions. In the homoskedastic case, things get significantly simplified. For all $\mathbf{x} \in R^d$,

$$L^* = L^*(\mathbf{x}) = \mathrm{Var}(\varepsilon) = \sigma^2 \quad \text{(homoskedastic case)}. \tag{11.11}$$

Notice also that f^* is precisely the function f in the decomposition of Theorem 11.1.

Example 11.2. Continuing Example 11.1 in the light of the previous discussion, we realize that an MMSE regression function in this problem is $f^*(x) = \sin(x)$, for $x \in [0, 2\pi]$, with

$$L^*(x) = \text{Var}(\varepsilon \mid X = x) = A(2\pi x - x^2), \quad x \in [0, 2\pi]. \tag{11.12}$$

In particular, $L^*(0) = L^*(2\pi) = 0$, so that the problem is deterministic at the extremes of the interval, while $L^*(x)$ is maximal, and the prediction problem most difficult, at the center of the interval $x = \pi$. To obtain a performance metric that is independent of x, we compute

$$L^* = E[L^*(X)] = A(2\pi E[X] - E[X^2]) = \frac{2\pi^2 A}{3}, \tag{11.13}$$

where we used the facts that $E[X] = \pi$ and $E[X^2] = \frac{4}{3}\pi^2$ for the uniform random variable X on the interval $[0, 2\pi]$. We observe that L^*, and the overall difficulty of prediction, increases (linearly) with the noise amplitude A. This confirms what was observed in Figure 11.1. ◇

Finally, notice that the error of any regression function $f \in F$ can be decomposed as follows:

$$
\begin{aligned}
L[f] &= E[(Y - f(\mathbf{X}))^2] = E[E[(Y - f(\mathbf{X}))^2 \mid \mathbf{X}]] = E[E[(f^*(\mathbf{X}) + \varepsilon - f(\mathbf{X}))^2 \mid \mathbf{X}]] \\
&= \underbrace{E[(f^*(\mathbf{X}) - f(\mathbf{X}))^2]}_{\text{reducible error}} + \underbrace{E[E[\varepsilon^2 \mid \mathbf{X}]]}_{L^*} + 2E[\underbrace{E[\varepsilon \mid \mathbf{X}]}_{=0}(f^*(\mathbf{X}) - f(\mathbf{X}))] \\
&= L^* + \bar{L}[f]
\end{aligned}
\tag{11.14}
$$

The reducible error $\bar{L}[f] = E[(f^*(\mathbf{X}) - f(\mathbf{X}))^2]$ is the excess error over the optimal regression; it is zero if and only if $f(\mathbf{X}) = f^*(\mathbf{X})$ with probability one. Notice that this derivation does not require homoskedasticity.

11.2   Sample-Based Regression

In practice, the joint feature-target distribution $P_{\mathbf{X},Y}$ is not know, or is only partially known, so that an optimal regression function f^* is not available, and an estimate f_n must be estimated from sample data S_n. The sample-based MSE of f_n is

$$L_n = L[f_n] = E[(Y - f_n(\mathbf{X}))^2 \mid S_n] \tag{11.15}$$

Since the data are random, f_n is random, and so is L_n. Taking expectation produces the data-independent *expected MSE* $E[L_n]$.

From (11.14), it follows that

$$
\begin{aligned}
E[L_n] &= L^* + E[\bar{L}[f_n]] \\
&= L^* + E[(f^*(\mathbf{X}) - f_n(\mathbf{X}))^2] \\
&= L^* + E[(f^*(\mathbf{X}) - f_n(\mathbf{X}))]^2 + \mathrm{Var}(f^*(\mathbf{X}) - f_n(\mathbf{X})) \\
&= L^* + \mathrm{Bias}(f_n)^2 + \mathrm{Variance}(f_n)
\end{aligned}
\tag{11.16}
$$

where the identity $E[Z^2] = E[Z]^2 + \mathrm{Var}(Z)$ was used. The "bias" term has to do with how far f_n is from the optimal regression f^* on average; it is zero if $E[f_n(\mathbf{X})] = f^*(\mathbf{X})$ over a region of probability one. The "variance" term measures the sensitivity of f_n is to the data; it is small if f_n changes little with different data, while it is large if small perturbation in the data create large changes in f_n. From (11.14), we see that both bias and variance should be made as small as possible. However, small variance usually is associated with large bias, and vice-versa; hence, the bias-variance trade-off. The terminology is borrowed from parameter estimation in classical statistics, where there is a trade-offbetween bias and variance.

Example 11.3. We illustrate the bias-variance trade-offby applying least-squares polynomial regression (this regression algorithm will be described in detail in the next Section) to the data in Figure 11.1(b). The trade-offcan be seen in the series of regression results using increasing polynomial order in Figure 11.2. The optimal regression is displayed for reference. At very low order (0 and 1), we can see that there is underfitting, with large bias. At very high order (24), there is obvious overfitting and a large variance. The best result seems to be achieved by a polynomial of order 3. The low optimal order reflects the simplicity of the sinusoidal data, despite the large variance of the noise. ◇

11.3 Parametric Regression

Parametric regression is the counterpart of parametric classification, which was discussed in Chapter 4. In parametric regression, the general model in Theorem 11.1 takes the form:

$$
Y = f(\mathbf{X}; \boldsymbol{\theta}) + \varepsilon,
\tag{11.17}
$$

where $\boldsymbol{\theta} \in \Theta \subseteq R^m$ is a parameter vector. It is desirable, but not necessary, for the optimal regression f^* to be a member of the family $\{f(\mathbf{X}; \boldsymbol{\theta}) \mid \boldsymbol{\theta} \in \Theta\}$.

Parametric regression seeks a sample-based estimate $\hat{\boldsymbol{\theta}}_n$ such that the plug-in sample-based regression $f_n(\mathbf{x}) = f(\mathbf{x}; \hat{\boldsymbol{\theta}}_n)$ has small regression error. The classical solution to this problem is contained in the quote by Gauss, in the epigraph to this chapter, which marks the origin of the *least-squares regression* approach (see the Bibliographical Notes). Given sample data $S_n = \{(\mathbf{X}_1, Y_1), \ldots, (\mathbf{X}_n, Y_n)\}$,

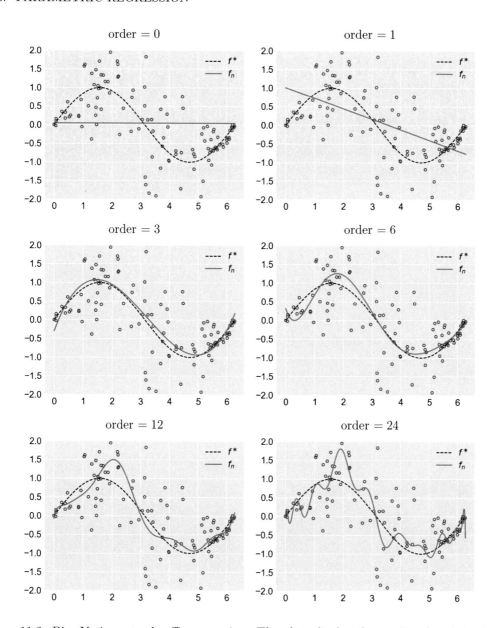

Figure 11.2: Bias-Variance trade-offin regression. The plots display the results of applying least-squares polynomial regression to the data of Figure 11.1(b), for increasing polynomial order. The optimal regression is displayed as well. As the order of the polynomial order increases, the bias decreases, but the variance increases. The best result appears to be obtained with a polynomial of order 3 (plots generated by `c11_poly.py`).

the *least-squares estimator* for parametric regression is given by

$$\hat{\boldsymbol{\theta}}_n^{\text{LS}} = \arg\min_{\boldsymbol{\theta} \in \Theta} \sum_{i=1}^{n} (Y_i - f(\mathbf{X}_i; \boldsymbol{\theta}))^2 \qquad (11.18)$$

The quantity being minimized is known as the *residual sum of squares* (RSS).

In the next section, we describe the application of the least-squares approach to the most common form of parametric regression, namely, the *linear regression model*.

11.3.1   Linear Regression

The basic form of the linear regression model, also known as *multivariate linear regression* in statistics, (11.17) takes the form:

$$Y = a_0 + a_1 X_1 + \cdots + a_d X_d + \varepsilon, \qquad (11.19)$$

with $\mathbf{X} = (X_1, \ldots, X_d) \in R^d$ and $\boldsymbol{\theta} = (a_0, a_1, \ldots, a_d) \in R^{d+1}$.

The name linear regression is somewhat misleading, since the previous model can be immediately extended to the *basis-function linear regression* model:

$$\begin{aligned} Y &= \theta_0 \phi_0(\mathbf{X}) + \theta_1 \phi_1(\mathbf{X}) + \cdots + \theta_k \phi_M(\mathbf{X}) + \varepsilon \\ &= \boldsymbol{\Phi}(\mathbf{X})^T \boldsymbol{\theta} + \varepsilon, \end{aligned} \qquad (11.20)$$

where $\phi_i : R^d \to R$ are basis functions and $\boldsymbol{\Phi} = (\phi_1, \ldots, \phi_M)$. Here $M + 1$ is the order of the model and $\boldsymbol{\theta} = (\theta_0, \theta_1, \ldots, \theta_M) \in R^{M+1}$. Notice that $M \neq d$, in general. The basis functions are quite general and do not need to be linear. The key point is that the model is still linear *in the parameters*. The standard linear model in (11.19) is of course a special case, with $\phi_0(\mathbf{X}) = 1$, $\phi_1(\mathbf{X}) = X_1$, ..., $\phi(\mathbf{X}) = X_d$, and $\theta_i = a_i$, for $i = 1, \ldots, d$.

An important example of basis-function linear regression is *polynomial regression*, already encountered in Example 11.3. In the univariate case, it takes the form:

$$Y = a_0 + a_1 X + a_2 X^2 + \cdots + a_k X^k + \varepsilon. \qquad (11.21)$$

Here the basis functions are $\phi_0(X) = 1$, $\phi_1(X) = X$, $\phi_2(X) = X^2$, ..., $\phi_k(X) = X^k$, and $\theta_i = a_i$, for $i = 1, \ldots, k$.

Next, we describe the application of least-squares parameter estimation to linear regression. Given the training data $S_n = \{(\mathbf{X}_1, Y_1), \ldots, (\mathbf{X}_n, Y_n)\}$, write one equation for each data point:

$$
\begin{aligned}
Y_1 &= \theta_0 \phi_0(\mathbf{X}_1) + \theta_1 \phi_1(\mathbf{X}_1) + \cdots + \theta_k \phi_k(\mathbf{X}_1) + \varepsilon_1, \\
Y_2 &= \theta_0 \phi_0(\mathbf{X}_2) + \theta_1 \phi_1(\mathbf{X}_2) + \cdots + \theta_k \phi_d(\mathbf{X}_2) + \varepsilon_2, \\
&\vdots \\
Y_n &= \theta_0 \phi_0(\mathbf{X}_n) + \theta_1 \phi_1(\mathbf{X}_n) + \cdots + \theta_k \phi_k(\mathbf{X}_n) + \varepsilon_n,
\end{aligned}
\tag{11.22}
$$

where we assume that $n > k$. This can be written in matrix form as

$$
\mathbf{Y}_{n \times 1} = H_{n \times k}(\mathbf{X}_1, \ldots, \mathbf{X}_n) \, \boldsymbol{\theta}_{k \times 1} + \boldsymbol{\varepsilon}_{n \times 1}, \tag{11.23}
$$

where $\mathbf{Y} = (Y_1, \ldots, Y_n)$, $\boldsymbol{\theta} = (\theta_1, \ldots, \theta_k)$, $\boldsymbol{\varepsilon} = (\varepsilon_1, \ldots, \varepsilon_n)$, and

$$
H(\mathbf{X}_1, \ldots, \mathbf{X}_n) = \begin{bmatrix} \phi_0(\mathbf{X}_1) & \cdots & \phi_k(\mathbf{X}_1) \\ \vdots & \ddots & \vdots \\ \phi_0(\mathbf{X}_n) & \cdots & \phi_k(\mathbf{X}_n) \end{bmatrix}. \tag{11.24}
$$

For simplicity, we shall omit the dependency on $(\mathbf{X}_1, \ldots, \mathbf{X}_n)$ and write H in the sequel.

From (11.17) and (11.20), we gather that $f(\mathbf{X}; \boldsymbol{\theta}) = \boldsymbol{\Phi}(\mathbf{X})^T \boldsymbol{\theta}$. With $\mathbf{f}(\mathbf{X}) = (f(\mathbf{X}_1), \ldots, f(\mathbf{X}_n))$, we can write $\mathbf{f}(\mathbf{X}) = H\hat{\boldsymbol{\theta}}$. The least-squares estimator in (11.18) can be written as

$$
\begin{aligned}
\hat{\boldsymbol{\theta}}_n^{\mathrm{LS}} &= \arg\min_{\boldsymbol{\theta} \in \Theta} \|\mathbf{Y} - \mathbf{f}(\mathbf{X})\|^2 \\
&= \arg\min_{\boldsymbol{\theta} \in \Theta} (\mathbf{Y} - H\hat{\boldsymbol{\theta}})^T (\mathbf{Y} - H\hat{\boldsymbol{\theta}}).
\end{aligned}
\tag{11.25}
$$

Assuming that H has full column rank ($n > k$ is usually sufficient for that) so that $H^T H$ is invertible, the solution is unique and given by (see Exercise 11.3):

$$
\hat{\boldsymbol{\theta}}_n^{\mathrm{LS}} = H^L \mathbf{Y} = (H^T H)^{-1} H^T \mathbf{Y} \tag{11.26}
$$

where $H^L = (H^T H)^{-1} H^T$ is the *left-inverse* of full-rank matrix H. The least-squares regression function at point $\mathbf{x} \in R^d$ is the plug-in estimate $f^{\mathrm{LS}}(\mathbf{x}) = \boldsymbol{\Phi}(\mathbf{x})^T \hat{\boldsymbol{\theta}}^{\mathrm{LS}}$.

A simple, but important, case that illustrates the general approach is that of univariate linear regression model:

$$
Y = \theta_0 + \theta_1 X + \varepsilon, \tag{11.27}
$$

where θ_0 and θ_1 are the *intercept* and *slope*, respectively. Given data $S_n = \{(X_1, Y_1), \ldots, (X_n, Y_n)\}$, we can write (11.23) as

$$
\begin{bmatrix} Y_1 \\ \cdots \\ Y_n \end{bmatrix} = \begin{bmatrix} 1 & X_1 \\ \cdots & \cdots \\ 1 & X_n \end{bmatrix} \begin{bmatrix} \theta_0 \\ \theta_1 \end{bmatrix} + \begin{bmatrix} \varepsilon_1 \\ \cdots \\ \varepsilon_n \end{bmatrix} \tag{11.28}
$$

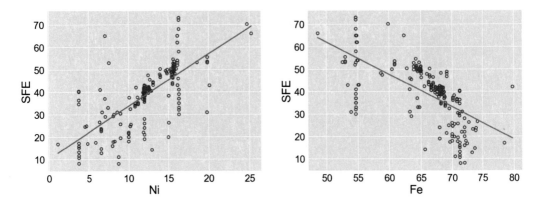

Figure 11.3: Linear regression example using the stacking fault energy (SFE) data set. We observe that SFE tends to increase with an increasing Nickel (Ni) content, but it decreases if the steel alloy contains more Iron (Fe) (plots generated by c11_SFE.py).

As can be verified by applying (11.26), the least-squares estimators of θ_0 and θ_1 are

$$\hat{\theta}_{0,n}^{\mathrm{LS}} = \overline{Y} - \hat{\theta}_{1,n}^{\mathrm{LS}} \, \overline{X} \,,$$
$$\hat{\theta}_{1,n}^{\mathrm{LS}} = \frac{\sum_{i=1}^n (X_i - \overline{X})(Y_i - \overline{Y})}{\sum_{i=1}^n (X_i - \overline{X})^2} \,, \tag{11.29}$$

where $\overline{X} = \frac{1}{n}\sum_{i=1}^n X_i$ and $\overline{Y} = \frac{1}{n}\sum_{i=1}^n Y_i$. Notice that $\hat{\theta}_{1,n}^{\mathrm{LS}} = S_{XY}/S_{XX}$, the ratio of the sample estimates of $\mathrm{Cov}(X,Y)$ and $\mathrm{Var}(X)$, respectively. The optimal regression line is given by $f^{\mathrm{LS}}(x) = \hat{\theta}_{0,n}^{\mathrm{LS}} + \hat{\theta}_{1,n}^{\mathrm{LS}} x$, for $x \in \mathbb{R}$.

Example 11.4. We apply least-squares linear regression to the stacking fault energy (SFE) data set (see Section A8.4), already used in Chapters 1 and 4 in a classification context. Here no quantization is applied on the SFE response. After preprocessing the data set, we obtain a data matrix containing 123 sample points and 7 features. Figure 11.3 displays the least-squares regression of the SFE response on the Fe and Ni atomic features, separately. From these plots, we may conclude that the stacking fault energy of the steel tends to decrease with a higher Iron (Fe) content, while the behavior is the opposite in the case of Nickel (Ni). This agrees with what was observed in Example 1.2. ◇

11.3.2 Gauss-Markov Theorem

Least-squares regression is a purely deterministic procedure, as the error vector $\boldsymbol{\varepsilon}$ in (11.23) represents simply a deviation and has no statistical properties. As a result, the least-squares method has nothing to say about the uncertainty of the fitted model.

However, if the noise ε is considered a random vector, then the model becomes stochastic, and one can talk about the statistical properties of the estimator $\hat{\boldsymbol{\theta}}$. In this section, we state and prove the Gauss-Markov Theorem, a classical result (with one of the most impressive names in mathematics) that shows that, under minimal distributional assumptions, the least-squares estimator in the previous section is unbiased and minimum-variance estimator over the class of all linear estimators.

First, given the model (11.23), repeated here for convenience,

$$\mathbf{Y}_{n \times 1} = H_{n \times k} \boldsymbol{\theta}_{k \times 1} + \boldsymbol{\varepsilon}_{n \times 1}, \tag{11.30}$$

where H is a function of $\mathbf{X}_1, \ldots, \mathbf{X}_n$, as before. We define a linear estimator as $\hat{\boldsymbol{\theta}} = B\mathbf{Y}$, where B is a $k \times n$ matrix. The estimator is *unbiased* if $E[\hat{\boldsymbol{\theta}}] = \boldsymbol{\theta}$. In addition, it is *minimum-variance* unbiased if $\text{Trace}(E[(\hat{\boldsymbol{\theta}} - \boldsymbol{\theta})(\hat{\boldsymbol{\theta}} - \boldsymbol{\theta})^T])$ is minimum among all estimators $\hat{\boldsymbol{\theta}}$ under consideration. Notice that $E[(\hat{\boldsymbol{\theta}} - \boldsymbol{\theta})(\hat{\boldsymbol{\theta}} - \boldsymbol{\theta})^T]$ is the covariance matrix of the unbiased estimator $\hat{\boldsymbol{\theta}}$, so that its trace is the sum of the variances of the individual estimators $\hat{\theta}_i$, for $i = 1, \ldots, k$. An estimator that is linear, unbiased, and minimum variance is known as a *best linear unbiased estimator* (BLUE).

Theorem 11.3. *(Gauss-Markov Theorem) If $E[\varepsilon] = 0$ and $E[\varepsilon\varepsilon^T] = \sigma^2 I_n$ (zero-mean uncorrelated noise), then the least-squares estimator*

$$\hat{\boldsymbol{\theta}}^{\text{LS}} = (H^T H)^{-1} H^T \mathbf{Y}, \tag{11.31}$$

is best linear unbiased.

Proof. The least-squares estimator is clearly linear, with $\hat{\boldsymbol{\theta}}^{\text{LS}} = B_0 \mathbf{Y}$, where $B_0 = H^T H^{-1} H^T$. Next we show that $\hat{\boldsymbol{\theta}}^{\text{LS}}$ is unbiased. First note that

$$
\begin{aligned}
\hat{\boldsymbol{\theta}}^{\text{LS}} &= (H^T H)^{-1} H^T \mathbf{Y} = (H^T H)^{-1} H^T (H\boldsymbol{\theta} + \varepsilon) \\
&= (H^T H)^{-1} (H^T H)\boldsymbol{\theta} + (H^T H)^{-1} H^T \varepsilon \\
&= \boldsymbol{\theta} + B_0 \varepsilon
\end{aligned}
\tag{11.32}
$$

Hence, $E[\hat{\boldsymbol{\theta}}^{\text{LS}}] = E[\boldsymbol{\theta} + B_0 \varepsilon] = \boldsymbol{\theta}$, given the assumption that $E[\varepsilon] = 0$. Now consider an linear estimator $\hat{\boldsymbol{\theta}} = B\mathbf{Y}$. If this estimator is to be unbiased, then the expectation

$$E[\hat{\boldsymbol{\theta}}] = E[B\mathbf{Y}] = E[B(H\boldsymbol{\theta} + \varepsilon)] = BH\boldsymbol{\theta} \tag{11.33}$$

must equal $\boldsymbol{\theta}$, so a linear estimator $\boldsymbol{\theta} = B\mathbf{Y}$ is unbiased if and only $BH = I$. Next note that, following the same derivation as in (11.32), we have $\hat{\boldsymbol{\theta}} - \boldsymbol{\theta} = B\varepsilon$. Hence, the covariance matrix of the unbiased estimator $\hat{\boldsymbol{\theta}} = B\mathbf{Y}$ is

$$E[(\hat{\boldsymbol{\theta}} - \boldsymbol{\theta})(\hat{\boldsymbol{\theta}} - \boldsymbol{\theta})^T] = E[B\varepsilon\varepsilon^T B^T] = BE[\varepsilon\varepsilon^T]B^T = \sigma^2 BB^T, \tag{11.34}$$

from the assumption that $E[\varepsilon\varepsilon^T] = \sigma^2 I$. Using the fact that $BH = I$, by unbiasedness, and the definition of B_0, it is easy to verify that:

$$B_0 B_0^T = B_0 B^T = B_0 B^T = (H^T H)^{-1}. \tag{11.35}$$

Hence, we can write

$$\begin{aligned} BB^T &= BB^T + 2B_0 B_0^T - B_0 B^T - BB_0^T \\ &= B_0 B_0^T + (B - B_0)(B - B_0)^T. \end{aligned} \tag{11.36}$$

Using now the fact that $\mathrm{Trace}(AA^T) \geq 0$ with equality only if $A \equiv 0$, we obtain $\mathrm{Trace}(BB^T) \geq \mathrm{Trace}(B_0 B_0^T)$, with equality only if $B = B_0$, proving the claim. $\diamond$

The Gauss-Markov Theorem makes minimal distributional assumptions, namely, that the noise ε is zero-mean and uncorrelated (there is even an essentially distribution-free version where ε is allowed to be correlated, see Exercise 11.5). If one is willing to assume further that the noise ε is Gaussian, then we will be able to show next that the least-squares estimator is not only BLUE, but also the maximum-likelihood solution to the model.

Assume that $\varepsilon \sim \mathcal{N}(0, \sigma^2 I_n)$ (zero-mean, uncorrelated, Gaussian noise). Then, the conditional distribution of $\mathbf{Y}$ given $\mathbf{X}_1, \ldots, \mathbf{X}_n$ is:

$$\mathbf{Y} = H\boldsymbol{\theta} + \varepsilon \sim \mathcal{N}(H\boldsymbol{\theta}, \sigma^2 I_n) \tag{11.37}$$

With fixed σ and $\hat{Y}_i = (H\boldsymbol{\theta})_i$, for $i = 1, \ldots, n$, we can write the conditional likelihood function for this model as

$$L(\boldsymbol{\theta}) = p(\mathbf{Y} \mid \boldsymbol{\theta}, \mathbf{X}_1, \ldots, \mathbf{X}_n) = \prod_{i=1}^{n} \frac{1}{\sqrt{2\pi}\sigma} \exp\left(-\frac{(Y_i - \hat{Y}_i)^2}{2\sigma^2}\right), \tag{11.38}$$

which leads to the log-likelihood

$$\ln L(\boldsymbol{\theta}) = \mathrm{const} - \sum_{i=1}^{n} \frac{(Y_i - \hat{Y}_i)^2}{2\sigma^2}. \tag{11.39}$$

For fixed σ, maximizing this log-likelihood is equivalent to minimizing the sum of squares $\sum_{i=1}^{n}(Y_i - \hat{Y}_i)^2$, and the MLE and least-squares estimators are the same.

To find the MLE of σ, one needs to maximize the likelihood

$$L(\sigma^2) = p(\mathbf{Y} \mid \sigma^2, \boldsymbol{\theta}^{\mathrm{LS}}, \mathbf{X}_1, \ldots, \mathbf{X}_n) = \prod_{i=1}^{n} \frac{1}{\sqrt{2\pi}\sigma} \exp\left(-\frac{(Y_i - \hat{Y}_i^{\mathrm{LS}})^2}{2\sigma^2}\right), \tag{11.40}$$

where $\hat{Y}_i^{\mathrm{LS}} = (H\hat{\boldsymbol{\theta}}^{\mathrm{LS}})_i$, for $i = 1, \ldots, n$, are the values predicted by the least-squares (and MLE) regression. It can be shown that (11.40) is maximized at

$$\hat{\sigma}_{\mathrm{MLE}}^2 = \frac{1}{n}\sum_{i=1}^{n}(Y_i - \hat{Y}_i^{\mathrm{LS}})^2 \tag{11.41}$$

The right hand side is just the normalized RSS. Therefore, to obtain an MLE estimate of the noise variance, one need only divide the RSS by the number of data points.

Even though $\boldsymbol{\theta}^{\text{LS}}$ is unbiased (as shown by the Gauss-Markov Theorem), the MLE estimator $\hat{\sigma}^2_{\text{MLE}}$ is not: it can be shown that, for a linear model with k parameters,

$$E[\hat{\sigma}^2_{\text{MLE}}] = \frac{n-k}{n}\sigma^2 \tag{11.42}$$

(however, this is still asymptotically unbiased, which is a property of all MLEs). To obtain an unbiased estimator, it is common in practice to use:

$$\hat{\sigma}^2_{\text{unbiased}} = \frac{n}{n-k}\hat{\sigma}^2_{\text{MLE}} = \frac{1}{n-k}\sum_{i=1}^{n}(Y_i - \hat{Y}_i)^2 = \frac{\text{RSS}}{n-k}. \tag{11.43}$$

For example, for regression with a line, one would estimate the error variance by $\text{RSS}/(n-2)$. Unless n is quite small, the estimators (11.41) and (11.43) yield very similar values.

Example 11.5. Continuing Example 11.4, the previous results state that an unbiased estimates of the noise variances are $21162.86/(211-2) = 101.26 \; 17865.18/(211-2) = 85.48$ for regressing SFE on Fe and Ni, respectively. This would correspond to standard deviations of roughly of 10.06 and 9.25, respectively. $\diamond$

11.3.3 Penalized Least Squares

In some cases, it is desirable to introduce constraints on the coefficients of parametric regression models in order to avoid overfitting. This is called penalized least squares or *ridge regression*.

In ridge regression, one adds a penalty term to the least-squares criterion,

$$\hat{\boldsymbol{\theta}}_n^{\text{RIDGE}} = \arg\min_{\boldsymbol{\theta}\in\Theta} ||\mathbf{Y} - H\boldsymbol{\theta}||^2 + \lambda||\boldsymbol{\theta}||^2 \tag{11.44}$$

where $\lambda > 0$ is a tunable parameter. The penalty term $\lambda||\boldsymbol{\theta}||^2$ forces the parameter vector to have small magnitude. In statistics, this process is called *shrinkage*. The solution to the minimization problem in (11.44) is given by (see Exercise 11.3):

$$\hat{\boldsymbol{\theta}}_n^{\text{RIDGE}} = (H^T H + \lambda I)^{-1} H^T \mathbf{Y} \tag{11.45}$$

Notice that this adds $\lambda > 0$ to each eigenvalue of $H^T H$, making this matrix better conditioned and the solution more stable. Of course, the case $\lambda = 0$ reduces to ordinary least-squares regression. The larger λ is, the closer to zero the regression coefficients are forced to be. However, they do not quite become zero. In Section 11.7.3, we discuss alternative penalized least squares techniques that can drive the regression coefficients to zero, producing a sparse solution.

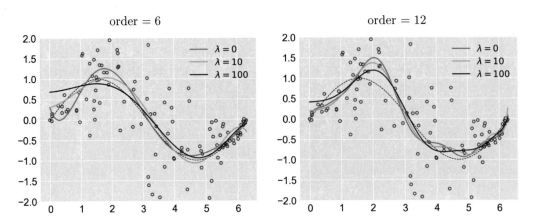

Figure 11.4: Ridge regression example. Polynomials of degree 6 and 12 are fitted to the data of Figure 11.1(b). Larger values of λ produce "flatter" curves, corresponding to the shrinkage of the polynomial coefficients to zero. The optimal regression is displayed as a dashed blue curve (plots generated by c11_poly_ridge.py).

Example 11.6. Figure 11.4 displays the ridge regression curves obtained by fitting polynomials of degree 6 and 12 to the data in Figure 11.1(b), for three different values of λ, including $\lambda = 0$ (no regularization), for comparison purposes. The optimal regression is also displayed for reference. We can observe that larger values of λ produce "flatter" curves, corresponding to the shrinkage of the polynomial coefficients to zero. $\diamond$

11.4   Nonparametric Regression

Let us recall the general model for regression, introduced in Theorem 11.1:

$$Y = f(\mathbf{X}) + \varepsilon. \tag{11.46}$$

As as done in Chapter 5 for classification, we can change the inference perspective by attempting to estimate the function f directly, rather than relying on a parametric model and plugging in parameter estimates. Alternatively, one may want to try the estimate the optimal MMSE regression $f^*(\mathbf{x}) = E[Y \mid \mathbf{X} = \mathbf{x}]$. In either case, the idea is to apply smoothing to the training data, as was the case in nonparametric classification. In this Section we consider two widely methods for nonparametric regression: one very classical, based on kernel smoothing, and the other a more modern approach using Bayesian inference and Gaussian processes.

11.4.1 Kernel Regression

As defined in Section 5.4, a kernel is a nonnegative function $K : R^d \to R$, and a radial-basis function (RBF) kernel is a monotonically decreasing function of $||\mathbf{x}||$. Several examples of kernels were given in that section.

Given data $S_n = \{(\mathbf{X}_1, Y_1), \ldots, (\mathbf{X}_n, Y_n)\}$, consider the following kernel estimates of the joint density $p(\mathbf{x}, y)$ and the marginal density $p(\mathbf{x})$:

$$
\begin{aligned}
p_n(\mathbf{x}, y) &= \frac{1}{n} \sum_{i=1}^{n} K_h(\mathbf{x} - \mathbf{X}_i) K_h(y - Y_i), \\
p_n(\mathbf{x}) &= \frac{1}{n} \sum_{i=1}^{n} K_h(\mathbf{x} - \mathbf{X}_i),
\end{aligned}
\tag{11.47}
$$

where h is the kernel bandwidth. (It may be illustrative to compare $p_n(\mathbf{x}, y)$ to the estimate $p_n^\diamond(\mathbf{x}, y)$ in (7.37), for the case $Y \in \{0, 1\}$.) By defining $p_n(y \mid \mathbf{x}) = p_n(\mathbf{x}, y)/p_n(\mathbf{x})$, one can define a nonparametric estimator of the optimal MMSE regression $f^*(\mathbf{x}) = E[Y \mid \mathbf{X} = \mathbf{x}]$:

$$
\begin{aligned}
f_n(\mathbf{x}) &= E_n[Y \mid \mathbf{X} = \mathbf{x}] = \int y p_n(y \mid \mathbf{x}) dy = \int y \frac{p_n(\mathbf{x}, y)}{p_n(\mathbf{x})} dy \\
&= \int y \frac{\sum_{i=1}^{n} K_h(\mathbf{x} - \mathbf{X}_i) K_h(y - Y_i)}{\sum_{i=1}^{n} K_h(\mathbf{x} - \mathbf{X}_i)} dy \\
&= \frac{\sum_{i=1}^{n} K_h(\mathbf{x} - \mathbf{X}_i) \int y K_h(y - Y_i) dy}{\sum_{i=1}^{n} K_h(\mathbf{x} - \mathbf{X}_i)} \\
&= \frac{\sum_{i=1}^{n} K_h(\mathbf{x} - \mathbf{X}_i) Y_i}{\sum_{i=1}^{n} K_h(\mathbf{x} - \mathbf{X}_i)}
\end{aligned}
\tag{11.48}
$$

where $\int y K_h(y - Y_i) dy = Y_i$ follows from the RBF assumption. This estimator is also known as the *Nadaraya-Watson kernel regression* estimator.

11.4.2 Gaussian Process Regression

The Gaussian process approach to nonparametric regression performs Bayesian inference directly on the space of functions f using a Gaussian stochastic process prior. Although derived in a completely different way, Gaussian process regression is a kernel approach, which is related to the Nadaraya-Watson kernel regression estimator of the previous section (see Exercise 11.8).

A *stochastic process* is an ensemble of real-valued random functions $\{f(\mathbf{x}, \xi); \mathbf{x} \in R^d, \xi \in S\}$, where S is a sample space in a probability space $(S, \mathcal{F}, P)$. For each fixed $\xi \in S$, $f(\mathbf{x}, \xi)$ is an ordinary function of $\mathbf{x}$, called a *sample function* of the process. Likewise, for each fixed $\mathbf{x} \in R^d$, $f(\mathbf{x}, \xi)$

is a random variable. A stochastic process is completely characterized by the distributions of the random vectors $\mathbf{f} = [f(\mathbf{x}_1), \ldots, f(\mathbf{x}_k)]$ for all finite sets of points $\mathbf{x}_1, \ldots, \mathbf{x}_k \in R^d$, $k \geq 1$. If all such random vectors have multivariate Gaussian distributions, then the stochastic process is a *Gaussian process*. Due to the nature of the multivariate Gaussian distribution, a Gaussian stochastic process depends only on the *mean function*

$$m(\mathbf{x}) = E[f(\mathbf{x})], \quad \mathbf{x} \in R^d, \tag{11.49}$$

and the *covariance function of Gaussian process* or *kernel*

$$k(\mathbf{x}, \mathbf{x}') = E[f(\mathbf{x})f(\mathbf{x}')] - m(\mathbf{x})m(\mathbf{x}'), \quad \mathbf{x}, \mathbf{x}' \in R^d. \tag{11.50}$$

We may thus denote a GP by $f(\mathbf{x}) \sim \mathcal{GP}(m(\mathbf{x}), k(\mathbf{x}, \mathbf{x}'))$.

A stochastic process is called *stationary* if the distribution of $\mathbf{f} = [f(\mathbf{x}_1), \ldots, f(\mathbf{x}_k)]$ is the same as that of $\mathbf{f_u} = [f(\mathbf{x}_1 + \mathbf{u}), \ldots, f(\mathbf{x}_k + \mathbf{u})]$, for all $\mathbf{u} \in R^d$ and finite sets of points $\mathbf{x}_1, \ldots, \mathbf{x}_k \in R^d$, $k \geq 1$. That is, the finite distributions of the process are all translation-invariant. The covariance function of a stationary process can only be a function of $\mathbf{x}' - \mathbf{x}'$ (see Exercise 11.6). By analogy, a a covariance function is called *stationary* if it is a function only of $\mathbf{x} - \mathbf{x}'$. Notice that having a stationary covariance function does not make a stochastic process stationary (it is only a necessary condition). The *variance function* $v(\mathbf{x}) = k(\mathbf{x}, \mathbf{x})$ is the variance of each random variable $f(\mathbf{x})$, for $\mathbf{x} \in R^d$. If the covariance function is stationary, then the variance function is constant: $v(\mathbf{x}) = \sigma_k^2 = k(\mathbf{x}, \mathbf{x})$, for any $\mathbf{x} \in R^d$. Finally, a stationary covariance function is *isotropic* if it is a function only of $||\mathbf{x} - \mathbf{x}'||$. By using the general fact that $k(\mathbf{x}, \mathbf{x}') = k(\mathbf{x}', \mathbf{x})$ we can see that, in the univariate case, a stationary covariance function is automatically isotropic.

Two important examples of isotropic stationary covariance functions are the squared exponential or "Gaussian"

$$k_{\text{SE}}(\mathbf{x}, \mathbf{x}') = \sigma_k^2 \exp\left(-\frac{||\mathbf{x} - \mathbf{x}||^2}{2\ell^2}\right), \tag{11.51}$$

and the absolute exponential

$$k_{\text{AE}}(\mathbf{x}, \mathbf{x}') = \sigma_k^2 \exp\left(-\frac{||\mathbf{x} - \mathbf{x}'||}{\ell}\right), \tag{11.52}$$

where in both cases, ℓ is the process *length-scale*. In the univariate case, the absolute exponential is the double-exponential covariance function $k(\tau) = \sigma_k^2 \exp(-|\tau|/\ell)$, with $\tau = x - x'$, hence the name "absolute exponential." The Gaussian and absolute exponential kernels can be seen as extremes in a family of isotropic stationary *Matérn* covariance functions:

$$k_{\text{MAT}}^{\nu}(\mathbf{x}, \mathbf{x}') = \sigma_k^2 \frac{2^{1-\nu}}{\Gamma(\nu)} \left(\frac{\sqrt{2\nu}\,||\mathbf{x} - \mathbf{x}'||}{\ell}\right)^{\nu} K_{\nu}\left(\frac{\sqrt{2\nu}\,||\mathbf{x} - \mathbf{x}'||}{\ell}\right), \tag{11.53}$$

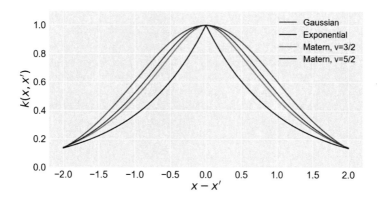

Figure 11.5: Univariate kernels used in Gaussian Process regression (plots generated by c11_GPkern.py).

where $\nu > 0$ is the order of the kernel, and K_ν is the incomplete Bessel function of the second kind. It is possible to show that $\nu = 1/2$ leads to the absolute exponential kernel, while $\nu \gg 1$ approximates very closely the Gaussian kernel (in fact, it converges to the Gaussian kernel as $\nu \to \infty$). Two other cases of interest in Gaussian process regression, for which (11.53) takes a simple form, are the case $\nu = 3/2$:

$$k_{\text{MAT}}^{\nu=3/2}(\mathbf{x}, \mathbf{x}') = \sigma_k^2 \left(1 + \frac{\sqrt{3}\,||\mathbf{x} - \mathbf{x}'||}{\ell} \right) \exp\left(-\frac{\sqrt{3}\,||\mathbf{x} - \mathbf{x}'||}{\ell} \right) \tag{11.54}$$

and $\nu = 5/2$:

$$k_{\text{MAT}}^{\nu=5/2}(\mathbf{x}, \mathbf{x}') = \sigma_k^2 \left(1 + \frac{\sqrt{5}\,||\mathbf{x} - \mathbf{x}'||}{\ell} + \frac{5\,||\mathbf{x} - \mathbf{x}'||^2}{3\ell^2} \right) \exp\left(-\frac{\sqrt{5}\,||\mathbf{x} - \mathbf{x}'||}{\ell} \right). \tag{11.55}$$

These covariance functions are plotted in Figure 11.5, in the univariate case.

Example 11.7. Sample functions of zero-mean, unit variance Gaussian processes for the squared exponential, Matérn ($\nu = 3/2$), and absolute exponential kernels, with $\sigma_k^2 = 1$ and different length-scales ℓ, are displayed in Figure 11.6. We can observe that the length scale ℓ controls the horizontal scale of the sample functions (the variance σ_k^2 controls the vertical scale). More importantly, we can see that the covariance function determines the smoothness of the sample functions, with the Gaussian kernel producing smooth functions and the absolute exponential kernel producing rough ones, while the Matérn kernel produces an intermediate result. ◇

The reason for the different smoothness properties produced by the different kernels in Figure 11.6 can actually be seen in the plot of the covariance functions in Figure 11.5. First, we observe in that plot that the Gaussian kernel produces the largest correlation at all distances $\mathbf{x} - x'$, while the

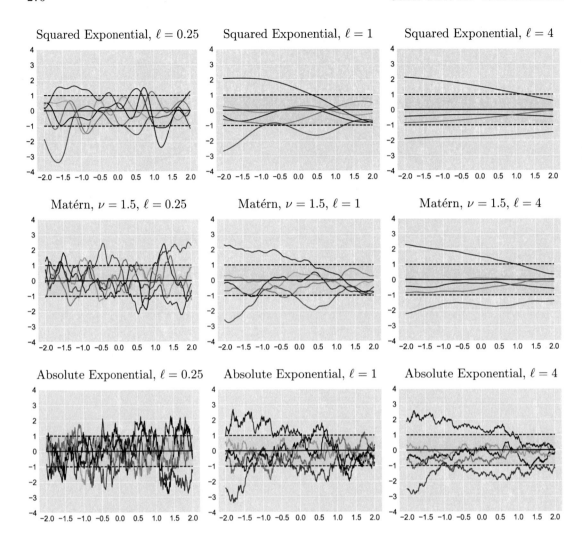

Figure 11.6: Sample functions from unit-variance, zero-mean Gaussian processes. The constant mean and variance functions are also displayed (plots generated by c11_GPsamp.py).

absolute exponential kernel produces the smallest, and the difference is more accentuated at short distances. If neighboring values $f(x)$ and $f(x')$ are more correlated, then the sample functions are more likely to be smooth. Second, though this is a sophisticated point that would get us too far afield to describe precisely (see the Bibliographical Notes), the average (in the mean-square sense) continuity and differentiability of a general stochastic process has to do with the continuity and differentiability of the covariance function $k(\mathbf{x}, \mathbf{x}')$ at the values $\mathbf{x} = \mathbf{x}'$. For a stationary kernel, this means continuity and differentiability at the origin $\mathbf{x} - \mathbf{x}' = 0$. The Gaussian kernel is infinitely differentiable at the origin, and as a consequence the associated process is infinitely differentiable everywhere in the mean-square sense, leading to very smooth sample functions. On the other hand, the absolute exponential kernel is not differentiable at all at the origin, so that the associated process is not mean-square differentiable anywhere, leading to very rough sample functions. It can be shown that a stochastic process with the Matérn covariance function of order ν is mean-square differentiable $\lceil \nu \rceil - 1$ times, producing an intermediate behavior; e.g. the Gaussian processes with $\nu = 3/2$ and $\nu = 5/2$ are once and twice differentiable, respectively, in the mean-square sense.

In practice, sample functions from a Gaussian process, such as the ones displayed in Figure 11.6 are simulated by picking a uniformly-spaced finite set of *testing points*

$$X^* = (\mathbf{x}_1^*, \ldots, \mathbf{x}_m^*), \tag{11.56}$$

drawing a sample $\mathbf{f}^* = (f(\mathbf{x}_1^*), \ldots, f(\mathbf{x}_m^*))$ from the multivariate Gaussian distribution

$$\mathbf{f}^* \sim \mathcal{N}(m(X^*), K(X^*, X^*)), \tag{11.57}$$

where

$$m(X^*) = (m(\mathbf{x}_1^*), \ldots, m(\mathbf{x}_m^*))^T, \tag{11.58}$$

and

$$K(X^*, X^*) = \begin{bmatrix} k(\mathbf{x}_1^*, \mathbf{x}_1^*) & k(\mathbf{x}_1^*, \mathbf{x}_2^*) & \cdots & k(\mathbf{x}_1^*, \mathbf{x}_m^*) \\ k(\mathbf{x}_2^*, \mathbf{x}_1^*) & k(\mathbf{x}_2^*, \mathbf{x}_2^*) & \cdots & k(\mathbf{x}_2^*, \mathbf{x}_m^*) \\ \vdots & \vdots & \ddots & \vdots \\ k(\mathbf{x}_m^*, \mathbf{x}_1^*) & k(\mathbf{x}_m^*, \mathbf{x}_2^*) & \cdots & k(\mathbf{x}_m^*, \mathbf{x}_m^*) \end{bmatrix}. \tag{11.59}$$

and then applying linear interpolation (connecting the dots with lines). The underlying point is that, although in Gaussian process regression we are talking about a stochastic process defined on Euclidean space R^d, in practice we deal only with a finite number of multivariate Gaussian random vectors.

The previous discussion about simulation segues naturally into the real objective of Gaussian process regression, which is to use training observations to predict the value of the unknown function f at a given arbitrary set of test points. Consider a set of training points $X = (\mathbf{x}_1, \ldots, \mathbf{x}_n)$ and corresponding set of responses $\mathbf{Y} = \mathbf{f} + \boldsymbol{\varepsilon}$, where $\mathbf{f} = (f(\mathbf{x}_1), \ldots, f(\mathbf{x}_n))$ and $\boldsymbol{\varepsilon} = (\varepsilon_1, \ldots, \varepsilon_n)$. We will consider here only the homoskedastic zero-mean Gaussian noise case, where $\boldsymbol{\varepsilon} \sim \mathcal{N}(0, \sigma^2 I)$ and $\mathcal{N}$ is independent of X. (One should be careful not to confuse the kernel variance σ_k^2 with the noise variance σ^2.) Given testing points $X^* = (\mathbf{x}_1^*, \ldots, \mathbf{x}_m^*)$, we would like to *predict* the value of $\mathbf{f}^* = (f(\mathbf{x}_1^*), \ldots, f(\mathbf{x}_m^*))$ in a way that is consistent with the training data. The Bayesian paradigm is to determine the *posterior distribution*, i.e., the conditional distribution of the vector $\mathbf{f}^*$ given $\mathbf{Y}$, and then make inferences on that — in this context, the *prior distribution* of $\mathbf{f}^*$ is described by (11.57)–(11.59). For Gaussian processes, this conditional distribution is Gaussian and can be written in closed form, as follows. If $\mathbf{X}$ and $\mathbf{X}'$ are jointly distributed Gaussian vectors, with multivariate distribution

$$\begin{bmatrix} \mathbf{X} \\ \mathbf{X}' \end{bmatrix} \sim \mathcal{N}\left(\begin{bmatrix} \boldsymbol{\mu}_{\mathbf{X}} \\ \boldsymbol{\mu}_{\mathbf{X}}' \end{bmatrix}, \begin{bmatrix} A & C \\ C^T & B \end{bmatrix} \right), \tag{11.60}$$

then $\mathbf{X} \mid \mathbf{X}'$ has a multivariate Gaussian distribution, given by

$$\mathbf{X} \mid \mathbf{X}' \sim \mathcal{N}\left(\boldsymbol{\mu}_{\mathbf{X}} + CB^{-1}(\mathbf{X}' - \boldsymbol{\mu}_{\mathbf{X}'}), A - CB^{-1}C^T \right). \tag{11.61}$$

Now, the joint distribution of the vectors $\mathbf{Y}$ and $\mathbf{f}(X^*)$ is multivariate Gaussian:

$$\begin{bmatrix} \mathbf{f}^* \\ \mathbf{Y} \end{bmatrix} \sim \mathcal{N}\left(\mathbf{0}, \begin{bmatrix} K(X^*, X^*) & K(X^*, X) \\ K(X^*, X)^T & K(X, X) + \sigma^2 I \end{bmatrix} \right) \tag{11.62}$$

where $K(X^*, X^*)$ is given by (11.59),

$$K(X^*, X) = \begin{bmatrix} k(\mathbf{x}_1^*, \mathbf{x}_1) & k(\mathbf{x}_1^*, \mathbf{x}_2) & \cdots & k(\mathbf{x}_1^*, \mathbf{x}_n) \\ k(\mathbf{x}_2^*, \mathbf{x}_1) & k(\mathbf{x}_2^*, \mathbf{x}_2) & \cdots & k(\mathbf{x}_2^*, \mathbf{x}_n) \\ \vdots & \vdots & \ddots & \vdots \\ k(\mathbf{x}_m^*, \mathbf{x}_1) & k(\mathbf{x}_m^*, \mathbf{x}_2) & \cdots & k(\mathbf{x}_m^*, \mathbf{x}_n) \end{bmatrix}, \tag{11.63}$$

and

$$K(X, X) = \begin{bmatrix} k(\mathbf{x}_1, \mathbf{x}_1) & k(\mathbf{x}_1, \mathbf{x}_2) & \cdots & k(\mathbf{x}_1, \mathbf{x}_n) \\ k(\mathbf{x}_2, \mathbf{x}_1) & k(\mathbf{x}_2, \mathbf{x}_2) & \cdots & k(\mathbf{x}_2, \mathbf{x}_n) \\ \vdots & \vdots & \ddots & \vdots \\ k(\mathbf{x}_n, \mathbf{x}_1) & k(\mathbf{x}_n, \mathbf{x}_2) & \cdots & k(\mathbf{x}_n, \mathbf{x}_n) \end{bmatrix}. \tag{11.64}$$

Applying (11.61), we gather that the posterior distribution is

$$\mathbf{f}^* \mid \mathbf{Y} \sim \mathcal{N}(\bar{\mathbf{f}}^*, \mathrm{Var}(\mathbf{f}^*)), \tag{11.65}$$

with posterior mean vector and posterior covariance matrix given by

$$\begin{aligned} \bar{\mathbf{f}}^* &= K(X^*, X)[K(X, X)^{-1} + \sigma^2 I_n]^{-1} \mathbf{Y} \\ \mathrm{Var}(\mathbf{f}^*) &= K(X^*, X^*) - K(X^*, X)[K(X, X) + \sigma^2 I]^{-1} K(X, X^*) \end{aligned} \tag{11.66}$$

Clearly, even if the prior Gaussian process is zero mean and has a stationary covariance function, this is no longer the case, in general, for the posterior Gaussian process.

In Gaussian process regression, we estimate the value of $\mathbf{f}^*$ at the test points by the conditional mean $\bar{\mathbf{f}}^*$, and the conditional regression error at each test point by the corresponding element in the diagonal of $\mathrm{Var}(\mathbf{f}^*)$. Notice that, in practice, σ^2 is rarely known, so the value used in (11.66) becomes a parameter σ_p^2 to be selected. This is all that is required for estimating the value of the unknown f at the given set of test points. However, if desired, an estimate of the entire function f

can be obtained by interpolating the conditional mean values (and the conditional variance values) over a dense set of test points. In the univariate case, this can be done by joining the estimated values with lines, as was done when simulating sample functions from the prior in Figure 11.6. In a multivariate setting, more advanced interpolation methods are required.

Example 11.8. We apply Gaussian process regression to the first 10 points of the data in Figure 11.1(b). Figure 11.7 displays the results of fitting a GP using the squared exponential, Matérn ($\nu = 3/2$), and absolute exponential kernels, under different length scales (same values as in Example 11.7). In all cases, $\sigma_k^2 = 0.3$ and $\sigma_p^2 = 0.1$. The training data is represented by red circles, while the estimated regression is represented by a black solid curve, which is obtained by linear interpolation of the posterior mean $\bar{f}^*$ over a dense test point mesh. The optimal regression is displayed for reference. A one standard-deviation confidence band is displayed, the boundaries of which are obtained by linear interpolation of $\bar{\mathbf{f}}^* \pm \sqrt{\mathrm{Var}(\mathbf{f}^*)}$ over the test point mesh. We observe that $\ell = 0.25$ undersmooths the data, while $\ell = 4$ oversmooths it, with $\ell = 1$ producing the best results. As expected, the confidence bands tend to be tighter near the training points and wider over intervals without data, particularly if there is not enough smoothing. The best result is achieved by RBF kernel with the intermediate length scale, $\ell = 1$, in which case the GP regression line follows the optimal sinusoid even in areas without data. The regression produced by the other kernels are too jagged and not appropriate for this problem, where the optimal regression is smooth. It is illustrative to compare these results, obtained using only 10 points, to those obtained in Figures 11.2 and 11.4. ◇

In the previous example, the parameters σ_k^2, ℓ, and σ_p^2 were set in ad-hoc fashion. The problem of how to select these parameters in a principled way is a model selection question (see Section 11.8). Next, we describe a popular model selection method for GP regression, which seeks the parameter values that maximize the so-called *marginal likelihood*. The latter is Bayesian terminology for the conditional distribution $p(\mathbf{Y} \mid X, \boldsymbol{\theta})$, where $\boldsymbol{\theta} = (\sigma_p^2, \sigma_k^2, \ell)$ is the vector of *hyperparameters*. In general, obtaining the marginal likelihood requires a complex computation, but in the case of the Gaussian process model, it can be obtained immediately from (11.62): $p(\mathbf{Y} \mid X, \boldsymbol{\theta}) = \mathcal{N}(0, K(X, X) + \sigma^2 I)$. The log-likelihood is thus given by:

$$\ln p(\mathbf{Y} \mid X, \boldsymbol{\theta}) = -\frac{1}{2}\mathbf{Y}^T(K(X, X) + \sigma^2 I)^{-1}\mathbf{y} - \frac{1}{2}\ln|K(X, X) + \sigma^2 I| - \frac{n}{2}\ln 2\pi. \quad (11.67)$$

The various terms in the right-hand side of (11.67) have model selection interpretations: the first one represents the empirical fit to the data, the second one is a complexity penalty term, while the last term is a normalization constant. Numerical maximization of (11.67) by gradient descent produces the required value of the hyperparameters (with the usual caveat about the need for multiple random restarts to deal with local maxima).

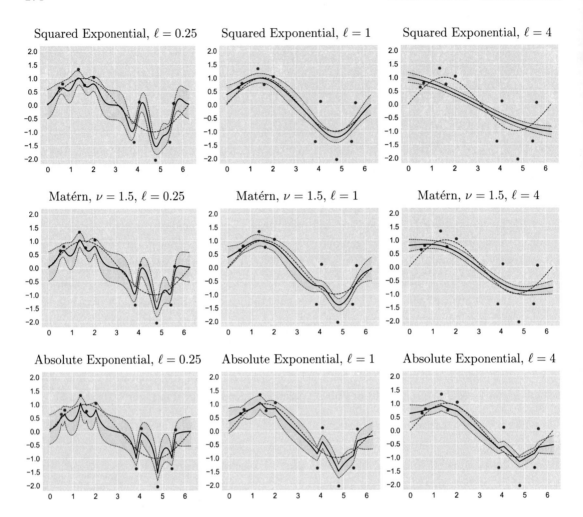

Figure 11.7: Gaussian process regression example. In all cases, $\sigma_k^2 = 0.3$ and $\sigma_p^2 = 0.1$. The training data is represented by red circles, the estimated regression is represented by a black solid curve, and a one standard-deviation confidence band is displayed as well. The optimal regression is displayed as a dashed blue curve. Smoothness of the regression line decreases from the Gaussian kernel at the top to the exponential kernel at the bottom. The confidence bands are tighter around the training points and wider away from them. The best result is achieved by the RBF kernel with the intermediate length scale, $\ell = 1$, when the GP regression line follows the optimal sinusoid even in areas where the training data set is very sparse (plots generated by c11_GPfit.py).

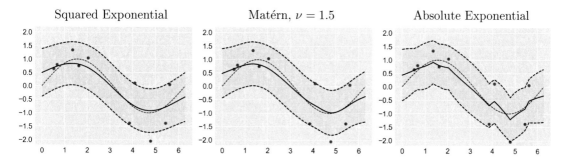

Figure 11.8: Gaussian process regression results for the same data and kernels in Figure 11.7, but using hyperparameter values selected by maximization of the marginal likelihood (plots generated by `c11_GPfitML.py`).

Example 11.9. Maximization of the marginal likelihood for the same data and kernels used in Example 11.8 produces the following values: $\sigma_k^2 = 0.63, \ell = 1.39, \sigma_p^2 = 0.53$ for the squared exponential kernel, $\sigma_k^2 = 0.65, \ell = 1.51, \sigma_p^2 = 0.53$ for the Matérm kernel, and $\sigma_k^2 = 0.74, \ell = 1.45, \sigma_p^2 = 0.45$ for the absolute exponential kernel. The corresponding regression results are displayed in Figure 11.8. We observe that the estimated length scale values are all very close to the intermediate length scale $\ell = 1$, in Example 11.8, and therefore the plots in Figure 11.8 resemble those in the middle column of Figure 11.7. The results achieved by the squared exponential and Matérn kernel are very similar, but not identical. ◇

11.5 Function-Approximation Regression

Most of the classification rules discussed in Chapter 6 can be easily adapted to regression. For example, by simply removing the output threshold nonlinearity, one gets a *neural network regressor*. The error criterion can be the square difference between network output and the training point response, and training proceeds as before.

In the case of CART, the modification is also very simple. Node splitting proceeds as before, but this time each node is assigned the average response of the training points inside it. The best splits selected for each node are those that minimize the RSS between the average response and the training responses in the children nodes. As in the case of classification, overfitting is quite common, which may be avoided by the use of regularization techniques, such as stopped splitting or pruning. Random forest regression can be obtained by perturbing the data and averaging the results, as in the case of classification. Perturbation may be applied to both $\mathbf{X}$ and Y.

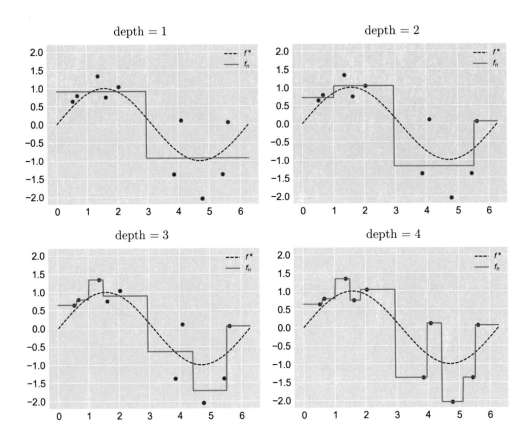

Figure 11.9: CART Regression example. Underfitting occurs at small depths, while overfitting is visible at large depths. The optimal regression is displayed as a dashed blue curve (plots generated by c11_CART.py).

Example 11.10. We apply CART regression to to the first 10 points of the data in Figure 11.1(b). A simple regularization method is applied, which consists of limiting the maximum depth of the tree. Figure 11.9 displays the results, for maximum depths from 1 to 4. The optimal regression is displayed for reference. We can see clear underfitting at small depths and overfitting at large depths. ◇

The extension of SVMs to regression is less straightforward. We briefly describe the idea next. The same idea of margin reappears, but now the points are supposed to be *within* a margin around the regression line. Slack variables are associated to outliers that break the margin criterion. As in the case of classification, only support vectors (margin and outlier vectors) determine regression curve. Nonlinear regression can be achieved by using the kernel trick, as before.

11.6 Error Estimation

A theory of regression error estimation, similar to the one for classification in Chapter 7, can be developed. Here we only outline the main aspects of regression error estimation.

Given a test sample $S_m = \{(\mathbf{X}_i^t, Y_i^t); i = 1, \ldots, m\}$, which is not used in training the regression function f_n, we can define a test-set MSE estimator by

$$\hat{L}_{n,m} = \frac{1}{m} \sum_{i=1}^{m} (Y_i^t - f_n(\mathbf{X}_i^t))^2. \tag{11.68}$$

Clearly, this test-set error estimator is unbiased in the sense that

$$E[\hat{L}_{n,m} \mid S_n] = L_n. \tag{11.69}$$

from which it follows that $E[\hat{L}_{n,m}] = E[L_n]$. From (11.69) and the Law of Large Numbers (see Theorem A.12) we also have that, given the training data S_n, $\hat{L}_{n,m} \to L_n$ with probability 1 as $m \to \infty$, regardless of the feature-label distribution.

As before, satisfactory performance of the test-set estimator depends on having both n and m sufficiently large. If that is not the case, then MSE estimation must proceed on the training data S_n only. The resubstitution estimator is

$$\hat{L}_n = \frac{1}{n} \sum_{i=1}^{n} (Y_i - f_n(\mathbf{X}_i)^2. \tag{11.70}$$

This is simply the (normalized) RSS of the previous sections. As in the case of classification, the RSS tends to be optimistically biased, and more so for more "flexible" algorithms. Another issue is that, unlike in the case of classification, the RSS is not bounded in the interval $[0, 1]$. A popular alternative is the R^2 *statistic*, which is always between 0 and 1. First, define the following estimator of Y without using $\mathbf{X}$:

$$\bar{Y} = \frac{1}{n} \sum_{i=1}^{n} Y_i. \tag{11.71}$$

The RSS of this estimator is

$$\text{TSS} = \sum_{i=1}^{n} (Y_i - \bar{Y})^2, \tag{11.72}$$

where TSS stands for *total sum of squares*. The R^2 statistic is the relative improvement in RSS by using $\mathbf{X}$ to predict Y over no using it:

$$R^2 = \frac{\text{TSS} - \text{RSS}}{\text{TSS}} = 1 - \frac{\text{RSS}}{\text{TSS}}. \tag{11.73}$$

This is also called, in some contexts, the *coefficient of determination*.

Resampling estimators of the MSE, such as cross-validation and bootstrap, can also be easily defined. As in the case of classification, subsamples of the training data are used to fit the regression, and the remaining data are used to form a test error. The process is repeated several times and the results are averaged.

11.7   Variable Selection

Just as in the case of classification, dimensionality reduction (see Chapter 9) is needed for improving prediction accuracy and reducing computational costs. Here we focus on variable (feature) selection as the most common form of dimensionality reduction in regression There are three main ways to perform variable selection in regression: wrapper search, statistical testing of coefficients, and sparsification methods.

11.7.1   Wrapper Search

Wrapper search algorithms in regression are entirely analogous to their classification counterparts. First some empirical criterion is defined, e.g., the resubstitution estimate (RSS) or the R^2 score, both defined in the previous section (In fact, minimizing the RSS is equivalent to maximizing R^2, as can be easily checked). The best feature of a given a size, according to the selected criterion, can be searched for exhaustively or greedily. In the latter case, sequential forward/backward and floating searches can be employed, just as in the case of classification.

However, if one does not know the optimal number of features, which is usually the case, and tries to perform wrapper selection to find it, one will usually end up with an overfit model with all variables. This is because the R^2 score will typically only increase if more variables are added. The most common way to avoid this issue is to employ the *adjusted R^2 statistic*:

$$\text{Adjusted } R^2 = 1 - \frac{\text{RSS}/(n - d - 1)}{\text{TSS}/(n - 1)} \tag{11.74}$$

This will penalize models with too many variables. Notice that this is not unlike structural risk minimization and other complexity penalization methods mentioned in Chapter 8. Another penalized criterion for regression is *Mallows' C_p*. In the case of a least-squares regression, this is given by

$$C_p = \frac{\text{RSS}}{n} + \frac{2d}{n}\hat{\sigma}^2 . \tag{11.75}$$

where $\hat{\sigma}^2$ is the MLE variance estimator in (11.41).

11.7.2 Statistical Testing

Statistical testing of regression coefficients is the standard approach in classical statistics. For example, in the generalized linear model of (11.20), one can test the hypothesis that each coefficient θ_i is nonzero, for $i = 1, \ldots, k$, and discard those for which the null hypothesis cannot be rejected. Stepwise search algorithms, including backward elimination, are common. This approach has some issues in complex models with many parameters, due to the issue of *multiple testing*. Statistical testing provides a counterpart to filter selection in classification.

11.7.3 LASSO and Elastic Net

Sparsification methods in parametric regression seek to shrink the model coefficients to zero, generating sparse feature vectors that contain a small number of nonzero elements. Recall that ridge regression for parametric models, discussed in Section 11.3.3, performs shrinkage, but the coefficients will usually not decrease all the way to zero. If one replaces the L^2 norm in (11.44) by the L^1 norm, on obtains the LASSO (Least Absolute Shrinkage and Selection Operator) estimator:

$$\hat{\boldsymbol{\theta}}_n^{\text{LASSO}} = \arg\min_{\boldsymbol{\theta} \in \Theta} \, ||\mathbf{Y} - H\boldsymbol{\theta}||^2 + \lambda ||\boldsymbol{\theta}||_1 \tag{11.76}$$

where $||\hat{\boldsymbol{\theta}}||_1 = |\hat{\theta}_1| + \cdots + |\hat{\theta}_k|$ is the L^1 norm of $\hat{\boldsymbol{\theta}}$. LASSO, unlike ridge regression, can drive coefficient values to zero.

Elastic Net is another kind of penalized least-squares estimator, which combines both L^1 and L^2 penalty terms

$$\hat{\boldsymbol{\theta}}_n^{\text{ENet}} = \arg\min_{\boldsymbol{\theta} \in \Theta} \, ||\mathbf{Y} - H\boldsymbol{\theta}||^2 + \lambda_1 ||\hat{\boldsymbol{\theta}}||_1 + \lambda_2 ||\hat{\boldsymbol{\theta}}||^2 \,, \tag{11.77}$$

where $\lambda_1, \lambda_2 > 0$, and thus produces a result that is intermediate between ridge regression and LASSO.

11.8 Model Selection

Just as in classification, one is often faced with selecting the values of free parameters in regression algorithms. Examples include the order of polynomial regression, the parameter λ in ridge regression, the bandwidth in kernel regression, and the kernel hyperparameters in Gaussian process regression, as well as the dimensionality (number of variables) in the model.

The simplest approach is to perform a grid search minimizing the residual sum of the squares of the regression, i.e., the empirical error on the training data. As in the case of minimizing the empirical error in classification, this approach will work if the ratio of complexity to sample size is small, where complexity includes the number of parameters, the flexibility of the regression model, and the dimensionality of the problem. Otherwise, overfitting and poor performance on future data are likely to occur.

If the ratio complexity to sample size is not favorable, some sort of complexity penalization must be considered. This topic was already discussed in Section 11.7.1, where minimization of the adjusted R^2 score and Mallow's Cp were proposed for choosing the number of variables for dimensionality reduction (which is also a model selection problem). Additional complexity penalization methods include minimizing the *Akaike information criterion* (AIC):

$$\text{AIC} = \frac{1}{n\hat{\sigma}^2}(\text{RSS} + 2d\hat{\sigma}^2) \tag{11.78}$$

and the *Bayesian information criterion* (BIC):

$$\text{BIC} = \frac{1}{n\hat{\sigma}^2}(\text{RSS} + \ln(n)d\hat{\sigma}^2) \tag{11.79}$$

where $\hat{\sigma}^2$ is the MLE variance estimator in (11.41) and d is the number of parameters in the linear model (11.20), which coincides with the number of variables in the multivariate linear regression model (11.19). Notice that in the cases considered here, AIC and Mallow's Cp are linearly related.

There is a theory of Structural Risk Minimization for regression. Consider a nested sequence $\{\mathcal{S}_k\}$ of spaces of real-valued functions associated with different regression algorithms, for $k = 1, \ldots, N$. Furthermore, assume that te VC dimensions $V_{\mathcal{S}_k}$ are finite for $k = 1, \ldots, N$. SRM proceeds by selecting the class (algorithm) $\mathcal{S}_{k^*}$ with the smallest value of the regression error bound

$$\text{MSE}_{\mathcal{S}_k} \leq \frac{\text{RSS}_{\mathcal{S}_k}}{n}\left(1 - \sqrt{\frac{V_{\mathcal{S}_k}}{n}\left(1 - \ln\frac{V_{\mathcal{S}_k}}{n}\right) + \frac{\ln n}{2n}}\right)^{-1} \tag{11.80}$$

where the bound is set to zero if negative. (Contrast this with the SRM procedure for classification in Section 8.3.3). Notice that the ratio between VC dimension and sample size features prominently, as was the case in classification. A small ratio helps make the bound in (11.80) smaller and is thus favored. The problem of estimating the VC dimension of a set of real-valued functions can be challenging, just as it is in the case of a set of indicator functions in classification. However, for the general linear model in (11.20), the VC dimension is finite and equal to $M + 1$.

Finally, as in classification, looking for the first local minimum RSS on a validation set and minimizing the cross-validated RSS are also used in model selection for regression.

11.9 Bibliographical Notes

The invention of both regression and the least-squares method (and some might say, statistical inference) is credited to Gauss — although Legendre published it first — who developed it in order to predict the position of planetary bodies with great accuracy on a nightly basis, using only incomplete and noisy observations [Stigler, 1981]. Before Gauss, Kepler had developed his Laws of Planetary Motion using an empirical ad-hoc approach, while Newton was able to establish all of Kepler's laws of planetary motions mechanistically, using only his Law of Universal Gravitation. Gauss appears to have been the first to notice, at the beginning of the 19th century, the limitation in these models: they were inaccurate, due to the influence of unobserved variables, and did not deal in a principled way with the irreducible uncertainty introduced by noise in the measurements.

Much of the literature on regression, especially in statistics, focuses on the multivariate linear regression model of (11.19). The classical statistics view of regression is summarized in Chapters 11 and 12 of Casella and Berger [2002]. The derivations of (11.41) and (11.42) for regression with a line can be found in Section 11.3 of the latter reference.

For a detailed discussion of the bias-variance decomposition in regression, see James et al. [2013]. Figure 11.2 is similar to Figure 1.4 in Bishop [2006].

Our proof of the Gauss-Markov Theorem follows that in Section 6.6 of Stark and Woods [1986].

The RSS is the empirical quadratic-loss error on the training data, and hence least-squares estimation is an *empirical risk minimization* (ERM) approach, similar to the ERM approach to classification, discussed in Exercise 8.5.

Excellent references on stochastic processes from an Engineering perspective are Jazwinski [2007] and Stark and Woods [1986]. In these references, mean-square continuity and differentiability of stochastic processes, and their relationship to the covariance function, are described in very readable fashion. It important to note that stochastic continuity and differentiability, while providing a general indication of smoothness, cannot guarantee that the sample functions are continuous and differentiable in all cases. A classic counter-example is afforded by discrete-valued stochastic processes with exponentially-distributed transition times between the values — e.g., any continuous-time Markov Chain, such as the Poisson process [Ross, 1995]. Such processes are continuous with probability one and in the mean-square sense at each point, but none of their sample functions is continuous at all points. What stochastic continuity and differentiability guarantee, roughly, is that are no "fixed" points at which the sample functions are discontinuous or nondifferentiable with positive probability. Thus the discontinuities of a Poisson process are "spread out" finely.

The standard reference on Gaussian Process regression is Rasmussen and Williams [2006]. Our discussion of that topic is based mostly on that reference, being also informed by Section 6.4 of Bishop. Despite the recent surge of interest, GP regression is an old subject. For instance. in the geostatistics literature, Gaussian process regression is known as Kriging [Cressie, 1991; Stein, 2012]. In the latter reference, Stein argues that the extreme smoothness of the sample functions produced by the squared-exponential kernel is unrealistic in natural processes, which lends support to the use of the Matérn kernel of appropriate order (usually $\nu = 3/2$ or $\nu = 5/2$, as larger values tend to produce results very similar to the squared-exponential kernel).

For a detailed account of the SVM algorithm for regression, see Section 7.2.1 of Bishop [2006].

The LASSO and the Elastic Net were introduced in Tibshirani [1996] and Zou and Hastie [2005], respectively.

Mallow's Cp method appeared in Mallows [1973]. There is some variance in the literature about the definitions of Mallow's CP, AIC, and BIC; the definitions adopted here are the ones in James et al. [2013].

The VC theory for regression, including the VC dimension of functions and complexity penalization, is covered in detail in Vapnik [1998]. For a readable review of the model selection problem in regression, including the VC theory, see Cherkassky and Ma [2003]. The theory of structural risk minimization (SRM) for regression, including the derivation of (11.80), is found in Cherkassky et al. [1999].

11.10   Exercises

11.1. Suppose that the optimal regression is sought in a subset of functions G, which does not necessarily contain the optimal quadratic-loss regression $f^*(\mathbf{x}) = E[Y \mid \mathbf{X} = \mathbf{x}]$:

$$f_G^* = \arg\min_{f \in G} L[f] = \arg\min_{f \in G} \int (y - f(\mathbf{x}))^2 p(\mathbf{x}, y) \, dx dy. \qquad (11.81)$$

Show that f_G^* minimizes the reducible error in (11.14), i.e., the L^2 distance to f^* among all functions $f \in G$, that is,

$$f_G^* = \arg\min_{f \in G} E[(f^*(\mathbf{X}) - f(\mathbf{X}))^2]. \qquad (11.82)$$

Hint: follow steps similar to those in the proof of Theorem 11.2.

11.2. Let (X, Y) be jointly Gaussian with $E[X] = \mu_X$, $\mathrm{Var}(X) = \sigma_X^2$, $E[Y] = \mu_Y$, $\mathrm{Var}(Y) = \sigma_Y^2$. Also let ρ be the correlation coefficient between X and Y.

(a) Show that the optimal MMSE regression is a line $f^*(x) = \theta_0 + \theta_1 x$, with parameters

$$\theta_0 = \mu_Y - \theta_1 \mu_X \quad \text{and} \quad \theta_1 = \rho \frac{\sigma_Y}{\sigma_X}. \tag{11.83}$$

(b) Show that the conditional and unconditional optimal regression errors are

$$L^*(x) = L^* = \sigma_Y^2 (1 - \rho^2) \tag{11.84}$$

regardless of the value of x.

(c) Show that least-squares regression is *consistent* in this case, in the sense that $\hat{\theta}_{0,n}^{\text{LS}} \to \theta_0$ and $\hat{\theta}_{1,n}^{\text{LS}} \to \theta_1$ in probability as $n \to \infty$.

(d) Plot the optimal regression line for the case $\sigma_x = \sigma_y$, $\mu_x = 0$, fixed μ_y and a few values of ρ. What do you observe as the correlation ρ changes? What happens for the case $\rho = 0$?

11.3. For the general linear model (11.23), with invertible $H^T H$:

(a) Show that the least-squares estimator for the general linear model is unique and given by (11.26).

Hint: Differentiate the quadratic function (11.25) and set the derivative to zero. Use the vectorial differentiation formulas:

$$\frac{\partial}{\partial \mathbf{u}} \mathbf{a}^T \mathbf{u} = \mathbf{a} \quad \text{and} \quad \frac{\partial}{\partial \mathbf{u}} \mathbf{u}^Y A \mathbf{u} = 2A\mathbf{u}. \tag{11.85}$$

(b) Show that the ridge estimator for the general linear model is unique and given by (11.45)

Hint: Modify the derivation in the previous item.

11.4. (Regression with a line through the origin.) Given data $S_n = \{(X_1, Y_1), \ldots, (X_n, Y_n)\}$, derive the least-squares estimator of the slope in the univariate model

$$Y = \theta X + \varepsilon. \tag{11.86}$$

Hint: Follow similar steps as in the derivation of (11.29).

11.5. (Gauss-Markov Theorem for correlated noise.) Consider the linear model $\mathbf{Y} = H\boldsymbol{\theta} + \varepsilon$, where H is full rank. Extend Theorem 11.3 by showing that if $E[\varepsilon] = 0$ and $E[\varepsilon\varepsilon^T] = K$, where K is a symmetric positive-definite matrix, then the estimator

$$\hat{\boldsymbol{\theta}} = (H^T K^{-1} H)^{-1} H^T K^{-1} \mathbf{Y} \tag{11.87}$$

is best linear unbiased.

11.6. Show that if a stochastic process is stationary, then its mean function $m(\mathbf{x})$ is constant and its covariance function $k(\mathbf{x}, \mathbf{x}')$ is only a function of $\mathbf{x} - \mathbf{x}'$. A process with only these properties is called *wide-sense stationary*.

11.7. In Gaussian process regression, if the responses can be assumed to be noiseless, i.e., $\mathbf{Y} = \mathbf{f} = (f(\mathbf{x}_1), \ldots, f(\mathbf{x}_n))$, the inference is based on the conditional distribution $\mathbf{f}^* \mid \mathbf{f}$. Derive the posterior mean and variance functions for this case. What happens if the test data is equal to the training data, i.e., $X^* = X$?

11.8. Verify that the Nadaraya-Watson kernel regression estimator in (11.48) can be written as a finite linear combination of kernels:

$$f_n(\mathbf{x}) = \sum_{i=1}^{n} a_{n,i} K_h(\mathbf{x} - \mathbf{X}_i), \tag{11.88}$$

where the coefficients $a_{n,i}$ are a function of the training data $S_n = \{(\mathbf{X}_1, Y_1), \ldots, (\mathbf{X}_n, Y_n)\}$. Now show that the GP regression estimator in (11.66) evaluated at a single test point $\mathbf{x}$ can be written similarly:

$$\bar{f}^*(\mathbf{x}) = \sum_{i=1}^{n} b_{n,i} k(\mathbf{x}, \mathbf{X}_i). \tag{11.89}$$

Obtain an expression for $b_{n,i}$ in terms of $k(\cdot, \cdot)$, σ_p^2, and the training data S_n.

11.9. Show that Akaike's Information Criterion for regression can be written as:

$$\text{AIC} = \frac{1 + d/n}{1 - d/n} \text{RSS}. \tag{11.90}$$

In this form, the AIC is also known as the final prediction error (FPE) [Akaike, 1970]

11.11   Python Assignments

11.10. Regarding the regression functions in Figures 11.2, 11.4, 11.7, 11.8, and 11.9.

(a) Rank the regression functions according normalized RSS. What do you observe?

(b) Use numerical integration to rank the regression functions according to reducible error $\bar{L}[f] = E[(f^*(\mathbf{X}) - f(\mathbf{X}))^2]$. What do you observe?

(c) The reducible error for the polynomial regressions in Figures 11.2 and 11.4 can be computed analytically in terms of the regression coefficients. Compare the numerical and analytical results for these regression functions.

Hint: run the python code for each figure to obtain the regression functions.

11.11. Apply linear regression to the stacking fault energy (SFE) data set.

(a) Modify `c11_SFE.py` to fit a univariate linear regression model (with intercept) separately to each of the seven variables remaining after preprocessing (two of these were already done in Example 11.4. List the fitted coefficients, the normalized RSS, and the R^2 statistic for each model. Which one of the seven variables is the best predictor of SFE, according to R^2? Plot the SFE response against each of the seven variables, with regression lines superimposed. How do you interpret these results?

(b) Perform multivariate linear regression with a wrapper search (for 1 to 5 variables) using the R^2 statistic as the search criterion. List the normalized RSS, the R^2 statistic, and the *adjusted* R^2 statistic for each model. Which would be the most predictive model according to adjusted R^2? How do you compare these results with those of item (a)?

11.12. Apply penalized least-squares multivariate linear regression to the SFE data set.

(a) Apply ridge regression to the entire data matrix, with regularization parameter $\lambda = 50, 30, 15, 7, 3, 1, 0.30, 0.10, 0.03, 0.01$. Do not apply any normalization or scaling to the data. List the regression coefficients for each value of λ.

(b) The "coefficient path" plot displays the values of each coefficient in a penalized least-squares multivariate linear regression as a function of the regularization parameter. Obtain the coefficient path plots for ridge regression and for the LASSO. Verify that the LASSO produces sparse solutions while ridge regression does not. Which atomic features produce the last two nonzero coefficients in the LASSO coefficient path? Does this agree with the results in Assignment 11.11?

(c) With $\lambda = 50$, the LASSO should produce an empty model (all coefficients equal to zero, and the intercept equal to the mean SFE), while with $\lambda = 30$, the lasso should produce a model with only one predictor. Plot the SFE response against this predictor with LASSO-regressed line and the corresponding ordinary regressed line superimposed. How do you interpret these results?

11.13. Apply Gaussian process regression to the SFE data set.

(a) Plot the posterior mean function with a one standard deviation confidence band, as in Figures 11.7 and 11.8, for the squared exponential, Matérn ($\nu = 1.5$), and absolute exponential kernels, with hyperparameters obtained by maximization of the conditional likelihood.

(b) Obtain the RSS and the best value of the conditional likelihood in each case, and compare the results.

11.14. Ordinary least-squares regression minimizes the sum of the squared *vertical* distances of each point (x_i, y_i) to their *vertical* projection (x_i', y_i') on the regression line. This assignment concerns the regression line that minimizes the sum of the squared distances of each point to their *horizontal* and *orthogonal* projections (x_i'', y_i'') and $(\bar{x}_i, \bar{y}_i)$, as in the figure below.

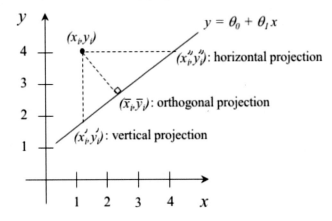

(a) Using (11.29) only (no new derivation is necessary), determine the θ_0 and θ_1 parameters that minimize the sum of squared distances to the horizontal projections.
Hint: Reason about the roles of X and Y.

(b) Apply this "horizontal least-squares" regression to the SFE data in Problem 11.10. Plot the results of regressing SFE on each of the seven filtered atomic features and record the residual sums of squares in each case. What do you observe?

(c) Determine the θ_0 and θ_1 parameters that minimize the sum of squared distances to the orthogonal projections.
Hint: Determine each orthogonal distance as a function of x_i, y_i, θ_0 and θ_1 using basic geometry. Find the value of θ_0 by comparison to the ordinary regression case, then find θ_1 using univariate optimization.

(d) Apply this "orthogonal least-squares" regression to the data in Figure 11.3. Plot the results of regressing SFE on each of the seven filtered atomic features and record the residual sums of squares in each case. What do you observe?

(e) It can be shown that the orthogonal regression line always lies between the vertical and horizontal ones. Confirm this by plotting all three regression lines in the same plot for each of the seven atomic features of the filtered stacking fault energy data set.

Appendix

A1 Probability Theory

The modern formulation of probability theory is due to Kolmogorov [1933]. In that 60-page monograph, Kolmogorov introduced the notion of probability spaces, the axiomatic definition of probability, the modern definition of random variables, and more. For an excellent review of Kolmogorov's fundamental contribution, see Nualart [2004]. In this Appendix, we review concepts of probability theory at the graduate level, including many concepts that are needed in the book. The language of measure theory is used, although measure-theoretical concepts are only needed in the book in the starred additional topics sections. For excellent book-length treatments of probability theory, the reader is referred to Billingsley [1995], Chung [1974], Loève [1977], Cramér [1999], and Rosenthal [2006], while a thorough elementary non-measure-theoretical introduction is provided by Ross [1994].

A1.1 Sample Space and Events

A *sample space* S is the set of all outcomes of an experiment. A σ-*algebra* is a collection $\mathcal{F}$ of subsets of S that is closed under complementation, (countable) intersection, and (countable) union. Each set E in $\mathcal{F}$ is called an *event*. Hence, complementation of events are events, and (countable) unions and intersections of events are events.

Event E is said to *occur* if it contains the outcome of the experiment. Whenever $E \subseteq F$ for two events E and F, the occurrence of E implies the occurrence of F. The complement event E^c is an event, which occurs iff(if and only if) E does not occur. The union $E \cup F$ is an event, which occurs iff E, F, or both E and F occur. On the other hand, the intersection $E \cap F$ is also an event, which occurs iffboth E and F occur. Finally, if $E \cap F = \emptyset$ (the latter is called the *impossible event*), then E or F may occur but not both.

© Springer Nature Switzerland AG 2020
U. Braga-Neto, *Fundamentals of Pattern Recognition and Machine Learning*,
https://doi.org/10.1007/978-3-030-27656-0

For example, if the experiment consists of flipping two coins, then

$$S = \{(H, H), (H, T), (T, H), (T, T)\}. \tag{A.1}$$

In this case, the σ-algebra contains all subsets of S (any subset of S is an event); e.g., event E that the first coin lands tails is: $E = \{(T, H), (T, T)\}$. Its complement E^c is the event that the first coin lands heads: $E^c = \{(H, H), (H, T)\}$. The union of these two events is the entire sample space S: one or the other must occur. The intersection is the impossible event: the coin cannot land both heads and tails on the first flip.

If on the other hand, the experiment consists in measuring the lifetime of a lightbulb, then

$$S = \{t \in R \mid t \geq 0\}. \tag{A.2}$$

Here, for reasons that will be described later, it is not desirable to consider all possible subsets in S as events. Instead, we consider the smallest σ-algebra that contains all intervals in S; this is called the *Borel σ-algebra* in S, and the events in it are called *Borel sets*; e.g., the event that the lightbulb will fail at or earlier than t time units is the Borel set $E = [0, t]$. The entire sample space is the countable union $\bigcup_{t=1}^{\infty} E_t$, where $\{E_t; t = 1, 2, \ldots\}$ is called an increasing *sequence of events*. Borel sets can be quite complicated (e.g., the famous Cantor set is a Borel set). There are sets of real numbers that are not Borel sets, but these are quite exotic and of no real interest. Generalizing, the Borel σ-algebra $\mathcal{B}^d$ of R^d is the smallest σ-algebra of subsets of R^d that contains all rectangular volumes in R^d. If $d = 1$, we write $\mathcal{B}^1 = \mathcal{B}$.

Limiting events are defined as follows. Given any sequence $\{E_n; n = 1, 2, \ldots\}$ of events, the *lim sup* is defined as:

$$\limsup_{n \to \infty} E_n = \bigcap_{n=1}^{\infty} \bigcup_{i=n}^{\infty} E_i. \tag{A.3}$$

We can see that $\limsup_{n \to \infty} E_n$ occurs iff E_n occurs for an infinite number of n, that is, E_n *occurs infinitely often*. This event is also denoted by $[E_n \, i.o.]$. On the other hand, the *lim inf* is defined as:

$$\liminf_{n \to \infty} E_n = \bigcup_{n=1}^{\infty} \bigcap_{i=n}^{\infty} E_i. \tag{A.4}$$

We can see that $\liminf_{n \to \infty} E_n$ occurs iff E_n occurs for all but a finite number of n, that is, E_n *eventually occurs for all n*. Clearly, $\liminf_{n \to \infty} E_n \subseteq \limsup_{n \to \infty} E_n$. If the two limiting events coincide, then we define

$$\lim_{n \to \infty} E_n = \liminf_{n \to \infty} E_n = \limsup_{n \to \infty} E_n. \tag{A.5}$$

Notice that, if $E_1 \subseteq E_2 \subseteq \ldots$ (an increasing sequence), then

$$\lim_{n \to \infty} E_n = \bigcup_{n=1}^{\infty} E_n, \tag{A.6}$$

whereas, if $E_1 \supseteq E_2 \supseteq \ldots$ (a decreasing sequence), then

$$\lim_{n \to \infty} E_n = \bigcap_{n=1}^{\infty} E_n . \tag{A.7}$$

A *measurable space* $(S, \mathcal{F})$ is a pair consisting of a set S and a σ-algebra defined on it. For example, $(R^d, \mathcal{B}^d)$ is the standard *Borel-measurable space*. A *measurable* function between two measurable spaces $(S, \mathcal{F})$ and $(T, \mathcal{G})$ is defined to be a mapping $f : S \to T$ such that for every $E \in \mathcal{G}$, the pre-image

$$f^{-1}(E) = \{x \in S \mid f(x) \in E\} \tag{A.8}$$

belongs to $\mathcal{F}$. A function $f : R^d \to R^k$ is said to be *Borel-measurable* if it is a measurable function between $(R^d, \mathcal{B}^d)$ and $(R^k, \mathcal{B}^k)$. A Borel-measurable function is a very general function. For our purposes, it can be considered to be an arbitrary function. In this book, all functions (including classifiers and regressions) are assumed to be Borel-measurable.

A1.2 Probability Measure

A *measure* on $(S, \mathcal{F})$ is a real-valued function μ defined on each $E \in \mathcal{F}$ such that

A1. $0 \leq \mu(E) \leq \infty$,

A2. $\mu(\emptyset) = 0$,

A3. Given any sequence $\{E_n; n = 1, 2, \ldots\}$ in $\mathcal{F}$ such that $E_i \cap E_j = \emptyset$ for all $i \neq j$,

$$\mu \left(\bigcup_{i=1}^{\infty} E_i \right) = \sum_{i=1}^{\infty} \mu(E_i) \quad (\sigma\text{-}additivity) . \tag{A.9}$$

The triple $(S, \mathcal{F}, \mu)$ is called a *measure space*. A *probability measure* P is a measure such that $P(S) = 1$. A *probability space* is a triple $(S, \mathcal{F}, P)$, consisting of a sample space S, a σ-algebra $\mathcal{F}$ containing all the events of interest, and a probability measure P. A probability space is a model for a stochastic experiment; the properties of the latter are completely determined once a probability space is specified.

Lebesgue measure on $(R^d, \mathcal{B}^d)$ is a measure λ that agrees with the usual definition of length of intervals in R, $\lambda([a, b]) = b - a$, area of rectangles in R^2, $\lambda([a, b] \times [c, d]) = (b - a)(d - c)$, and so on for higher-dimensional spaces, and uniquely *extends* it to complicated (Borel) sets. Notice that $\lambda(\{x\}) = 0$, for all $\mathbf{x} \in R^d$, since a point has no spatial extension (it follows that it makes no difference whether intervals and rectangles are open, closed, or half-open). By σ-additivity, any countable subset of R^d has Lebesgue measure zero, and there are uncountable sets that have Lebesgue measure zero as well (e.g., the Cantor set in R). Sets of Lebesgue measure zero are very sparse; any property that holds in R^d outside of such a set is said to hold *almost everywhere* (a.e.). The measure space $(R^d, \mathcal{B}^d, \lambda)$ provides the standard setting for mathematical analysis.

Lebesgue measure restricted to $([0,1], \mathcal{B}_0)$, where $\mathcal{B}_0$ is the σ-algebra containing all Borel subsets of $[0,1]$, is a probability measure, since $\lambda([0,1]) = 1$. The probability space $([0,1], \mathcal{B}_0.\lambda)$ provides a model for the familiar uniform distribution on $[0,1]$. A famous impossibility theorem states that there does not exist a probability measure defined on $([0,1], 2^{[0,1]})$, where $2^{[0,1]}$ denotes the σ-algebra of all subsets of $[0,1]$, such that $P(\{x\}) = 0$ for all $x \in [0,1]$ [Billingsley, 1995, p. 46]. Therefore, λ cannot be extended to all subsets of $[0,1]$. This shows the need to restrict attention to the σ-algebra of Borel sets, where a unique extension of λ exists. (Lebesgue measure can be uniquely extended to even more general sets, but this is not of interest here.)

The following properties of a probability measure are straightforward consequences of axioms A1–A3 plus the requirement that $P(S) = 1$:

P1. $P(E^c) = 1 - P(E)$.

P2. If $E \subseteq F$ then $P(E) \leq P(F)$.

P3. $P(E \cup F) = P(E) + P(F) - P(E \cap F)$.

P4. (Union Bound) For any sequence of events $E_1, E_2, \ldots$

$$P\left(\bigcup_{n=1}^{\infty} E_n\right) \leq \sum_{n=1}^{\infty} P(E_n). \tag{A.10}$$

P5. (Continuity from below.) If $\{E_n; n = 1, 2, \ldots\}$ is an increasing sequence of events, then

$$P(E_n) \uparrow P\left(\bigcup_{n=1}^{\infty} E_n\right) \tag{A.11}$$

P6. (Continuity from above.) If $\{E_n; n = 1, 2, \ldots\}$ is an decreasing sequence of events, then

$$P(E_n) \downarrow P\left(\bigcap_{n=1}^{\infty} E_n\right) \tag{A.12}$$

Using P5 and P6 above, it is easy to show that

$$P\left(\liminf_{n \to \infty} E_n\right) \leq \liminf_{n \to \infty} P(E_n) \leq \limsup_{n \to \infty} P(E_n) \leq P\left(\limsup_{n \to \infty} E_n\right). \tag{A.13}$$

From this, the general *continuity of probability measure* property follows: for any sequence of events $\{E_n; n = 1, 2, \ldots\}$,

$$P\left(\lim_{n \to \infty} E_n\right) = \lim_{n \to \infty} P(E_n). \tag{A.14}$$

In some cases, it can be easy to determine the probability of limsup and liminf events. For example, it follows from (A.13) that mere convergence of $P(E_n)$ to 1 or 0 as $n \to \infty$ implies that $P(\limsup_{n \to \infty} E_n) = 1$ and $P(\liminf_{n \to \infty} E_n) = 0$, respectively. In the general case, it may not be simple to determine the value of these probabilities. The *Borel-Cantelli Lemmas* give sufficient conditions for the probability of limsup to be 0 and 1 (through the identity $P(\liminf E_n) = 1 - P(\limsup E^c)$, corresponding results on the probability of liminf can be derived).

Theorem A.1. *(First Borel-Cantelli Lemma.) For any sequence of events* $E_1, E_2, \ldots$

$$\sum_{n=1}^{\infty} P(E_n) < \infty \;\Rightarrow\; P([E_n \; i.o.]) = 0. \tag{A.15}$$

Proof. Continuity of probability measure and the union bound allow one to write

$$P([E_n \; i.o.]) = P\left(\bigcap_{n=1}^{\infty} \bigcup_{i=n}^{\infty} E_i\right) = P\left(\lim_{n \to \infty} \bigcup_{i=n}^{\infty} E_i\right) = \lim_{n \to \infty} P\left(\bigcup_{i=n}^{\infty} E_i\right) \leq \lim_{n \to \infty} \sum_{i=n}^{\infty} P(E_i). \tag{A.16}$$

But if $\sum_{n=1}^{\infty} P(E_n) < \infty$ then the last limit must be zero, proving the claim. $\diamond$

The converse to the First Lemma holds if the events are independent.

Theorem A.2. *(Second Borel-Cantelli Lemma.) For an* independent *sequence of events* $E_1, E_2, \ldots$,

$$\sum_{n=1}^{\infty} P(E_n) = \infty \;\Rightarrow\; P([E_n \; i.o.]) = 1 \tag{A.17}$$

Proof. By continuity of probability measure,

$$P([E_n \; i.o.]) = P\left(\bigcap_{n=1}^{\infty} \bigcup_{i=n}^{\infty} E_i\right) = P\left(\lim_{n \to \infty} \bigcup_{i=n}^{\infty} E_i\right) = \lim_{n \to \infty} P\left(\bigcup_{i=n}^{\infty} E_i\right) = 1 - \lim_{n \to \infty} P\left(\bigcap_{i=n}^{\infty} E_i^c\right), \tag{A.18}$$

where the last equality follows from DeMorgan's Law. Now, by independence,

$$P\left(\bigcap_{i=n}^{\infty} E_i^c\right) = \prod_{i=n}^{\infty} P(E_i^c) = \prod_{i=n}^{\infty} (1 - P(E_i)) \tag{A.19}$$

From the inequality $1 - x \leq e^{-x}$ we obtain

$$P\left(\bigcap_{i=n}^{\infty} E_i^c\right) \leq \prod_{i=1}^{\infty} \exp(-P(E_i)) = \exp\left(-\sum_{i=n}^{\infty} P(E_i)\right) = 0 \tag{A.20}$$

since, by assumption, $\sum_{i=n}^{\infty} P(E_i) = \infty$, for all n. From (A.18) and (A.20), $P([E_n \; i.o.]) = 1$, as required. $\diamond$

A1.3   Conditional Probability and Independence

Given that an event F has occurred, for E to occur, $E \cap F$ has to occur. In addition, the sample space gets *restricted* to those outcomes in F, so a normalization factor $P(F)$ has to be introduced. Therefore, assuming that $P(F) > 0$,

$$P(E \mid F) = \frac{P(E \cap F)}{P(F)}. \tag{A.21}$$

For simplicity, it is usual to write $P(E \cap F) = P(E, F)$ to indicate the *joint probability* of E and F. From (A.21), one then obtains

$$P(E, F) = P(E \mid F)P(F), \tag{A.22}$$

which is known as the *multiplication rule*. One can also condition on multiple events:

$$P(E \mid F_1, F_2, \ldots, F_n) = \frac{P(E \cap F_1 \cap F_2 \cap \ldots \cap F_n)}{P(F_1 \cap F_2 \cap \ldots \cap F_n)}. \tag{A.23}$$

This allows one to generalize the multiplication rule thus:

$$P(E_1, E_2, \ldots, E_n) = P(E_n \mid E_1, \ldots, E_{n-1})P(E_{n-1} \mid E_1, \ldots, E_{n-2}) \cdots P(E_2 \mid E_1)P(E_1). \tag{A.24}$$

The *Law of Total Probability* is a consequence of axioms of probability and the multiplication rule:

$$P(E) = P(E, F) + P(E, F^c) = P(E \mid F)P(F) + P(E \mid F^c)(1 - P(F)). \tag{A.25}$$

This property allows one to compute a hard unconditional probability in terms of easier conditional ones. It can be extended to multiple conditioning events via

$$P(E) = \sum_{i=1}^{n} P(E, F_i) = \sum_{i=1}^{n} P(E \mid F_i)P(F_i), \tag{A.26}$$

for pairwise disjoint F_i such that $\bigcup F_i \supseteq E$.

One of the most useful results of probability theory is *Bayes Theorem*:

$$P(E \mid F) = \frac{P(F \mid E)P(E)}{P(F)} = \frac{P(F \mid E)P(E)}{P(F \mid E)P(E) + P(F \mid E^c)(1 - P(E)))} \tag{A.27}$$

Bayes Theorem can be interpreted as a way to (1) "invert" the probability $P(F \mid E)$ to obtain the probability $P(E \mid F)$; or (2) "update" the "prior" probability $P(E)$ to obtain the "posterior" probability $P(E \mid F)$.

Events E and F are independent if the occurrence of one does not carry information as to the occurrence of the other. That is, assuming that all events have nonzero probability,

$$P(E \mid F) = P(E) \text{ and } P(F \mid E) = P(F). \tag{A.28}$$

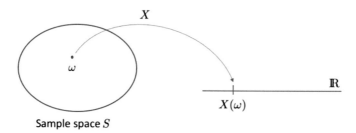

Figure A.1: A real-valued random variable.

It is easy to see that this is equivalent to the condition

$$P(E, F) = P(E)P(F). \tag{A.29}$$

If E and F are independent, so are the pairs (E, F^c), (E^c, F), and (E^c, F^c). However, E being independent of F and G does not imply that E is independent of $F \cap G$. Furthermore, three events E, F, G are independent if $P(E, F, G) = P(E)P(F)P(G)$ and each pair of events is independent. This can be extended to independence of any number of events, by requiring that the joint probability factor and that all subsets of events be independent.

Finally, we remark that $P(\cdot|F)$ is a probability measure, so that it satisfies all properties mentioned previously. In particular, it is possible to define the notion of conditional independence of events.

A1.4   Random Variables

A random variable can be thought of roughly as a "random number." Formally, a random variable X defined on a probability space $(S, \mathcal{F}, P)$ is a measurable function X between $(S, \mathcal{F})$ and $(R, \mathcal{B})$ (see Section A1.1 for the required definitions). Thus, a random variable X assigns to each outcome $\omega \in S$ a real number $X(\omega)$ — see Figure A.1 for an illustration.

By using properties of the inverse set function, it is easy to see that the set function

$$P_X(B) = P(X \in B) = P(X^{-1}(B)), \quad \text{for } B \in \mathcal{B}, \tag{A.30}$$

is a probability measure on $(R, \mathcal{B})$, called the *distribution* or *law* of X. (Note that P_X is well defined, since X is assumed measurable, and thus $X^{-1}(B)$ is an event in $\mathcal{F}$, for each $B \in \mathcal{B}$.) If $P_X = P_Y$ then X and Y are *identically distributed*. This does not mean they are identical: take X and Y to be uniform over $[0, 1]$ with $Y = 1 - X$. In this case, $P_X = P_Y$ but $P(X = Y) = 0$. On the other hand, if $P(X = Y) = 1$, then X and Y are identically distributed.

An alternative characterization of a random variable X is provided by the *cumulative distribution function* (CDF) $F_X : R \to [0, 1]$, defined by

$$F_X(x) = P_X((-\infty, x]) = P(X \le x), \quad x \in R. \tag{A.31}$$

It can be seen that the CDF has the following properties:

F1. F_X is non-decreasing: $x_1 \le x_2 \Rightarrow F(x_1) \le F(x_2)$.

F2. $\lim_{x \to -\infty} F_X(x) = 0$ and $\lim_{x \to +\infty} F_X(x) = 1$.

F3. F_X is right-continuous: $\lim_{x \to x_0^+} F_X(x) = F_X(x_0)$.

The following remarkable theorem states that the information in the *set function* P_X is equivalent to the information in the *point function* F_X; for a proof, see [Rosenthal, 2006, Prop. 6.0.2].

Theorem A.3. *Let X and Y be two random variables (possibly defined on two different probability spaces). Then $P_X = P_Y$ if and only if $F_X = F_Y$.*

Furthermore, it can be shown that given a probability measure P_X on $(R, \mathcal{B})$, there is a random variable X defined on some probability space that has P_X for its distribution; and equivalently, given any function F_X satisfying properties F1-F3 above, there is an X that has F_X as its CDF [Billingsley, 1995, Thm 14.1].

If $X_1, \ldots, X_n$ are *jointly-distributed* random variables (i.e., defined on the same probability space) then they are said to be independent if

$$P(\{X_1 \in B_1\} \cap \ldots \cap \{X_n \in B_n\}) = P_{X_1}(B_1) \cdots P_{X_n}(B_n), \tag{A.32}$$

for any Borel sets $B_1, \ldots, B_n$. Equivalently, they are independent if

$$P(\{X_1 \le x_1\} \cap \ldots \cap \{X_n \le x_n\}) = F_{X_1}(x_1) \cdots F_{X_n}(x_n), \tag{A.33}$$

for any points $x_1, \ldots x_n \in R$. If in addition $P_{X_1} = \cdots = P_{X_n}$, or equivalently, $F_{X_1} = \cdots = F_{X_n}$, then $X_1, \ldots, X_n$ are *independent and identically distributed* (i.i.d.) random variables.

Discrete Random Variables

If the distribution of a random variable X is concentrated on a countable number of points $x_1, x_2, \ldots$, i.e., $P_X(\{x_1, x_2, \ldots\}) = 1$, then X is said to be a *discrete* random variable. For example, let X be the numerical outcome of the roll of a six-sided. Then P_X is concentrated on the set $\{1, 2, 3, 4, 5, 6\}$.

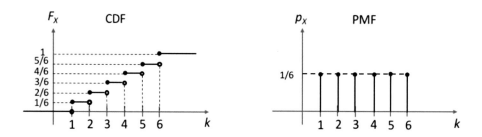

Figure A.2: The CDF and PMF of a uniform discrete random variable.

The CDF F_X for this example can be seen in Figure A.2. As seen in this plot, F_X is a "staircase" function, with "jumps" located at the points masses in P_X. This is a general fact for any discrete random variable X.

A discrete random variable X can thus be completely specified by the location and size of the jumps in F_X (since that specifies F_X). In other words, a discrete random variable X is specified by its *probability mass function* (PMF), defined by

$$p_X(x_k) \ = \ P(X = x_k) \ = \ F_X(x_k) - F_X(x_k-) \,, \tag{A.34}$$

at all points $x_k \in R$ such that $P_X(\{x_k\}) > 0$. See Figure A.2 for the PMF in the previous die-rolling example.

Clearly, discrete random variables $X_1, \ldots, X_n$ are independent if

$$P(\{X_1 = x_{k_1}\} \cap \ldots \cap \{X_n = x_{k_n}\}), \ = \ p_{X_1}(x_{k_1}) \cdots p_{X_n}(x_{k_n}) \tag{A.35}$$

at all sets of points where the corresponding PMFs are defined.

Useful discrete random variables include the already mentioned uniform r.v. over a finite set of numbers K with PMF

$$p_X(x_k) \ = \ \frac{1}{|K|} \,, \quad k \in K \,, \tag{A.36}$$

the Bernoulli with parameter $0 < p < 1$, with PMF

$$\begin{aligned} p_X(0) &= 1 - p \,, \\ p_X(1) &= p \,, \end{aligned} \tag{A.37}$$

the Binomial with parameters $n \in \{1, 2, \ldots\}$ and $0 < p < 1$, such that

$$p_X(x_k) \ = \ \binom{n}{k} p^k (1-p)^{n-k}, \quad k = 0, 1, \ldots, n \,, \tag{A.38}$$

the Poisson with parameter $\lambda > 0$, such that

$$p_X(x_k) = e^{-\lambda}\frac{\lambda^k}{k!}, \quad k = 0, 1, \ldots \tag{A.39}$$

and the Geometric with parameter $0 < p < 1$ such that

$$p_X(x_k) = (1-p)^{k-1}p, \quad k = 1, 2, \ldots \tag{A.40}$$

A binomial r.v. with parameters n and p has the distribution of a a sum of n i.i.d. Bernoulli r.v.s with parameter p.

Continuous Random Variables

The transition from discrete to continuous random variables is nontrivial. A continuous random variable X should have the following two smoothness properties:

C1. F_X is continuous, i.e., it contains no jumps; i.e., $P(X = x) = 0$ for all $x \in R$.

C2. There is a nonnegative function p_X such that

$$P(a \leq X \leq b) = F_X(b) - F_X(a) = \int_a^b p_X(x)\,dx, \tag{A.41}$$

for $a, b \in R$, with $a \leq b$. In particular, $\int_{-\infty}^{\infty} p_X(x)\,dx = 1$.

It follows from the properties of the integral that C2 implies C1. However, it is one of the surprising facts of probability theory that C1 does not imply C2: there are continuous CDFs that do not satisfy C2. The counterexamples are admittedly exotic. For instance, the *Cantor function* is a continuous increasing function defined on the interval $[0, 1]$, which has derivative equal to zero on the complement of the Cantor set, i.e., almost everywhere, but grows continuously from 0 to 1. The Cantor function is constant almost everywhere, but manages to grow continuously, without jumps. Such functions are called *singular* (or "devil staircases" in the popular literature). The Cantor function (suitably extended outside the interval $[0, 1]$) defines a continuous CDF that cannot satisfy C2. Such exotic examples can be ruled out if one requires the CDF to have a smoothness property known as *absolute continuity* (which is more stringent than simple continuity). In fact, it can be shown that absolute continuity of F_X is *equivalent* to C2. It is also equivalent to the requirement that $P(X \in B) = 0$ for any Borel set B of measure zero, not simply on isolated points, as in C1, or countable set of points. It can indeed be shown that any CDF can be decomposed uniquely into a sum of a discrete, singular, and absolute continuous components.[1]

---

[1] For proofs and more details, the reader is referred to Sections 31 and 32 of Billingsley [1995] and Chapter 1 of Chung [1974]. The construction of the Cantor function is described in Chapter 7 of Schroeder [2009].

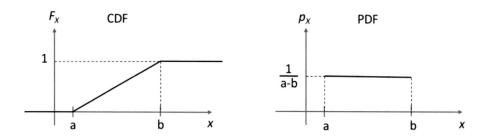

Figure A.3: The CDF and PDF of a uniform continuous random variable.

The definition of a continuous random variable X requires F_X to be absolutely continuous, not simply continuous, in which case C2 is satisfied, and p_X is called a *probability density function* (PDF). (Perhaps it would be more appropriate to call these *absolutely continuous* random variables, but the terminology "continuous random variable" is entrenched.) See Figure A.3 for an illustration of the CDF and PDF of a uniform continuous random variable. The CDF of a continuous random variable does not have to be differentiable everywhere (in this example, it fails to be differentiable at a and b). But where it is differentiable, $dF_X(x)/dx = p_X(x)$ (the density can take arbitrary values where F_X is not differentiable, and this happens at most over a set of Lebesgue measure zero).

Useful continuous random variables include the already mentioned uniform r.v. over the interval $[a, b]$, with density

$$p_X(x) = \frac{1}{b - a}, \quad a < x < b, \tag{A.42}$$

the univariate Gaussian r.v. with parameters μ and $\sigma > 0$, such that

$$p_X(x) = \frac{1}{\sqrt{2\pi\sigma^2}} \exp\left(-\frac{(x - \mu)^2}{2\sigma^2}\right), \quad x \in R, \tag{A.43}$$

the exponential r.v. with parameter $\lambda > 0$, such that

$$p_X(x) = \lambda e^{-\lambda x}, \quad x \geq 0 \tag{A.44}$$

the gamma r.v. with parameters $\lambda, t > 0$, such that

$$p_X(x) = \frac{\lambda e^{-\lambda x}(\lambda x)^{t-1}}{\Gamma(t)}, \quad x \geq 0, \tag{A.45}$$

where $\Gamma(t) = \int_0^\infty e^{-u} u^{t-1} du$, and the beta r.v. with parameters $a, b > 0$, such that:

$$p_X(x) = \frac{1}{B(a, b)} x^{a-1}(1 - x)^{b-1}, \quad 0 < x < 1, \tag{A.46}$$

where $B(a, b) = \Gamma(a + b)/\Gamma(a)\Gamma(b)$. Among these, the Gaussian is the only one defined over the entire real line; the exponential and gamma are defined over the nonnegative real numbers, while the uniform and beta have bounded support. In fact, the uniform r.v. over $[0, 1]$ is just a beta with $a = b = 1$, while an exponential r.v. is a gamma with $t = 1$.

General Random Variables

There are random variables that are neither continuous nor discrete. Of course, an example of that is afforded by a mixture of a discrete random variable and a continuous random variable. The CDF of such a mixed random variable has jumps, but it is not a staircase function. However, there are more general random variables that are *not* mixtures of this kind; e.g., the random variable corresponding to the Cantor CDF.

A1.5   Joint and Conditional Distributions

The *joint CDF* of two jointly-distributed random variables X and Y is a function $F_{XY} : R \times R \to [0, 1]$ defined by

$$F_{XY}(x, y) = P(\{X \leq x\} \cap \{Y \leq y\}) = P(X \leq x, Y \leq y), \quad x, y \in R. \tag{A.47}$$

This is the probability of the "lower-left quadrant" with corner at (x, y). Note that $F_{XY}(x, \infty) = F_X(x)$ and $F_{XY}(\infty, y) = F_Y(y)$. These are called the *marginal CDFs*.

If X and Y are jointly-distributed continuous random variables, then we define the *joint density*

$$p_{XY}(x, y) = \frac{\partial^2 F_{XY}(x, y)}{\partial x \partial y} \quad x, y \in R, \tag{A.48}$$

at all points where the derivative is defined. The joint density function $p_{XY}(x, y)$ integrates to 1 over R^2. The *marginal densities* are given by

$$\begin{aligned}
p_X(x) &= \int_{-\infty}^{\infty} p_{XY}(x, y) \, dy, \quad x \in R, \\
p_Y(y) &= \int_{-\infty}^{\infty} p_{XY}(x, y) \, dx, \quad y \in R,
\end{aligned} \tag{A.49}$$

The random variables X and Y are *independent* if $p_{XY}(x, y) = p_X(x)p_Y(y)$, for all $x, y \in R$. It can be shown that if X and Y are independent and $Z = X + Y$ then

$$p_Z(z) = \int_{-\infty}^{\infty} p_X(x)p_Y(z - x) \, dx, \quad z \in R, \tag{A.50}$$

with a similar expression in the discrete case for the corresponding PMFs. The above integral is known as the *convolution integral*.

If $p_Y(y) > 0$, the *conditional density* of X given $Y = y$ is defined by:

$$p_{X|Y}(x \mid y) = \frac{p_{XY}(x, y)}{p_Y(y)}, \quad x \in R. \tag{A.51}$$

For an event E, the conditional probability $P(E \mid Y = y)$ needs care if Y is a continuous random variable, as $P(Y = y) = 0$. But as long as $p_Y(y) > 0$, this probability can be defined (the details are outside of the scope of this review):

$$P(E \mid Y = y) = \int_E p_{X|Y}(x \mid y) \, dx. \tag{A.52}$$

The "Law of Total Probability" for random variables is a generalization of (A.26):

$$P(E) = \int_{-\infty}^{\infty} P(E \mid Y = y) \, p_Y(y) \, dy. \tag{A.53}$$

The concepts of joint PMF, marginal PMFs, and conditional PMF can defined in a similar way. For conciseness, this is omitted in this review.

A1.6 Expectation

The expectation of a random variable has several important interpretations: 1) its average value (weighted by the probabilities); 2) a summary of its distribution (sometimes referred to as a "location parameter"); 3) a prediction of its future value. The latter meaning is the most important one for pattern recognition and machine learning.

Expectation can be formalized by using the notion of integration, which we briefly review next. For a measure space $(S, \mathcal{F}, \mu)$ and a Borel-measurable function $f : S \to R$, one defines the integral

$$\int f \, d\mu = \int f(\omega) \, \mu(d\omega) \tag{A.54}$$

as a number in $R \cup \{-\infty, \infty\}$, as follows. First, if $f = I_A$ is the indicator of a set $A \in \mathcal{F}$, then $\int f \, d\mu = \mu(A)$, i.e., integrating a constant "1" over a set produces just the measure of that set. Next, if $f = \sum_{i=1}^n x_i I_{A_i}$, where the $x_i \in R$ and the A_i are measurable sets that partition S, then

$$\int f \, d\mu = \sum_{i=1}^n x_i \mu(A_i). \tag{A.55}$$

Such a function f is called *simple*, as it takes on a finite number of values $x_1, \ldots, x_n$, with $f^{-1}(\{x_i\}) = A_i$, for $i = 1, \ldots, n$. Next, for general nonnegative function f, one defines its integral as

$$\int f \, d\mu = \sup \left\{ \int g \, d\mu \mid g : S \to R \text{ is simple and } g \le f \right\}. \tag{A.56}$$

Finally, for general f, define nonnegative functions $f^+(\omega) = f(\omega) I_{f(\omega)>0}$ and $f^-(\omega) = -f(\omega) I_{f(\omega)\le 0}$. Clearly, $f = f^+ - f^-$, so the integral of f is defined as

$$\int f \, d\mu = \int f^+ \, d\mu - \int f^- \, d\mu, \tag{A.57}$$

provided that at least one of $\int f^+ \, d\mu$ and $\int f^- \, d\mu$ is finite. If both are finite, then $-\infty < \int f \, d\mu < \infty$, and f is said to be *integrable* with respect to measure μ. Since $|f| = f^+ + f^-$, f is integrable if and only if $\int |f| \, d\mu < \infty$. If $\int f^+ \, d\mu = \int f^- \, d\mu = \infty$, then the integral of f is not defined at all.

The integral ignores everything that happens over sets of measure zero: if $f = g$ outside a set of measure zero, then $\int f \, d\mu = \int g \, d\mu$. Hence, if $f = 0$ a.e., then $\int f \, d\mu = 0$, and the integral of nonnegative f is positive if and only if $f > 0$ over a set of nonzero measure.

The integral of f over a set $A \in \mathcal{F}$ is defined as $\int_A f \, d\mu = \int I_A f \, d\mu$, if it exists. If f is nonnegative, then $\nu(A) = \int_A f \, d\mu$ defines a measure on $(S, \mathcal{F})$, and f is called a *density* of ν with respect to μ (densities are unique up to sets of μ-measure zero). It is clear that $\nu(A) = 0$ whenever $\mu(A) = 0$; any measure ν with this property is said to be *absolutely continuous* with respect to μ (this is a generalization of the previous definition, as we comment below). The following theorem can be proved by showing that it holds for indicators, simple functions, and then nonnegative functions through (A.56).

Theorem A.4. *If $g : S \to R$ is integrable and $f : S \to R$ is a density of ν with respect to μ, then*

$$\int g(\omega) \, \nu(d\omega) = \int g(\omega) f(\omega) \, \mu(d\omega) . \tag{A.58}$$

The general integral has all the properties with which one if familiar in Calculus, such as linearity: it can be shown that if f and g are integrable and a and b are constants, then

$$\int (af + bg) \, d\mu = a \int f \, d\mu + b \int g \, d\mu . \tag{A.59}$$

If the measure space is $(R, \mathcal{B}, \lambda)$ then the integral of a function $f : R \to R$,

$$\int f \, d\lambda = \int f(x) \, \lambda(dx) \tag{A.60}$$

is the *Lebesgue integral* of f, if it exists. It can be shown that the Lebesgue integral coincides with the usual Riemann integral, whenever the latter exists. But the full generality of the Lebesgue integral is needed to integrate complicated functions, or functions over complicated sets. The classical example is provided by the function $f : R \to R$ defined as $f(x) = 1$ if x is rational, and $f(x) = 0$, otherwise. Notice that $f = I_Q$, the indicator of the set of rationals Q. This function is extremely irregular (discontinuous and nondifferentiable at every point) and not Riemann-integrable. However, f is measurable and Lebesgue-integrable, with $\int f(x) \, \lambda(dx) = \lambda(Q) = 0$. All integrals mentioned before in this Appendix, including (A.41), should be considered to be Lebesgue integrals.

Now, given a random variable X defined on a probability space $(S, \mathcal{F}, P)$, the expectation $E[X]$ is simply the integral of X over S according to the probability measure P:

$$E[X] = \int X \, dP = \int X(\omega) \, P(d\omega) , \tag{A.61}$$

if it exists. So expectation is an integral, and all definitions and properties mentioned previously in this section apply; e.g., we get the familiar formulas $E[I_E] = P(E)$, for an event E, and

$$E[aX + bY] = aE[X] + b[Y], \tag{A.62}$$

for jointly-distributed integrable random variables X and Y and constants a and b, as in (A.59). This extends to any finite number of random variables, by induction. One of the most important results of probability theory is stated next, without proof.

Theorem A.5. *(Change of Variable Theorem.) If $g : R \to R$ is a measurable function, then*

$$E[g(X)] = \int_S g(X(\omega)) \, P(d\omega) = \int_{-\infty}^{\infty} g(x) \, P_X(dx), \tag{A.63}$$

where P_X is the distribution of X, defined in (A.30).

Hence, expectations can be computed by integration over the real line. The previous theory is entirely general, and applies equally well to continuous, discrete, and more general random variables.

If X is continuous, then it satisfies (A.41), where the integral should be interpreted as Lebesgue integral over the interval $[a, b]$. It can be shown then that p_X is a density for the distribution P_X with respect to Lebesgue measure. Combining Theorems A.4 and A.5 produces the familiar formula:

$$E[g(X)] = \int_{-\infty}^{\infty} g(x) \, p_X(x) \, dx, \tag{A.64}$$

where the integral is the Lebesgue integral, which reduces to the ordinary integral if the integrand is Riemann-integrable. If $g(x) = x$, one gets the usual definition $E[X] = \int x \, p_X(x) \, dx$.

On the other hand, if X is discrete, then P_X is concentrated on a countable number of points $x_1, x_2, \ldots$, and Thm A.5 produces

$$E[g(X)] = \sum_{k=1}^{\infty} g(x_k) \, p_X(x_k), \tag{A.65}$$

if the sum is well-defined. If $g(x) = x$, we get the familiar formula $E[X] = \sum_{k=1}^{\infty} x_k \, p_X(x_k)$.

From now on we assume that random variables are integrable. If $f : R \to R$ is Borel-measurable and *concave* (i.e., f lies at or above a line joining any of its points) then *Jensen's Inequality* is:

$$E[f(X)] \leq f(E[X]). \tag{A.66}$$

It can be shown that X and Y are independent if and only if $E[f(X)g(Y)] = E[f(X)]E[g(Y)]$ for all Borel-measurable functions $f, g : R \to R$. If this condition is satisfied for at least $f(X) = X$ and

$g(Y) = Y$, that is, if $E[XY] = E[X]E[Y]$, then X and Y are said to be *uncorrelated*. Of course, independence implies uncorrelatedness. The converse is only true in special cases; e.g. jointly Gaussian random variables.

Holder's Inequality states that, for $1 < r < \infty$ and $1/r + 1/s = 1$,

$$E[|XY|] \leq E[|X|^r]^{1/r} E[|Y|^s]^{1/s}. \tag{A.67}$$

The special case $r = s = 2$ results in the *Cauchy-Schwarz Inequality*:

$$E[|XY|] \leq \sqrt{E[X^2]E[Y^2]}. \tag{A.68}$$

The expectation of a random variable X is affected by its *probability tails*, given by $F_X(a) = P(X \leq a)$ and $1 - F_X(a) = P(X \geq a)$. If the probability tails on both sides fail to vanish sufficiently fast (X has "fat tails"), then X will not be integrable and $E[X]$ is undefined. The standard example is the Cauchy random variable, with density $p_X(x) = [\pi(1 + x^2)]^{-1}$. For a nonnegative random variable X, there is only one probability tail, the upper tail $P(X > a)$, and there is a simple formula relating $E[X]$ to it:

$$E[X] = \int_0^\infty P(X > x)\,dx. \tag{A.69}$$

A small $E[X]$ constrain the upper tail to be thin. This is guaranteed by *Markov's inequality:* if X is a nonnegative random variable,

$$P(X \geq a) \leq \frac{E[X]}{a}, \quad \text{for all } a > 0. \tag{A.70}$$

Finally, a particular result that if of interest to our purposes relates an exponentially-vanishing upper tail of a nonnegative random variable to a bound on its expectation.

Lemma A.1. *If X is a non-negative random variable such that $P(X > t) \leq ce^{-at^2}$, for all $t > 0$ and given $a, c > 0$, we have*

$$E[X] \leq \sqrt{\frac{1 + \ln c}{a}}. \tag{A.71}$$

Proof. Note that $P(X^2 > t) = P(X > \sqrt{t}) \leq ce^{-at}$. From (A.69) we get:

$$\begin{aligned}
E[X^2] &= \int_0^\infty P(X^2 > t)\,dt = \int_0^u P(X^2 > t)\,dt + \int_u^\infty P(X^2 > t)\,dt \\
&\leq u + \int_u^\infty ce^{-at}\,dt = u + \frac{c}{a}e^{-au}.
\end{aligned} \tag{A.72}$$

By direct differentiation, it is easy to verify that the upper bound on the right hand side is minimized at $u = (\ln c)/a$. Substituting this value back into the bound leads to $E[X^2] \leq (1 + \ln c)/a$. The result then follows from the fact that $E[X] \leq \sqrt{E[X^2]}$. $\diamond$

If the second moment exists, the variance $\mathrm{Var}(X)$ of a random variable X is a nonnegative quantity defined by:

$$\mathrm{Var}(X) = E[(X - E[X])^2] = E[X^2] - (E[X])^2. \tag{A.73}$$

The variance of a random variable can be interpreted as: 1) its "spread" around the mean; 2) a second summary of its distribution (the "scale parameter"); 3) the uncertainty in the prediction of its future value by its expectation.

The following property follows directly from the definition:

$$\mathrm{Var}(aX + c) = a^2 \mathrm{Var}(X). \tag{A.74}$$

A small variance constrains the random variable to be close to its mean with high probability. This follows from *Chebyshev's Inequality*:

$$P(|X - E[X]| \geq \tau) \leq \frac{\mathrm{Var}(X)}{\tau^2}, \quad \text{for all } \tau > 0. \tag{A.75}$$

Chebyshev's inequality follows directly from the application of Markov's Inequality (A.70) to the random variable $|X - E[X]|^2$ with $a = \tau^2$.

Expectation has the linearity property, so that, given any pair of jointly distributed random variables X and Y, it is always true that $E[X + Y] = E[X] + E[Y]$ (provided that all expectations exist). However, it is not always true that $\mathrm{Var}(X + Y) = \mathrm{Var}(X) + \mathrm{Var}(Y)$. In order to investigate this issue, it is necessary to introduce the *covariance* between X and Y:

$$\mathrm{Cov}(X, Y) = E[(X - E[X])(Y - E[Y])] = E[XY] - E[X]E[Y]. \tag{A.76}$$

If $\mathrm{Cov}(X, Y) > 0$ then X and Y are *positively correlated*; otherwise, they are *negatively correlated*. Clearly, X and Y are uncorrelated if and only if $\mathrm{Cov}(X, Y) = 0$. Clearly, $\mathrm{Cov}(X, X) = \mathrm{Var}(X)$. In addition, $\mathrm{Cov}(\sum_{i=1}^{n} X_i, \sum_{j=1}^{m} Y_j) = \sum_{i=1}^{n} \sum_{j=1}^{m} \mathrm{Cov}(X_i, Y_j)$.

Now, it follows directly from the definition of variance that

$$\mathrm{Var}(X_1 + X_2) = \mathrm{Var}(X_1) + \mathrm{Var}(X_2) + 2\mathrm{Cov}(X_1, X_2). \tag{A.77}$$

This can be extended to any number of random variables by induction:

$$\mathrm{Var}\left(\sum_{i=1}^{n} X_i\right) = \sum_{i-1}^{n} \mathrm{Var}(X_i) + 2\sum_{i<j} \mathrm{Cov}(X_i, X_j). \tag{A.78}$$

Hence, the variance is distributive over sums if all variables are *pairwise uncorrelated*. o It follows directly from the Cauchy-Schwarz Inequality (A.68) that $|\mathrm{Cov}(X, Y)| \leq \sqrt{\mathrm{Var}(X)\mathrm{Var}(Y)}$. Therefore, the covariance can be normalized to be in the interval $[-1, 1]$ thus:

$$\rho(X, Y) = \frac{\mathrm{Cov}(X, Y)}{\sqrt{\mathrm{Var}(X)\mathrm{Var}(Y)}}, \tag{A.79}$$

with $-1 \leq \rho(X, Y) \leq 1$. This is called the correlation coefficient between X and Y. The closer $|\rho|$ is to 1, the tighter is the relationship between X and Y. The limiting case where $\rho(X, Y) = \pm 1$ occurs if and only if $Y = a \pm bX$, i.e., X and Y are perfectly related to each other through a linear (affine) relationship. For this reason, $\rho(X, Y)$ is sometimes called the linear correlation coefficient between X and Y; it does not respond to nonlinear relationships.

Conditional expectation allows the prediction of the value of a random variable given the *observed* value of the other, i.e., prediction given data, while conditional variance yields the uncertainty of that prediction.

If X and Y are jointly continuous random variables and the conditional density $p_{X|Y}(x \mid y)$ is well defined for $Y = y$, then the conditional expectation of X *given* $Y = y$ is:

$$E[X \mid Y = y] = \int_{-\infty}^{\infty} x\, p_{X|Y}(x \mid y)\, dx \tag{A.80}$$

with a similar definition for discrete random variables using conditional PMFs.

The conditional variance of X given $Y = y$ is defined using conditional expectation as:

$$\text{Var}(X \mid Y = y) = E[(X - E[X \mid Y = y])^2 \mid Y = y] = E[X^2 \mid Y = y] - (E[X \mid Y = y])^2. \tag{A.81}$$

Most of the properties of expectation and variance apply without modification to conditional expectations and conditional variances, respectively. For example, $E[\sum_{i=1}^{n} X_i \mid Y = y] = \sum_{i=1}^{n} E[X_i \mid Y = y]$ and $\text{Var}(aX + c \mid Y = y) = a^2 \text{Var}(X \mid Y = y)$.

Now, both $E[X \mid Y = y]$ and $\text{Var}(X \mid Y = y)$ are deterministic quantities for each value of $Y = y$ (just as the ordinary expectation and variance are). But if the specific value $Y = y$ is not specified and allowed to vary, then we can look at $E[X \mid Y]$ and $\text{Var}(X \mid Y)$ as functions of the random variable Y, and therefore, random variables themselves. The reasons why these are valid random variables are nontrivial and beyond the scope of this review.

One can show that the expectation of the random variable $E[X \mid Y]$ is precisely $E[X]$:

$$E[E[X \mid Y]] = E[X]. \tag{A.82}$$

An equivalent statement is:

$$E[X] = \int_{-\infty}^{\infty} E[X \mid Y = y]\, p(y)\, dy, \tag{A.83}$$

with a similar expression in the discrete case. Paraphrasing the Law of Total Probability (A.26), the previous equation might be called the *Law of Total Expectation*.

On the other hand, it is not the case that $\text{Var}(X) = E[\text{Var}(X \mid Y)]$. The answer is slightly more complicated:

$$\text{Var}(X) = E[\text{Var}(X \mid Y)] + \text{Var}(E[X \mid Y]). \tag{A.84}$$

This is known as the *Conditional Variance Formula*. It is an "analysis of variance" formula, as it breaks down the total variance of X into a "within-rows" component and an "across-rows" component. One might call this the *Law of Total Variance*. This formula plays a key role in Chapter 7.

Now, suppose one is interested in predicting the value of a random variable Y using a predictor $\hat{Y}$. One would like $\hat{Y}$ to be optimal according to some criterion. The criterion most widely used is the *mean-square error:*

$$\text{MSE} = E[(Y - \hat{Y})^2]. \tag{A.85}$$

It can be shown easily that the minimum mean-square error (MMSE) estimator is simply the mean: $\hat{Y}^* = E[Y]$. This is a constant estimator, since no data are available. Clearly, the MSE of $\hat{Y}^*$ is simply the variance of Y. Therefore, the best one can do in the absence of any extra information is to predict the mean $E[Y]$, with an uncertainty equal to the variance $\text{Var}(Y)$.

If $\text{Var}(Y)$ is very small, i.e., if there were very small uncertainty in Y to begin with, then $E[Y]$ could actually be an acceptable estimator. In practice, this is rarely the case. Therefore, observations on an auxiliary random variable X (i.e., *data*) are sought to improve prediction. Naturally, it is known (or hoped) that X and Y are not independent, otherwise no improvement over the constant estimator is possible. One defines the *conditional MSE* of a data-dependent estimator $\hat{Y} = h(X)$ as

$$\text{MSE}(X) = E[(Y - h(X))^2 \mid X]. \tag{A.86}$$

By taking expectation over X, one obtains the unconditional MSE: $E[(Y - h(X))^2]$. The conditional MSE is often the most important one in practice, since it concerns the particular data at hand, while the unconditional MSE is data-independent and used to compare the performance of different predictors. Regardless, the MMSE estimator in *both cases* is the conditional mean $h^*(X) = E[Y \mid X]$, as shown in Chapter 11. This is one of the most important results in supervised learning. The *posterior-probability function* $\eta(x) = E[Y \mid X = x]$ is the optimal regression of Y on X. This is not in general the optimal estimator if Y is discrete; e.g., in the case of classification. This is because $\eta(X)$ may not be in the range of values taken by Y, so it does not define a valid estimator. It is shown in Chapter 2 that one needs to threshold $\eta(x)$ at $1/2$ to obtain the optimal estimator (optimal classifier) in the case $Y \in \{0, 1\}$.

A1.7 Vector Random Variables

The previous theory can be extended to vector random variables, or *random vectors*, defined on a probability space $(S, \mathcal{F}, P)$. A random vector is a Borel-measurable function $\mathbf{X} : S \to R^d$, with a probability distribution $P_{\mathbf{X}}$ defined on $(R^d, \mathcal{B}^d)$. The components of the random vector $\mathbf{X} = (X_1, \ldots, X_d)$ are jointly-distributed random variables X_i on $(S, \mathcal{F}, P)$, for $i = 1, \ldots, d$.

The expected value of $\mathbf{X}$ is the vector of expected values of the components, if they exist:

$$E[\mathbf{X}] = \begin{bmatrix} E[X_1] \\ \cdots \\ E[X_d] \end{bmatrix}. \tag{A.87}$$

The second moments of a random vector are contained in the $d \times d$ *covariance matrix*:

$$\Sigma = E[(\mathbf{X} - \boldsymbol{\mu})(\mathbf{X} - \boldsymbol{\mu})^T], \tag{A.88}$$

where $\Sigma_{ii} = \mathrm{Var}(X_i)$ and $\Sigma_{ij} = \mathrm{Cov}(X_i, X_j)$, for $i, j = 1, \ldots, d$, and the expectation of the matrix is defined as the matrix of the expected values of its components, assuming they exist. The covariance matrix is real symmetric and thus diagonalizable:

$$\Sigma = UDU^T, \tag{A.89}$$

where U is the orthogonal matrix of eigenvectors and D is the diagonal matrix of eigenvalues (a review of basic matrix theory facts is given in Section A2). All eigenvalues are nonnegative (Σ is *positive semi-definite*). In fact, except for "degenerate" cases, all eigenvalues are positive, and so Σ is invertible (Σ is said to be *positive definite* in this case).

It is easy to check that the random vector

$$\mathbf{Y} = \Sigma^{-\frac{1}{2}}(\mathbf{X} - \boldsymbol{\mu}) = D^{-\frac{1}{2}}U^T(\mathbf{X} - \boldsymbol{\mu}) \tag{A.90}$$

has zero mean and covariance matrix $\mathbf{I}_d$ (so that all components of $\mathbf{Y}$ are zero-mean, unit-variance, and uncorrelated). This is called *whitening or the Mahalanobis transformation*.

Given n *independent and identically-distributed* (i.i.d.) sample observations $\mathbf{X}_1, \ldots, \mathbf{X}_n$ of the random vector $\mathbf{X}$, then the maximum-likelihood estimator of $\boldsymbol{\mu} = E[\mathbf{X}]$, known as the *sample mean*, is

$$\hat{\boldsymbol{\mu}} = \frac{1}{n}\sum_{i=1}^{n}\mathbf{X}_i. \tag{A.91}$$

It can be shown that this estimator is *unbiased* (that is, $E[\hat{\boldsymbol{\mu}}] = \boldsymbol{\mu}$) and *consistent* (that is, $\hat{\boldsymbol{\mu}}$ converges in probability to $\boldsymbol{\mu}$ as $n \to \infty$; see Section A1.8 and Theorem A.12). On the other hand, the *sample covariance* estimator is given by:

$$\hat{\Sigma} = \frac{1}{n-1}\sum_{i=1}^{n}(\mathbf{X}_i - \hat{\boldsymbol{\mu}})(\mathbf{X}_i - \hat{\boldsymbol{\mu}})^T. \tag{A.92}$$

This is an unbiased and consistent estimator of Σ.

The multivariate Gaussian distribution is probably the most important probability distribution in Engineering and Science. The random vector $\mathbf{X}$ has a multivariate Gaussian distribution with mean $\boldsymbol{\mu}$ and covariance matrix Σ (assuming Σ invertible) if its density is given by

$$p(\mathbf{x}) = \frac{1}{\sqrt{(2\pi)^d \det(\Sigma)}} \exp\left(-\frac{1}{2}(\mathbf{x} - \boldsymbol{\mu})^T \Sigma^{-1}(\mathbf{x} - \boldsymbol{\mu})\right). \tag{A.93}$$

We write $\mathbf{X} \sim N_d(\boldsymbol{\mu}, \Sigma)$.

The multivariate Gaussian has ellipsoidal contours of constant density of the form

$$(\mathbf{x} - \boldsymbol{\mu})^T \Sigma^{-1}(\mathbf{x} - \boldsymbol{\mu}) = c^2, \quad c > 0. \tag{A.94}$$

The axes of the ellipsoids are given by the eigenvectors of Σ and the length of the axes are proportional to its eigenvalues. In the case $\Sigma = \sigma^2 I_d$, where I_d denotes the $d \times d$ identity matrix, the contours are spherical with center at $\boldsymbol{\mu}$. This can be seen by substituting $\Sigma = \sigma^2 I_d$ in (A.94), which leads to the following equation for the contours:

$$||\mathbf{x} - \boldsymbol{\mu}||^2 = r^2, \quad r > 0, \tag{A.95}$$

If $d = 1$, one gets the univariate Gaussian distribution $X \sim \mathcal{N}(\mu, \sigma^2)$. With $\mu = 0$ and $\sigma = 1$, the CDF of X is given by

$$P(X \leq x) = \Phi(x) = \int_{-\infty}^{x} \frac{1}{2\pi} e^{-\frac{u^2}{2}} du. \tag{A.96}$$

It is clear that the function $\Phi(\cdot)$ satisfies the property $\Phi(-x) = 1 - \Phi(x)$.

The following are useful properties of a multivariate Gaussian random vector $\mathbf{X} \sim \mathcal{N}(\mu, \Sigma)$:

G1. The density of each component X_i is univariate gaussian $\mathcal{N}(\mu_i, \Sigma_{ii})$.

G2. The components of $\mathbf{X}$ are independent *if and only if* they are uncorrelated, i.e., Σ is a diagonal matrix.

G3. The whitening transformation $\mathbf{Y} = \Sigma^{-\frac{1}{2}}(\mathbf{X} - \boldsymbol{\mu})$ produces a multivariate gaussian $\mathbf{Y} \sim \mathcal{N}(0, I_p)$ (so that all components of $\mathbf{Y}$ are zero-mean, unit-variance, and uncorrelated Gaussian random variables).

G4. In general, if $\mathbf{A}$ is a nonsingular $p \times p$ matrix and $\mathbf{c}$ is a p-vector, then $\mathbf{Y} = \mathbf{A}\mathbf{X} + \mathbf{c} \sim N_p(\mathbf{A}\boldsymbol{\mu} + \mathbf{c}, \mathbf{A}\Sigma\mathbf{A}^T)$.

G5. The random vectors $\mathbf{A}\mathbf{X}$ and $\mathbf{B}\mathbf{X}$ are independent iff $\mathbf{A}\Sigma\mathbf{B}^T = 0$.

G6. If $\mathbf{Y}$ and $\mathbf{X}$ are jointly multivariate Gaussian, then the distribution of $\mathbf{Y}$ given $\mathbf{X}$ is again multivariate Gaussian.

G7. The best MMSE predictor $E[\mathbf{Y} \mid \mathbf{X}]$ is a linear function of $\mathbf{X}$.

A1.8   Convergence of Random Sequences

It is often necessary in pattern recognition and machine learning to investigate the long-term behavior of random sequences, such as the sequence of true or estimated classification error rates indexed by sample size. In this section and the next, we review basic results about convergence of random sequences. We consider only the case of real-valued random variables, but nearly all the definitions and results can be directly extended to random vectors, with the appropriate modifications.

A *random sequence* $\{X_n; n = 1, 2, \ldots\}$ is a sequence of random variables. The standard modes of convergence for random sequences are:

1. *"Sure"* convergence: $X_n \to X$ surely if for all outcomes $\omega \in S$ in the sample space one has $\lim_{n\to\infty} X_n(\omega) = X(\omega)$.

2. *Almost-sure (a.s.) convergence* or *convergence with probability 1*: $X_n \xrightarrow{a.s.} X$ if pointwise converge fails only for an event of probability zero, i.e.:

$$P\left(\left\{\omega \in S \mid \lim_{n\to\infty} X_n(\omega) = X(\omega)\right\}\right) = 1 \,. \tag{A.97}$$

3. *L^p-convergence*: $X_n \to X$ in L^p, for $p > 0$, also denoted by $X_n \xrightarrow{L^p} X$, if $E[|X_n|^p] < \infty$ for $n = 1, 2, \ldots$, $E[|X|^p] < \infty$, and:

$$\lim_{n\to\infty} E[|X_n - X|^p] = 0 \,. \tag{A.98}$$

   The special case of L^2 convergence is also called *mean-square* (m.s.) convergence.

4. *Convergence in probability*: $X_n \to X$ in probability, also denoted by $X_n \xrightarrow{P} X$, if the "probability of error" converges to zero:

$$\lim_{n\to\infty} P(|X_n - X| > \tau) = 0 \,, \quad \text{for all } \tau > 0 \,. \tag{A.99}$$

5. *Convergence in Distribution* : $X_n \to X$ in distribution, also denoted by $X_n \xrightarrow{D} X$, if the corresponding CDFs converge:
$$\lim_{n\to\infty} F_{X_n}(a) = F_X(a) \,, \tag{A.100}$$
   at all points $a \in R$ where F_X is continuous.

We state, without proof, the relationships among the various modes of convergence:

$$\left.\begin{array}{c} \text{sure} \Rightarrow \text{almost-sure} \\ L^p \end{array}\right\} \Rightarrow \text{probability} \Rightarrow \text{distribution} \,. \tag{A.101}$$

Hence, sure convergence is the strongest mode of convergence and convergence in distribution is the weakest. However, sure convergence is unnecessarily demanding, and almost-sure convergence is the strongest mode of convergence employed. On the other hand, convergence is distribution is really convergence of CDFs, and does not have all the properties one expects from convergence. For example, it can be shown that convergence X_n to X and Y_n to Y in distribution does not imply in general that $X_n + Y_n$ converges to $X + Y$ in distribution, whereas this would be true for convergence almost surely, in L^p, and in probability [Chung, 1974].

To show consistency of parametric classification rules (see Chapters 3 and 4), an essential fact about convergence with probability 1 and in probability is that, similarly to ordinary convergence, they are preserved by application of continuous functions. The following result is stated without proof.

Theorem A.6. *(Continuous Mapping Theorem.)* *If $f : R \to R$ is continuous a.e. with respect to X, i.e. $P(X \in C) = 1$, where C is the set of points of continuity of f, then*

(i) $X_n \xrightarrow{a.s.} X$ *implies that* $f(X_n) \xrightarrow{a.s.} f(X)$.

(ii) $X_n \xrightarrow{P} X$ *implies that* $f(X_n) \xrightarrow{P} f(X)$.

(iii) $X_n \xrightarrow{D} X$ *implies that* $f(X_n) \xrightarrow{D} f(X)$.

A special case of interest is $X = c$, i.e., the distribution of X is a point mass at c. In this case, the continuous mapping theorem requires f to be merely continuous at c.

The following classical result is stated here without proof.

Theorem A.7. *(Dominated Convergence Theorem.)* *If there is an integrable random variable Y, i.e., $E[|Y|] < \infty$, with $P(|X_n| \le Y) = 1$, for $n = 1, 2, \ldots$, then $X_n \xrightarrow{P} X$ implies that $E[X_n] \to E[X]$.*

The next result provides a common way to show strong consistency (e.g., see Chapter 7). It is a consequence of the First Borel-Cantelli Lemma, and it indicates that converge with probability 1 is in a sense a "fast" form of convergence in probability.

Theorem A.8. *If, for all $\tau > 0$, $P(|X_n - X| > \tau) \to 0$ fast enough to obtain*

$$\sum_{n=1}^{\infty} P(|X_n - X| > \tau) < \infty, \tag{A.102}$$

then $X_n \xrightarrow{a.s.} X$.

Proof. First notice that a sample sequence $X_n(\omega)$ fails to converge to $X(\omega)$ if and only if there is a $\tau > 0$ such that $|X_n(\omega)) - X(\omega)| > \tau$ infinitely often as $n \to \infty$. Hence, $X_n \to X$ a.s. if and only if $P(|X_n - X| > \tau)$ *i.o.*$) = 0$, for all $\tau > 0$. The result then follows from the First Borel-Cantelli Lemma (see Thm. A.1). $\diamond$

The previous result implies that convergence in probability can produce convergence with probability 1 along a subsequence, obtained by "downsampling" the original sequence, as shown next.

Theorem A.9. *If* $X_n \xrightarrow{P} X$, *then there is an increasing sequence of indices* n_k *such that* $X_{n_k} \xrightarrow{a.s.} X$.

Proof. Since $P(|X_n - X| > \tau) \to 0$, for all $\tau > 0$, we can pick an increasing sequence of indices n_k such that $P(|X_{n_k} - X| > 1/k) \le 2^{-k}$. Given any $\tau > 0$, pick k_τ such that $1/k_\tau < \tau$. We have

$$\sum_{k=k_\tau}^{\infty} P(|X_{n_k} - X| > \tau) \le \sum_{k=k_\tau}^{\infty} P(|X_{n_k} - X| > 1/k) \le \sum_{k=k_\tau}^{\infty} 2^{-k} < \infty, \tag{A.103}$$

so that $X_{n_k} \xrightarrow{a.s.} X$ by Theorem A.8. $\diamond$

The previous theorem provides a criterion to *disprove* convergence $X_n \to X$ in probability: it is enough to show that there is no subsequence that converges to X with probability 1. This criterion is used in Chapter 4 (see Example 4.4).

Notice also that if X_n is monotone and $P(|X_n - X| > \tau) \to 0$, then $P(|X_n - X| > \tau)$ *i.o.*$) = 0$. Hence, if X_n is monotone, $X_n \to X$ in probability if and only if $X_n \to X$ with probability 1 (see the proof of Thm. A.8).

Stronger relations among the modes of convergence hold in special cases. In particular, we prove below that L^p convergence and convergence in probability are equivalent if the random sequence $\{X_n; n = 1, 2, \ldots\}$ is *uniformly bounded*, i.e., if there exists a finite $K > 0$, *which does not depend on* n, such that

$$|X_n| \le K, \text{with probability 1, for all } n = 1, 2, \ldots \tag{A.104}$$

meaning that $P(|X_n| < K) = 1$, for all $n = 1, 2, \ldots$ The classification error rate sequence $\{\varepsilon_n; n = 1, 2, \ldots\}$ is an example of uniformly bounded random sequence, with $K = 1$, therefore this is an important topic for our purposes. We have the following theorem.

Theorem A.10. *Let* $\{X_n; n = 1, 2, \ldots\}$ *be a uniformly bounded random sequence. The following statements are equivalent.*

(1) $X_n \xrightarrow{L^p} X$, *for some* $p > 0$.

(2) $X_n \xrightarrow{L^q} X$, *for all* $q > 0$.

(3) $X_n \xrightarrow{P} X$.

Proof. First note that we can assume without loss of generality that $X = 0$, since $X_n \to X$ if and only if $X_n - X \to 0$, and $X_n - X$ is also uniformly bounded, with $E[|X_n - X|^p] < \infty$. Showing that $(1) \Leftrightarrow (2)$ requires showing that $X_n \to 0$ in L^p, for some $p > 0$ implies that $X_n \to 0$ in L^q, for all $q > 0$. First observe that $E[|X_n|^q] \leq E[K^q] = K^q < \infty$, for all $q > 0$. If $q > p$, the result is immediate. Let $0 < q < p$. With $X = X_n^q$, $Y = 1$ and $r = p/q$, Holder's Inequality (A.67) yields

$$E[|X_n|^q] \leq E[|X_n|^p]^{q/p}. \tag{A.105}$$

Hence, if $E[|X_n|^p] \to 0$, then $E[|X_n|^q] \to 0$, proving the assertion. To show that $(2) \Leftrightarrow (3)$, first we show the direct implication by writing Markov's Inequality (A.70) with $X = |X_n|^p$ and $a = \tau^p$:

$$P(|X_n| \geq \tau) \leq \frac{E[|X_n|^p]}{\tau^p}, \quad \text{for all } \tau > 0. \tag{A.106}$$

The right-hand side goes to 0 by hypothesis, and thus so does the left-hand side, which is equivalent to (A.99) with $X = 0$. To show the reverse implication, write

$$E[|X_n|^p] = E[|X_n|^p I_{|X_n| < \tau}] + E[|X_n|^p I_{|X_n| \geq \tau}] \leq \tau^p + K^p P(|X_n| \geq \tau). \tag{A.107}$$

By assumption, $P(|X_n| \geq \tau) \to 0$, for all $\tau > 0$, so that $\lim E[|X_n|^p] \leq \tau^p$. Since $\tau > 0$ is arbitrary, this establishes the desired result. $\diamond$

The previous theorem implies that, for uniformly bounded random sequences, the relationships among the modes of convergence become:

$$\text{sure} \Rightarrow \text{almost-sure} \Rightarrow \left\{ \begin{matrix} L^p \\ \text{probability} \end{matrix} \right\} \Rightarrow \text{distribution} \tag{A.108}$$

As a simple corollary of Theorem A.10, we have the following useful result, which is also a corollary of Theorem A.7.

Theorem A.11. *(Bounded Convergence Theorem.) If $\{X_n; n = 1, 2, \ldots\}$ is a uniformly bounded random sequence and $X_n \xrightarrow{P} X$, then $E[X_n] \to E[X]$.*

Proof. From the previous theorem, $X_n \xrightarrow{L^1} X$, i.e., $E[|X_n - X|] \to 0$. But $|E[X_n - X]| \leq E[|X_n - X|]$, hence $|E[X_n - X]| \to 0$ and $E[X_n - X] \to 0$. $\diamond$

Example A.1. To illustrate these concepts, consider a sequence of independent binary random variables $X_1, X_2, \ldots$ that take on values in $\{0, 1\}$ such that

$$P(\{X_n = 1\}) = \frac{1}{n}, \quad n = 1, 2, \ldots \tag{A.109}$$

Then $X_n \xrightarrow{P} 0$, since $P(X_n > \tau) \to 0$, for every $\tau > 0$. By Theorem A.10, $X_n \xrightarrow{L^p} 0$ as well. However, X_n does not converge to 0 with probability 1. Indeed,

$$\sum_{n=1}^{\infty} P(\{X_n = 1\}) = \sum_{n=1}^{\infty} P(\{X_n = 0\}) = \infty, \tag{A.110}$$

and it follows from the 2nd Borel-Cantelli lemma that

$$P([\{X_n = 1\} \ i.o.]) = P([\{X_n = 0\} \ i.o.]) = 1, \tag{A.111}$$

so that X_n does not converge with probability 1. However, if convergence of the probabilities to zero is faster, e.g.

$$P(\{X_n = 1\}) = \frac{1}{n^2}, \quad n = 1, 2, \dots \tag{A.112}$$

then $\sum_{n=1}^{\infty} P(\{X_n = 1\}) < \infty$ and Theorem A.8 ensures that X_n converges to 0 with probability 1.
◇

In the previous example, note that, with $P(X_n = 1) = 1/n$, the probability of observing a 1 becomes infinitesimally small as $n \to \infty$, so the sequence consists, for all practice purposes, of all zeros for large enough n. Convergence in probability and in L^p of X_n to 0 agrees with this fact, but the lack of convergence with probability 1 does not. This is an indication that almost-sure convergence may be too stringent a criterion to be useful in practice, and convergence in probability and in L^p (assuming boundedness) may be enough. For example, this is the case in most signal processing applications, where L^2 is the criterion of choice. More generally, Engineering applications usually concern average performance and rates of failure.

A1.9   Asymptotic Theorems

The classical asymptotic theorems in probability theory are the Law of Large Numbers and the Central Limit Theorem, the proofs of which can be found, for example, in Chung [1974].

Theorem A.12. *(Law of Large Numbers.) Given an i.i.d. random sequence $\{X_n; n = 1, 2, \dots\}$ with common finite mean μ,*

$$\frac{1}{n} \sum_{i=1}^{n} X_i \xrightarrow{a.s.} \mu. \tag{A.113}$$

Theorem A.13. *(Central Limit Theorem.) Given an i.i.d. random sequence $\{X_n; n = 1, 2, \dots\}$ with common finite mean μ and common finite variance σ^2,*

$$\frac{1}{\sigma\sqrt{n}} \left(\sum_{i=1}^{n} X_i - n\mu \right) \xrightarrow{D} \mathcal{N}(0, 1). \tag{A.114}$$

The previous asymptotic theorems concern the behavior of a sum of n random variables as n approach infinity. It is also useful to have an idea of how partial sums differ from expected values for finite n. This issue is addressed by the so-called *concentration inequalities*, the most famous of which is Hoeffding's inequality, derived in Hoeffding [1963].

Theorem A.14. *(Hoeffding's Inequality.) Given independent (not necessarily identically-distributed) random variables $W_1, \ldots, W_n$ such that $P(a \leq W_i \leq b) = 1$, for $i = 1, \ldots, n$, the sum $Z_n = \sum_{i=1}^{n} W_i$ satisfies*

$$P(|Z_n - E[Z_n]| \geq \tau) \leq 2e^{-\frac{2\tau^2}{n(a-b)^2}}, \quad \text{for all } \tau > 0. \tag{A.115}$$

A2 Basic Matrix Theory

The material in this section is a summary of concepts and results from main matrix theory that are useful in the text. For an in-depth treatment, see Horn and Johnson [1990].

We assume that the reader is familiar with the concepts of vector, matrix, matrix product, transpose, determinant, and matrix inverse. We say that a set of vectors $\{\mathbf{x}_1, \ldots, \mathbf{x}_n\}$ is *linearly dependent* if the equation

$$a_1 \mathbf{x}_1 + \cdots + a_n \mathbf{x}_n = 0 \tag{A.116}$$

is satisfied for coefficients $a_1, \ldots, a_n$ that are not all zero. In other words, some of the vectors can be written as a linear combination of other vectors. If a set of vectors is not linearly dependent, then it is said to be *linearly independent*.

The *rank* of a matrix $A_{m \times n}$ is the largest number of columns of A that form a linearly independent set. This must be equal to the maximum number of rows that form a linearly independent set (row rank = column rank). A square matrix $A_{n \times n}$ is *nonsingular* if the inverse A^{-1} exists, or equivalently, the determinant $|A|$ is nonzero. The following are useful facts:

- $\text{rank}(A) = \text{rank}(A^T) = \text{rank}(AA^T) = \text{rank}(A^T A)$, where A^T denotes matrix transpose.

- $\text{rank}(A_{m \times n}) \leq \min\{m, n\}$. If equality is achieved, A is said to be *full-rank*.

- $A_{n \times n}$ is nonsingular if and only if $\text{rank}(A) = n$, i.e., A is full-rank. By the definition of rank, this means that the system of equations $A\mathbf{x} = 0$ has a unique solution $\mathbf{x} = 0$..

- If $B_{m \times m}$ is nonsingular then $\text{rank}(BA_{m \times n}) = \text{rank}(A)$ (multiplication by a nonsingular matrix preserves rank).

- $\text{rank}(A_{m \times n}) = \text{rank}(B_{m \times n})$ if and only if there are nonsingular matrices $X_{m \times m}$ and $Y_{n \times n}$ such that $B = XAY$.

- If $\text{rank}(A_{m \times n}) = k$, then there is a nonsingular matrix $B_{k \times k}$ and matrices $X_{m \times k}$ and $Y_{k \times n}$ such that $A = XBY$.

- As a corollary from the previous fact, $A_{m \times n}$ is a *rank-1 matrix* if A is a product of two vectors, $A = \mathbf{x}\mathbf{y}^T$, where the lengths of $\mathbf{x}$ and $\mathbf{y}$ are m and n, respectively.

An *eigenvalue* λ of a square matrix $A_{n \times n}$ is a solution of the equation

$$A\mathbf{x} = \lambda \mathbf{x}, \quad \mathbf{x} \neq 0, \tag{A.117}$$

in which case $\mathbf{x}$ is an *eigenvector* of A associated with λ. Complex λ and $\mathbf{x}$ are allowed. The following are useful facts:

- The eigenvalues of A and A^T are the same.

- If A is real symmetric, then all its eigenvalues are real.

- Since A is singular if and only if $A\mathbf{x} = 0$ with nonzero $\mathbf{x}$, we conclude that A is singular if and only if it has a zero eigenvalue.

From (A.117), λ is an eigenvalue if and only if $(A - \lambda I_n)\mathbf{x} = 0$ with nonzero $\mathbf{x}$. From previous facts, we conclude that $A - \lambda I_n$ is singular, that is, $|A - \lambda I_n| = 0$. But $p(\lambda) = |A - \lambda I_n|$ is a polynomial of degree n, which thus has exactly n roots (allowing for multiplicity), so we have proved the following useful fact.

Theorem A.15. *Any square matrix $A_{n \times n}$ has exactly n (possibly complex) eigenvalues $\{\lambda_1, \ldots, \lambda_n\}$, which are the roots of the* characteristic polynomial $p(\lambda) = |A - \lambda I_n|$.

If A is a diagonal matrix, then the eigenvalues are clearly the elements in its diagonal, so that $\text{Trace}(A) = \sum_{i=1}^n \lambda_i$ and $|A| = \prod_{i=1}^n \lambda_i$. It is a remarkable fact that it is still true that $\text{Trace}(A) = \sum_{i=1}^n \lambda_i$ and $|A| = \prod_{i=1}^n \lambda_i$ for any, not necessarily diagonal, square matrix A.

Matrix $B_{n \times n}$ is *similar* to matrix $A_{n \times n}$ if there is a nonsingular matrix $S_{n \times n}$ such that

$$B = S^{-1}AS. \tag{A.118}$$

It is easy to show that if A and B are similar, they have the same characteristic polynomial, and therefore the same set of eigenvalues (however, having the same set of eigenvalues is not sufficient for similarity).

Matrix A is said to be *diagonalizable* if it is similar to a diagonal matrix D. Since similarity preserves the characteristic polynomial, the eigenvalues of A are equal to the elements in the diagonal of D. The following theorem is not difficult to prove.

Theorem A.16. *A matrix $A_{n \times n}$ is diagonalizable if and only if it has a set of n linearly independent eigenvectors.*

A real-valued matrix $U_{n \times n}$ is said to be *orthogonal* if $U^T U = U U^T = I_n$, i.e., $U^{-1} = U^T$. Clearly, this happens if and only if the columns (and rows) of U are a set of unit-norm orthogonal vectors in R^n. Matrix $A_{n \times n}$ is said to be *orthogonally diagonalizable* if it is diagonalizable by an orthogonal matrix $U_{n \times n}$, i.e., $A = U^T D U$, where D is diagonal. Since

The following theorem, stated without proof, is one of the most important results in matrix theory.

Theorem A.17. *(Spectral Theorem.) If A is real symmetric, then it is orthogonally diagonalizable.*

Therefore, of A is real symmetric, we can write $A = U^T \Lambda U$ and $\Lambda = U A U^T$, where λ is a diagonal matrix containing the n eigenvalues of A on its diagonal. Furthermore, $U A = \Lambda U$, and thus the $i-the$ column of U is the eigenvector of A associated with the eigenvalue in the i-the position of the diagonal of Λ, for $i = 1, \ldots, n$.

A real symmetric matrix $A_{n \times n}$ is said to be *positive definite* if

$$\mathbf{x}^T A \mathbf{x} > 0, \quad \text{for all } \mathbf{x} \neq 0. \tag{A.119}$$

If the condition is relaxed to $\mathbf{x}^T A \mathbf{x} \geq 0$, then A is said to be *positive semi definite*. As we mentioned in the text, a covariance matrix is always at least positive semi-definite.

The following theorem is not difficult to prove.

Theorem A.18. *A real symmetric matrix A is positive definite if and only if all its eigenvalues are positive. It is positive semidefinite if and only if all eigenvalues are nonnegative.*

In particular, a positive definite matrix A is nonsingular. Another useful fact is that A is positive definite if and only if there is a nonsingular matrix C such that $A = C C^T$.

A3 Basic Lagrange-Multiplier Optimization

In this section we review results from Lagrange Multiplier theory that are needed in Section 6.1.1. For simplicity, we consider only minimization with inequality constraints, which is the case of the linear SVM optimization problems (6.6) and (6.20). Our presentation follows largely Chapter 5 of Boyd and Vandenberghe [2004], with some elements from Chapters 5 and 6 of Bertsekas [1995].

Consider the general (not necessarily convex) optimization problem:

$$\begin{aligned} \min \quad & f(\mathbf{x}) \\ \text{s.t.} \quad & g_i(\mathbf{x}) \leq 0, \quad i = 1,\ldots,n. \end{aligned} \tag{A.120}$$

where all functions are defined on R^d.

The *primal Lagrangian functional* is defined as

$$L_P(\mathbf{x},\boldsymbol{\lambda}) \;=\; f(\mathbf{x}) + \sum_{i=1}^{n} \lambda_i g_i(\mathbf{x}), \tag{A.121}$$

where λ_i is the *Lagrange multiplier* associated with constraint $g_i(\mathbf{x}) \leq 0$ and $\boldsymbol{\lambda} = (\lambda_1,\ldots,\lambda_n)$.

The *dual Lagrangian functional* is defined as:

$$L_D(\boldsymbol{\lambda}) \;=\; \inf_{\mathbf{x}\in R^d} L_P(\mathbf{x},\boldsymbol{\lambda}) \;=\; \inf_{\mathbf{x}\in R^d}\left(f(\mathbf{x}) + \sum_{i=1}^{n} \lambda_i g_i(\mathbf{x}) \right). \tag{A.122}$$

Using the properties of infimum, we have

$$\begin{aligned} L_D(\alpha\boldsymbol{\lambda}_1 + (1-\alpha)\boldsymbol{\lambda}_2) &= \inf_{\mathbf{x}\in R^d}\left(f(\mathbf{x}) + \sum_{i=1}^{n}(\alpha\lambda_{1,i} + (1-\alpha)\lambda_{2,i})g_i(\mathbf{x}) \right) \\ &= \inf_{\mathbf{x}\in R^d}\left(\alpha\left(f(\mathbf{x}) + \sum_{i=1}^{n}\lambda_{1,i}g_i(\mathbf{x}) \right) + (1-\alpha)\left(f(\mathbf{x}) + \sum_{i=1}^{n}\lambda_{2,i}g_i(\mathbf{x}) \right) \right) \\ &\geq \alpha \inf_{\mathbf{x}\in R^d}\left(f(\mathbf{x}) + \sum_{i=1}^{n}\lambda_{1,i}g_i(\mathbf{x}) \right) + (1-\alpha)\inf_{\mathbf{x}\in R^d}\left(f(\mathbf{x}) + \sum_{i=1}^{n}\lambda_{2,i}g_i(\mathbf{x}) \right) \\ &= \alpha L_D(\boldsymbol{\lambda}_1) + (1-\alpha)L_D(\boldsymbol{\lambda}_2), \end{aligned} \tag{A.123}$$

for all $\boldsymbol{\lambda}_1, \boldsymbol{\lambda}_2 \in R^n$ and $0 \leq \alpha \leq 1$. The dual Lagrangian functional $L_D(\boldsymbol{\lambda})$ is therefore a *concave* function. Furthermore, for all $\mathbf{x} \in F$, where F is the feasible region of (A.120), and $\boldsymbol{\lambda} \geq 0$,

$$L_P(\mathbf{x},\boldsymbol{\lambda}) \;=\; f(\mathbf{x}) + \sum_{i=1}^{n} \lambda_i g_i(\mathbf{x}) \;\leq\; f(\mathbf{x}), \tag{A.124}$$

since $g_i(\mathbf{x}) \leq 0$, for $i = 1,\ldots,n$. It follows that

$$L_D(\boldsymbol{\lambda}) \;=\; \inf_{\mathbf{x}\in R^d} L_P(\mathbf{x},\boldsymbol{\lambda}) \;\leq\; \inf_{\mathbf{x}\in F} f(\mathbf{x}) \;=\; f(\mathbf{x}^*), \quad \text{for all } \boldsymbol{\lambda} \geq 0, \tag{A.125}$$

showing that $L_D(\boldsymbol{\lambda})$ is a lower bound on $f(\mathbf{x}^*)$, whenever $\boldsymbol{\lambda} \geq 0$.

The natural next step is to maximize this lower bound. This leads to the *dual* optimization problem:

$$\begin{aligned} \max \quad & L_D(\boldsymbol{\lambda}) \\ \text{s.t.} \quad & \boldsymbol{\lambda} \geq 0. \end{aligned} \tag{A.126}$$

Since the cost $L_D(\boldsymbol{\lambda})$ is concave (as shown previously) and the feasible region is a convex set, this is a convex optimization problem, for which there are efficient solution methods. This is true whether or not the original problem (A.120) is convex.

If $\boldsymbol{\lambda}^*$ is a solution of (A.126), then it follows from (A.125) that $L_D(\boldsymbol{\lambda}^*) \leq f(\mathbf{x}^*)$, which is known as the *weak duality* property. If equality is achieved,

$$L_D(\boldsymbol{\lambda}^*) = f(\mathbf{x}^*), \tag{A.127}$$

then the problem is said to satisfy the *strong duality* property. This property is not always satisfied, but there are several sets of conditions, called *constraint qualifications*, that ensure strong duality. For convex optimization problems with affine constraints, such as the linear SVM optimization problems (6.6) and (6.20), a simple constraint qualification condition, known as *Slater's condition*, guarantees strong duality as long as the feasible region is nonempty.

The point $(\bar{\mathbf{w}}, \bar{\mathbf{z}})$, where $\bar{\mathbf{w}} \in W$ and $\bar{\mathbf{z}} \in Z$, is a *saddle point* of a function h defined on $W \times Z$ if

$$h(\bar{\mathbf{y}}, \bar{\mathbf{z}}) = \inf_{\mathbf{w} \in W} h(\mathbf{w}, \bar{\mathbf{z}}) \quad \text{and} \quad h(\bar{\mathbf{y}}, \bar{\mathbf{z}}) = \sup_{\mathbf{z} \in Z} h(\bar{\mathbf{w}}, \mathbf{z}). \tag{A.128}$$

Under strong duality,

$$
\begin{aligned}
f(\mathbf{x}^*) = L_D(\boldsymbol{\lambda}^*) = \inf_{\mathbf{x} \in R^d} L_P(\mathbf{x}, \boldsymbol{\lambda}^*) &= \inf_{\mathbf{x} \in R^d} \left(f(\mathbf{x}) + \sum_{i=1}^{n} \lambda_i^* g_i(\mathbf{x}) \right) \\
&\leq L_P(\mathbf{x}^*, \boldsymbol{\lambda}^*) = f(\mathbf{x}^*) + \sum_{i=1}^{n} \lambda_i^* g_i(\mathbf{x}^*) \leq f(\mathbf{x}^*).
\end{aligned}
\tag{A.129}
$$

The first inequality follows from the definition of inf, whereas the second inequality follows from the facts that $\lambda_i^* \geq 0$ and $g_i(\mathbf{x}^*) \leq 0$, for $i = 1, \ldots, n$. It follows from (A.129) that both inequalities hold with equality. In particular,

$$L_P(\mathbf{x}^*, \boldsymbol{\lambda}^*) = \inf_{\mathbf{x} \in R^d} L_P(\mathbf{x}, \boldsymbol{\lambda}^*). \tag{A.130}$$

On the other hand, it is always true that

$$\sup_{\boldsymbol{\lambda} \geq 0} L_P(\mathbf{x}^*, \boldsymbol{\lambda}) = \sup_{\boldsymbol{\lambda} \geq 0} \left(f(\mathbf{x}^*) + \sum_{i=1}^{n} \lambda_i^* g_i(\mathbf{x}^*) \right) = f(\mathbf{x}^*), \tag{A.131}$$

because $g_i(\mathbf{x}^*) \leq 0$, for $i = 1, \ldots, n$, so that $f(\mathbf{x}^*)$ maximizes $L_P(\mathbf{x}^*, \boldsymbol{\lambda})$ at $\boldsymbol{\lambda} = 0$. With the extra condition of strong duality, we have from (A.129) that $f(\mathbf{x}^*) = L_P(\mathbf{x}^*, \boldsymbol{\lambda}^*)$, so we obtain

$$L_P(\mathbf{x}^*, \boldsymbol{\lambda}^*) = \sup_{\boldsymbol{\lambda} \geq 0} L_P(\mathbf{x}^*, \boldsymbol{\lambda}). \tag{A.132}$$

It follows from (A.130) and (A.132) that strong duality implies that $(\mathbf{x}^*, \boldsymbol{\lambda}^*)$ is a saddle point of $L_p(\mathbf{x}, \boldsymbol{\lambda})$. It follows immediately from the general relations

$$f(\mathbf{x}^*) = \sup_{\boldsymbol{\lambda} \geq 0} L_P(\mathbf{x}^*, \boldsymbol{\lambda}) \quad \text{and} \quad L_D(\boldsymbol{\lambda}^*) = \inf_{\mathbf{x} \in R^d} L_P(\mathbf{x}, \boldsymbol{\lambda}^*) \tag{A.133}$$

that the converse is true: if $(\mathbf{x}^*, \boldsymbol{\lambda}^*)$ is a saddle point of $L_p(\mathbf{x}, \boldsymbol{\lambda})$ then strong duality holds.

An optimal point $(\mathbf{x}^*, \boldsymbol{\lambda}^*)$, under strong duality, simultaneously minimizes $L_P(\mathbf{x}, \boldsymbol{\lambda})$ with respect to $\mathbf{x}$ and maximizes $L_P(\mathbf{x}, \boldsymbol{\lambda})$ with respect to $\boldsymbol{\lambda}$. In particular, an optimal point $(\mathbf{x}^*, \boldsymbol{\lambda}^*)$ satisfies

$$\mathbf{x}^* = \arg \min_{\mathbf{x} \in R^d} L_P(\mathbf{x}, \boldsymbol{\lambda}^*). \tag{A.134}$$

Since this is an *unconstrained* minimization problem, necessary conditions for unconstrained minima apply. In particular, assuming that f and g_i are differentiable, for $i = 1, \ldots, n$, the general stationarity condition must be satisfied:

$$\nabla_{\mathbf{x}} L_P(\mathbf{x}^*, \boldsymbol{\lambda}^*) = \nabla_{\mathbf{x}} f(\mathbf{x}^*) + \sum_{i=1}^{n} \lambda_i^* \nabla_{\mathbf{x}} g_i(\mathbf{x}^*) = 0. \tag{A.135}$$

Another consequence of (A.129) is

$$f(\mathbf{x}^*) = f(\mathbf{x}^*) + \sum_{i=1}^{n} \lambda_i^* g_i(\mathbf{x}^*) \Rightarrow \sum_{i=1}^{n} \lambda_i^* g_i(\mathbf{x}^*) = 0, \tag{A.136}$$

from which the following important *complementary slackness* conditions follow:

$$\lambda_i^* g_i(\mathbf{x}^*) = 0, \quad i = 1, \ldots, n. \tag{A.137}$$

This means that if a constraint is inactive at the optimum, i.e., $g_i(\mathbf{x}^*) < 0$, then the corresponding optimal Lagrange multiplier λ_i^* must be zero. Conversely, $\lambda_i^* > 0$ implies that $g_i(\mathbf{x}^*) = 0$, i.e., the corresponding constraint is active (tight) at the optimum.

We can summarize all the previous results in the following classical theorem.

Theorem A.19. *(Karush-Kuhn-Tucker Conditions). Let $\mathbf{x}^*$ be a solution of the original optimization problem in (A.120), and let $\boldsymbol{\lambda}^*$ be a solution of the dual optimization problem in (A.126) such that strong duality is satisfied. Assume further that f and g_i are differentiable, for $i = 1, \ldots, n$. Then the following conditions must be satisfied:*

$$
\begin{aligned}
\nabla_{\mathbf{x}} L_P(\mathbf{x}^*, \boldsymbol{\lambda}^*) &= \nabla_{\mathbf{x}} f(\mathbf{x}^*) + \sum_{i=1}^{n} \lambda_i^* \nabla_{\mathbf{x}} g_i(\mathbf{x}^*) = 0, && \text{(stationarity)} \\
g_i(\mathbf{x}^*) &\leq 0, \quad i = 1, \ldots, n, && \text{(primal feasibility)} \\
\lambda_i^* &\geq 0, \quad i = 1, \ldots, n, && \text{(dual feasibility)} \\
\lambda_i^* g_i(\mathbf{x}^*) &= 0, \quad i = 1, \ldots, n. && \text{(complementary slackness)}
\end{aligned}
\tag{A.138}
$$

Furthermore, it can be shown that if the original optimization problem in (A.120) is convex with affine constraints, then the KKT conditions are also sufficient for optimality.

A4 Proof of the Cover-Hart Theorem

In this section we present proofs of Thm 5.1 and 5.3. The proof of Thm 5.1 follows the general structure of the original proof in Cover and Hart [1967], with some differences. This proof assumes existence and continuity almost everywhere of the class-conditional densities. In Stone [1977] a more general proof is given, which does not assume existence of densities (see also Chapter 5 of Devroye et al. [1996]).

Proof of Theorem 5.1

First, one has to show that the nearest neighbor $\mathbf{X}_n^{(1)}$ of a test point $\mathbf{X}$ converges to $\mathbf{X}$ as $n \to \infty$. The existence of densities makes this simple to show. First note that, for any $\tau > 0$,

$$P(||\mathbf{X}_n^{(1)} - \mathbf{X}|| > \tau) = P(||\mathbf{X}_i - \mathbf{X}|| > \tau; \, i = 1, \ldots, n) = (1 - P(||\mathbf{X}_1 - \mathbf{X}|| < \tau))^n. \quad \text{(A.139)}$$

If we can show that $P(||\mathbf{X}_1 - \mathbf{X}|| < \tau) > 0$, then it follows from (A.139) that $P(||\mathbf{X}_n^{(1)} - \mathbf{X}|| > \tau) \to 0$, so that $\mathbf{X}_n^{(1)} \to \mathbf{X}$ in probability. Since $\mathbf{X}_1$ and $\mathbf{X}$ are independent and identically distributed with density $p_{\mathbf{X}}$, $\mathbf{X}_1 - \mathbf{X}$ has a density $p_{\mathbf{X}_1 - \mathbf{X}}$, given by the classical convolution formula:

$$p_{\mathbf{X}_1 - \mathbf{X}}(\mathbf{x}) = \int_{R^d} p_{\mathbf{X}}(\mathbf{x} + \mathbf{u}) \, p_{\mathbf{X}}(\mathbf{u}) \, d\mathbf{u}. \quad \text{(A.140)}$$

From this, we have $p_{\mathbf{X}_1 - \mathbf{X}}(\mathbf{0}) = \int_{R^d} p_{\mathbf{X}}^2(\mathbf{x}) \, d\mathbf{u} > 0$. It follows, by continuity of the integral, that $p_{\mathbf{X}_1 - \mathbf{X}}$ must be nonzero in a neighborhood of $\mathbf{0}$, i.e., $P(||\mathbf{X}_1 - \mathbf{X}|| < \tau) > 0$, as was to be shown.

Now, let Y_n' denote the label of the nearest neighbor $\mathbf{X}_n^{(1)}$. Consider the conditional error rate

$$\begin{aligned} P(\psi_n(\mathbf{X}) \neq Y \mid \mathbf{X}, \mathbf{X}_1, \ldots, \mathbf{X}_n) &= P(Y_n' \neq Y \mid \mathbf{X}, \mathbf{X}_n^{(1)}) \\ &= P(Y = 1, Y_n' = 0 \mid \mathbf{X}, \mathbf{X}_n^{(1)}) + P(Y = 0, Y_n' = 1 \mid \mathbf{X}, \mathbf{X}_n^{(1)}) \\ &= P(Y = 1 \mid \mathbf{X})P(Y_n' = 0 \mid \mathbf{X}_n^{(1)}) + P(Y = 0 \mid \mathbf{X})P(Y_n' = 1 \mid \mathbf{X}_n^{(1)}) \\ &= \eta(\mathbf{X})(1 - \eta(\mathbf{X}_n^{(1)})) + (1 - \eta(\mathbf{X}))\eta(\mathbf{X}_n^{(1)}) \end{aligned} \quad \text{(A.141)}$$

where independence of $(\mathbf{X}_n^{(1)}, Y_n')$ and $(\mathbf{X}, Y)$ was used. We now use the assumption that the class-conditional densities exist and are continuous a.e., which implies that η is continuous a.e. We had established previously that $\mathbf{X}_n^{(1)} \to \mathbf{X}$ in probability. By the Continuous Mapping Theorem (see Theorem A.6), $\eta(\mathbf{X}_n^{(1)}) \to \eta(\mathbf{X})$ in probability and

$$P(\psi_n(\mathbf{X}) \neq Y \mid \mathbf{X}, \mathbf{X}_1, \ldots, \mathbf{X}_n) \to 2\eta(\mathbf{X})(1 - \eta(\mathbf{X})) \text{ in probability.} \quad \text{(A.142)}$$

Since all random variables are bounded in the interval $[0, 1]$, we can apply the Bounded Convergence Theorem (see Thm. A.11) to obtain

$$E[\varepsilon_n] = E[P(\psi_n(\mathbf{X}) \neq Y \mid \mathbf{X}, \mathbf{X}_1, \ldots, \mathbf{X}_n)] \to E[2\eta(\mathbf{X})(1 - \eta(\mathbf{X}))], \quad \text{(A.143)}$$

proving the first part of the theorem.

For the second part, let $r(\mathbf{X}) = \min\{\eta(\mathbf{X}), 1-\eta(\mathbf{X})\}$ and note that $\eta(\mathbf{X})(1-\eta(\mathbf{X})) = r(\mathbf{X})(1-r(\mathbf{X}))$. It follows that

$$
\begin{aligned}
\varepsilon_{\mathrm{NN}} &= E[2\eta(\mathbf{X})(1-\eta(\mathbf{X}))] = E[2r(\mathbf{X})(1-r(\mathbf{X}))] \\
&= 2E[r(\mathbf{X})]E[(1-r(\mathbf{X}))] + 2\mathrm{Cov}(r(\mathbf{X}), 1-r(\mathbf{X})) \\
&= 2\varepsilon^*(1-\varepsilon^*) - 2\mathrm{Var}(r(\mathbf{X})) \leq 2\varepsilon^*(1-\varepsilon^*) \leq 2\varepsilon^*,
\end{aligned}
\tag{A.144}
$$

as required.

Proof of Theorem 5.3

The proof of (5.13) and (5.14) follows the same structure as in the case $k = 1$. As before, the first step is to show that the ith-nearest neighbor $\mathbf{X}_n^{(i)}$ of $\mathbf{X}$, for $i = 1, \ldots, k$, converges to $\mathbf{X}$ in probability as $n \to \infty$. This is so because, for every $\tau > 0$,

$$
P(\|\mathbf{X}_n^{(i)} - \mathbf{X}\| > \tau) = P(\|\mathbf{X}_j - \mathbf{X}\| > \tau; j = k, \ldots, n) = (1 - P(\|\mathbf{X}_1 - \mathbf{X}\| < \tau))^{n-k-1} \to 0,
\tag{A.145}
$$

since $P(\|\mathbf{X}_1 - \mathbf{X}\| < \tau) > 0$, as shown in the previous proof. Next, let the label of the $i$the-nearest neighbor $\mathbf{X}_n^{(i)}$ of $\mathbf{X}$ by $Y_n^{(i)}$, and consider the conditional error rate

$$
\begin{aligned}
&P(\psi_n(\mathbf{X}) \neq Y \mid \mathbf{X}, \mathbf{X}_1, \ldots, \mathbf{X}_n) \\
&= P(Y = 1, \textstyle\sum_{i=1}^{k} Y_n^{(i)} < \frac{k}{2} \mid \mathbf{X}, \mathbf{X}_n^{(1)}, \ldots, \mathbf{X}_n^{(k)}) + P(Y = 0, \textstyle\sum_{i=1}^{k} Y_n^{(i)} > \frac{k}{2} \mid \mathbf{X}, \mathbf{X}_n^{(1)}, \ldots, \mathbf{X}_n^{(k)}) \\
&= P(Y = 1 \mid \mathbf{X})P(\textstyle\sum_{i=1}^{k} Y_n^{(i)} < \frac{k}{2} \mid \mathbf{X}_n^{(1)}, \ldots, \mathbf{X}_n^{(k)}) \\
&\quad + P(Y = 0 \mid \mathbf{X})P(\textstyle\sum_{i=1}^{k} Y_n^{(i)} > \frac{k}{2} \mid \mathbf{X}_n^{(1)}, \ldots, \mathbf{X}_n^{(k)}) \\
&= \eta(\mathbf{X})\textstyle\sum_{i=0}^{(k-1)/2} P(\textstyle\sum_{j=1}^{k} Y_n^{(j)} = i \mid \mathbf{X}_n^{(1)}, \ldots, \mathbf{X}_n^{(k)}) \\
&\quad + (1-\eta(\mathbf{X}))\textstyle\sum_{i=(k+1)/2}^{k} P(\textstyle\sum_{j=1}^{k} Y_n^{(j)} = i \mid \mathbf{X}_n^{(1)}, \ldots, \mathbf{X}_n^{(k)}),
\end{aligned}
\tag{A.146}
$$

where

$$
\begin{aligned}
P(\textstyle\sum_{j=1}^{k} Y_n^{(j)} = i \mid \mathbf{X}_n^{(1)}, \ldots, \mathbf{X}_n^{(k)}) &= \sum_{\substack{m_1,\ldots,m_k \in \{0,1\} \\ m_1 + \cdots + m_k = i}} \prod_{j=1}^{k} P(Y_n^{(j)} = m_j \mid \mathbf{X}_n^{(j)}) \\
&= \sum_{\substack{m_1,\ldots,m_k \in \{0,1\} \\ m_1 + \cdots + m_k = i}} \prod_{j=1}^{k} \eta(\mathbf{X}_n^{(j)})^{m_j} (1 - \eta(\mathbf{X}_n^{(j)}))^{1-m_j}.
\end{aligned}
\tag{A.147}
$$

Using the previously established fact that $\mathbf{X}_n^{(j)} \to \mathbf{X}$ in probability, for $i = 1, \ldots, k$, it follows from the assumption of continuity of the distributions a.e. and the Continuous Mapping Theorem (see Theorem A.6) that

$$P\left(\sum_{j=1}^{k} Y_n^{(j)} = i \mid \mathbf{X}_n^{(1)}, \ldots, \mathbf{X}_n^{(k)}\right) \xrightarrow{P} \sum_{\substack{m_1,\ldots,m_k \in \{0,1\} \\ m_1+\cdots+m_k=i}} \prod_{j=1}^{k} \eta(\mathbf{X})^{m_j} (1 - \eta(\mathbf{X}))^{1-m_j}$$

$$= \binom{k}{i} \eta(\mathbf{X})^i (1 - \eta(\mathbf{X}))^{k-i} \tag{A.148}$$

and

$$P(\psi_n(\mathbf{X}) \neq Y \mid \mathbf{X}, \mathbf{X}_1, \ldots, \mathbf{X}_n) \xrightarrow{P} \sum_{i=0}^{(k-1)/2} \eta(\mathbf{X})^{i+1}(1 - \eta(\mathbf{X}))^{k-i}$$
$$+ \sum_{i=(k+1)/2}^{k} \eta(\mathbf{X})^i (1 - \eta(\mathbf{X}))^{k+1-i}. \tag{A.149}$$

Since all random variables are bounded in the interval $[0, 1]$, we can apply the Bounded Convergence Theorem (see Thm. A.11) to obtain

$$E[\varepsilon_n] = E[P(\psi_n(\mathbf{X}) \neq Y \mid \mathbf{X}, \mathbf{X}_1, \ldots, \mathbf{X}_n)]$$
$$\to E\left[\sum_{i=0}^{(k-1)/2} \eta(\mathbf{X})^{i+1}(1 - \eta(\mathbf{X}))^{k-i} + \sum_{i=(k+1)/2}^{k} \eta(\mathbf{X})^i (1 - \eta(\mathbf{X}))^{k+1-i}\right], \tag{A.150}$$

establishing (5.13) and (5.14).

For the second part, as before, we let $r(\mathbf{X}) = \min\{\eta(\mathbf{X}), 1 - \eta(\mathbf{X})\}$ and note that $\eta(\mathbf{X})(1 - \eta(\mathbf{X})) = r(\mathbf{X})(1 - r(\mathbf{X}))$. By symmetry, it is easy to see that $\alpha_k(\eta(\mathbf{X})) = \alpha_k(r(\mathbf{X}))$. We seek an inequality $\alpha_k(r(\mathbf{X})) \leq a_k r(\mathbf{X})$, so that

$$\varepsilon_{\mathrm{kNN}} = E[\alpha_k(\eta(\mathbf{X}))] = E[\alpha_k(r(\mathbf{X}))] \leq a_k E[r(\mathbf{X})] = a_k \varepsilon^*, \tag{A.151}$$

where $a_k > 1$ is as small as possible. But as can be seen in Figure 5.8, a_k corresponds to the slope of the tangent line to $\alpha_k(p)$, in the range $p \in [0, \frac{1}{2}]$, through the origin, so it must satisfy (5.21).

A5 Proof of Stone's Theorem

In this section, we present a proof of Thm 5.4, which essentially follows the proof given by Devroye et al. [1996]. The original proof in Stone [1977] is more general, relaxing the nonnegativity and normalization assumptions (5.2) on the weights, while also showing that, under (5.2), the conditions on the weights given in the theorem are both necessary and sufficient for universal consistency.

Proof of Theorem 5.4

It follows from Lemma 5.1, and the comment following it, that it is sufficient to show that $E[(\eta_n(\mathbf{X}) - \eta(\mathbf{X}))^2] \to 0$, as $n \to \infty$. Introduce the smoothed posterior-probability function

$$\tilde{\eta}_n(\mathbf{x}) = \sum_{i=1}^{n} W_{n,i}(\mathbf{x})\eta(\mathbf{X}_i). \tag{A.152}$$

This is not a true estimator, since it is a function of $\eta(\mathbf{x})$. However, it allows one to break the problem down into two manageable parts:

$$
\begin{aligned}
E[(\eta_n(\mathbf{X}) - \eta(\mathbf{X}))^2] &= E[(\eta_n(\mathbf{X}) - \tilde{\eta}_n(\mathbf{X}) + \tilde{\eta}_n(\mathbf{X}) - \eta(\mathbf{X}))^2] \\
&\le 2E[(\eta_n(\mathbf{X}) - \tilde{\eta}_n(\mathbf{X}))^2] + 2E[(\tilde{\eta}_n(\mathbf{X}) - \eta(\mathbf{X}))^2],
\end{aligned}
\tag{A.153}
$$

where the inequality follows from the fact that $(a+b)^2 \le 2(a^2 + b^2)$. The rest of the proof consists in showing that $E[(\eta_n(\mathbf{X}) - \tilde{\eta}_n(\mathbf{X}))^2] \to 0$, and then showing that $E[(\tilde{\eta}_n(\mathbf{X}) - \eta(\mathbf{X}))^2] \to 0$.

For the first part, notice that

$$
\begin{aligned}
E[(\eta_n(\mathbf{X}) - \tilde{\eta}_n(\mathbf{X}))^2] &= E\left[\left(\sum_{i=1}^{n} W_{ni}(\mathbf{X})(Y_i - \eta(\mathbf{X}_i))\right)^2\right] \\
&= \sum_{i=1}^{n}\sum_{j=1}^{n} E\left[W_{ni}(\mathbf{X})W_{nj}(\mathbf{X})(Y_i - \eta(\mathbf{X}_i)(Y_j - \eta(\mathbf{X}_j)\right] \\
&= \sum_{i=1}^{n}\sum_{j=1}^{n} E\left[E\left[W_{ni}(\mathbf{X})W_{nj}(\mathbf{X})(Y_i - \eta(\mathbf{X}_i)(Y_j - \eta(\mathbf{X}_j) \mid \mathbf{X}, \mathbf{X}_1, \ldots, \mathbf{X}_n\right]\right]
\end{aligned}
\tag{A.154}
$$

Now, given $\mathbf{X}, \mathbf{X}_1, \ldots, \mathbf{X}_n$, $W_{ni}(\mathbf{X})$ and $W_{nj}(\mathbf{X})$ are constants, and $Y_i - \eta(\mathbf{X}_i)$ and $Y_j - \eta(\mathbf{X}_j)$ are zero-mean random variables. Furthermore, $Y_i - \eta(\mathbf{X}_i)$ and $Y_j - \eta(\mathbf{X}_j)$ are independent if $i \neq j$. Therefore, $E\left[W_{ni}(\mathbf{X})W_{nj}(\mathbf{X})(Y_i - \eta(\mathbf{X}_i)(Y_j - \eta(\mathbf{X}_j) \mid \mathbf{X}, \mathbf{X}_1, \ldots, \mathbf{X}_n\right] = 0$, for $i \neq j$, and we obtain

$$
\begin{aligned}
E[(\eta_n(\mathbf{X}) - \tilde{\eta}_n(\mathbf{X}))^2] &= \sum_{i=1}^{n} E\left[W_{ni}^2(\mathbf{X})(Y_i - \eta(\mathbf{X}_i)^2]\right] \\
&\le E\left[\sum_{i=1}^{n} W_{ni}^2(\mathbf{X})\right] \le E\left[\max_{i=1,\ldots,n} W_{n,i}(\mathbf{x})\sum_{i=1}^{n} W_{ni}(\mathbf{X})\right] = E\left[\max_{i=1,\ldots,n} W_{n,i}(\mathbf{x})\right] \to 0,
\end{aligned}
\tag{A.155}
$$

by condition (ii) of Stone's Theorem and the Bounded Convergence Theorem A.11.

The second part is more technical. First, given $\tau > 0$, find a function η^* such that $0 \le \eta^*(\mathbf{x}) \le 1$, η^* is $P_{\mathbf{X}}$-square-integrable, continuous, and has compact support, and $E[(\eta^*(\mathbf{X}) - \eta(\mathbf{X}))^2] < \tau$. Such a function exists, because $\eta(\mathbf{x})$ is $P_{\mathbf{X}}$-integrable (see Section 2.6.3), and therefore square-integrable, since $\eta^2(\mathbf{x}) \le \eta(\mathbf{x})$, and the set of continuous function with compact support is dense in the set of square-integrable functions. Now, write

$$
E[(\tilde{\eta}_n(\mathbf{X}) - \eta(\mathbf{X}))^2] = E\left[\left(\sum_{i=1}^{n} W_{ni}(\mathbf{X})(\eta(\mathbf{X}_i) - \eta(\mathbf{X}))\right)^2\right] \leq E\left[\sum_{i=1}^{n} W_{ni}(\mathbf{X})(\eta(\mathbf{X}_i) - \eta(\mathbf{X}))^2\right]
$$

$$
= E\left[\sum_{i=1}^{n} W_{ni}(\mathbf{X})\left((\eta(\mathbf{X}_i) - \eta^*(\mathbf{X}_i)) + (\eta^*(\mathbf{X}_i) - \eta^*(\mathbf{X})) + (\eta^*(\mathbf{X}) - \eta(\mathbf{X}))\right)^2\right]
$$

$$
\leq 3E\left[\sum_{i=1}^{n} W_{ni}(\mathbf{X})\left((\eta(\mathbf{X}_i) - \eta^*(\mathbf{X}_i))^2 + (\eta^*(\mathbf{X}_i) - \eta^*(\mathbf{X}))^2 + (\eta^*(\mathbf{X}) - \eta(\mathbf{X}))^2\right)\right]
$$

$$
\leq 3E\left[\sum_{i=1}^{n} W_{ni}(\mathbf{X})(\eta(\mathbf{X}_i) - \eta^*(\mathbf{X}_i))^2\right] + 3E\left[\sum_{i=1}^{n} W_{ni}(\mathbf{X})(\eta^*(\mathbf{X}_i) - \eta^*(\mathbf{X}))^2\right] + 3E\left[(\eta^*(\mathbf{X}) - \eta(\mathbf{X}))^2\right]
$$

$$
= I + II + III\,,
$$

(A.156)

where the first inequality follows from Jensen's Inequality, while the second inequality follows from the fact that $(a + b + c)^2 \leq 3(a^2 + b^2 + c^2)$. Now, by construction of η^* and condition (iii) of Stone's Theorem, it follows that $I < 3\tau$ and $III < 3c\tau$. To bound II, notice that η^*, being continuous on a compact support, is also uniformly continuous. Hence, given $\tau > 0$, there is a $\delta > 0$ such that $\|\mathbf{x}' - \mathbf{x}\| < \delta$ implies that $|\eta^*(\mathbf{x}') - \eta^*(\mathbf{x})| < \tau$, for all $\mathbf{x}', \mathbf{x} \in R^d$. Hence,

$$
II \leq 3E\left[\sum_{i=1}^{n} W_{n,i}(\mathbf{X})I_{\|\mathbf{X}_i - \mathbf{X}\| > \delta}\right] + 3E\left[\sum_{i=1}^{n} W_{n,i}(\mathbf{X})\tau\right] = 3E\left[\sum_{i=1}^{n} W_{n,i}(\mathbf{X})I_{\|\mathbf{X}_i - \mathbf{X}\| > \delta}\right] + 3\tau\,,
$$

(A.157)

where we used the fact that $|\eta^*(\mathbf{x}') - \eta^*(\mathbf{x})| \leq 1$. Using condition (i) of Stone's Theorem and the Bounded Convergence Theorem A.11, it follows that $\limsup_{n \to \infty} II \leq 3\tau$. Putting all together,

$$
\limsup_{n \to \infty} E[(\tilde{\eta}_n(\mathbf{X}) - \eta(\mathbf{X}))^2] \leq 3\tau + 3c\tau + 3\tau = 3(c + 2)\tau\,.
$$

(A.158)

Since τ is arbitrary, it follows that $E[(\tilde{\eta}_n(\mathbf{X}) - \eta(\mathbf{X}))^2] \to 0$ and the proof is complete.

A6   Proof of the Vapnik-Chervonenkis Theorem

In this section, we present a proof of Thm 8.2. Our proof combines elements of the proofs given by Pollard [1984] and Devroye et al. [1996], who credit Dudley [1978]. See also Castro [2020]. We prove a general version of the result and then specialize it to the classification case.

Consider a probability space $(R^p, \mathcal{B}^p, \nu)$, and n i.i.d. random variables $Z_1, \ldots, Z_n \sim \nu$. (For a review of probability theory, see Section A1.) Note that each Z_i is in fact a random vector, but we do not employ the usual boldface type here, so as not to encumber the notation. An *empirical measure* is a random measure on $(R^p, \mathcal{B}^P)$ that is a function of $Z_1, \ldots, Z_n$. The standard empirical measure ν_n puts mass $1/n$ over each Z_i, so that

$$\nu_n(A) = \frac{1}{n} \sum_{i=1}^{n} I_{Z_i \in A}, \tag{A.159}$$

for $A \in \mathcal{B}^p$. By the Law of Large Numbers (LLN), $\nu_n(A) \overset{a.s.}{\to} \nu(A)$, as $n \to \infty$, for any fixed A. In the VC theorem, one is interested instead in a *uniform* version of the LLN: $\sup_{A \in \mathcal{A}} |\nu_n(A) - \nu(A)| \overset{a.s.}{\to} 0$, for a suitably provided family of sets $\mathcal{A} \subset \mathcal{B}^p$. General conditions to ensure the measurability of $\sup_{A \in \mathcal{A}} |\nu_n(A) - \nu(A)|$ and of various other quantities in the proofs are discussed in Pollard [1984]; such will be assumed tacitly below.

Define a second (signed) empirical measure $\tilde{\nu}_n$, which puts mass $1/n$ or $-1/n$ randomly over each Z_i, i.e.,

$$\tilde{\nu}_n(A) = \frac{1}{n} \sum_{i=1}^{n} \sigma_i I_{Z_i \in A} \tag{A.160}$$

for $A \in \mathcal{A}$, where $\sigma_1, \ldots, \sigma_n$ are i.i.d. random variables with $P(\sigma_1 = 1) = P(\sigma_1 = -1) = 1/2$, independently of $Z_1, \ldots, Z_n$.

It turns out that the VC theorem, much as Theorem 8.1, can be proved by a direct application of the Union Bound (A.10) and Hoeffding's Inequality (8.8), with the addition of the next key lemma.

Lemma A.2. *(Symmetrization Lemma). Regardless of the measure ν,*

$$P\left(\sup_{A \in \mathcal{A}} |\nu_n(A) - \nu(A)| > \tau\right) \leq 4P\left(\sup_{A \in \mathcal{A}} |\tilde{\nu}_n(A)| > \frac{\tau}{4}\right), \quad \text{for all } \tau > 0 \text{ and } n \geq 2\tau^{-2}. \tag{A.161}$$

Proof. Consider a second sample $Z_1', \ldots, Z_n' \sim \nu$, independent of $Z_1, \ldots, Z_n$ and the signs $\sigma_1, \ldots, \sigma_n$. In the first part of the proof, one seeks to relate the tail probability of $\sup_{A \in \mathcal{A}} |\nu_n(A) - \nu(A)|$ in (A.161) to a tail probability of $\sup_{A \in \mathcal{A}} |\nu_n'(A) - \nu_n(A)|$, where

$$\nu_n'(A) = \frac{1}{n} \sum_{i=1}^{n} I_{Z_i' \in A}, \tag{A.162}$$

for $A \in \mathcal{A}$, and, in the second part, relate that to the tail probability of $\sup_{A \in \mathcal{A}} |\tilde{\nu}_n(A)|$ in (A.161).

Notice that, whenever $\sup_{A \in \mathcal{A}} |\nu_n(A) - \nu(A)| > \tau$, there is an $A^* \in \mathcal{A}$, which is a function of $Z_1, \ldots, Z_n$, such that $|\nu_n(A^*) - \nu(A^*)| > \tau$, with probability 1. In other words,

$$P\left(|\nu_n(A^*) - \nu(A^*)| > \tau \,\middle|\, \sup_{A \in \mathcal{A}} |\nu_n(A) - \nu(A)| > \tau\right) = 1, \tag{A.163}$$

which in turn implies that

$$P\left(|\nu_n(A^*) - \nu(A^*)| > \tau\right) \geq P\left(\sup_{A \in \mathcal{A}} |\nu_n(A) - \nu(A)| > \tau\right). \tag{A.164}$$

Now, conditioned on $Z_1, \ldots, Z_n$, $\mathcal{A}^*$ is fixed (nonrandom). Notice that $E[\nu'_n(A^*) \mid Z_1, \ldots, Z_n] = \nu(A^*)$ and $\mathrm{Var}(\nu'_n(A^*) \mid Z_1, \ldots, Z_n) = \nu(A^*)(1 - \nu(A^{**}))/n$. Hence, we can apply Chebyshev's Inequality (A.75) to get:

$$P\left(|\nu'_n(A^*) - \nu(A^*)| < \frac{\tau}{2} \;\Big|\; Z_1, \ldots, Z_n\right) \geq 1 - \frac{4\nu(A^*)(1 - \nu(A^*))}{n\tau^2} \geq 1 - \frac{1}{n\tau^2} \geq \frac{1}{2}, \quad (A.165)$$

for $n \geq 2\tau^{-2}$. Now,

$$P\left(\sup_{A \in \mathcal{A}} |\nu'_n(A) - \nu_n(A)| > \frac{\tau}{2} \;\Big|\; Z_1, \ldots, Z_n\right) \geq P\left(|\nu'_n(A^*) - \nu_n(A^*)| > \frac{\tau}{2} \;\Big|\; Z_1, \ldots, Z_n\right)$$

$$\geq I_{|\nu_n(A^*) - \nu(A^*)| > \tau}\, P\left(|\nu'_n(A^*) - \nu(A^*)| < \frac{\tau}{2} \;\Big|\; Z_1, \ldots, Z_n\right) \geq \frac{1}{2} I_{|\nu_n(A^*) - \nu(A^*)| > \tau}\,. \quad (A.166)$$

where the second inequality follows from the fact that $|a - c| > \tau$ and $|b - c| < \tau/2$ imply that $|a - b| > \tau/2$. Integrating (A.166) on both sides with respect to $Z_1, \ldots, Z_n$ and using (A.164) yields

$$P\left(\sup_{A \in \mathcal{A}} |\nu'_n(A) - \nu_n(A)| > \frac{\tau}{2}\right) \geq \frac{1}{2} P\left(\sup_{A \in \mathcal{A}} |\nu_n(A) - \nu(A)| > \tau\right), \quad (A.167)$$

which completes the first part of the proof. Next, define

$$\tilde{\nu}'_n(A) = \frac{1}{n} \sum_{i=1}^{n} \sigma_i I_{Z'_i \in A} \quad (A.168)$$

for $A \in \mathcal{A}$. The key observation at this point is that $\sup_{A \in \mathcal{A}} |\nu'_n(A) - \nu_n(A)|$ has the same distribution as $\sup_{A \in \mathcal{A}} |\tilde{\nu}'_n(A) - \tilde{\nu}_n(A)|$, which can be seen by conditioning on $\sigma_1, \ldots, \sigma_n$. Hence,

$$P\left(\sup_{A \in \mathcal{A}} |\nu'_n(A) - \nu_n(A)| > \frac{\tau}{2}\right) = P\left(\sup_{A \in \mathcal{A}} |\tilde{\nu}'_n(A) - \tilde{\nu}_n(A)| > \frac{\tau}{2}\right)$$

$$\leq P\left(\left\{\sup_{A \in \mathcal{A}} |\tilde{\nu}'_n(A)| > \frac{\tau}{4}\right\} \bigcup \left\{\sup_{A \in \mathcal{A}} |\tilde{\nu}_n(A)| > \frac{\tau}{4}\right\}\right) \quad (A.169)$$

$$\leq P\left(\sup_{A \in \mathcal{A}} |\tilde{\nu}'_n(A)| > \frac{\tau}{4}\right) + P\left(\sup_{A \in \mathcal{A}} |\tilde{\nu}_n(A)| > \frac{\tau}{4}\right) = 2P\left(\sup_{A \in \mathcal{A}} |\tilde{\nu}_n(A)| > \frac{\tau}{4}\right),$$

where the first inequality follows from the fact that $|a - b| > \tau/2$ implies that $|a| > \tau/4$ or $|b| > \tau/4$, while the second inequality is an application of the Union Bound (A.10). Combining (A.167) and (A.169) proves the lemma. $\diamond$

Equipped with the Symmetrization Lemma, the proof of the following theorem is fairly simple, but also quite instructive.

Theorem A.20. *(General Vapnik-Chervonenkis Theorem.) Regardless of the measure ν,*

$$P\left(\sup_{A \in \mathcal{A}} |\nu_n(A) - \nu(A)| > \tau\right) \leq 8\,\mathcal{S}(\mathcal{A}, n) e^{-n\tau^2/32}, \quad \text{for all } \tau > 0\,. \quad (A.170)$$

where $\mathcal{S}(\mathcal{A}, n)$ is the nth shatter coefficient of $\mathcal{A}$, defined in (8.14).

Proof. For fixed $Z_1 = z_1, \ldots, Z_n = z_n$, consider the binary vector $(I_{z_i \in A}, \ldots, I_{z_i \in A})$, as A ranges over $\mathcal{A}$. There are of course a maximum of 2^n distinct values that this vector can take on. But, for a given $\mathcal{A}$, this number may be smaller than 2^n. Indeed, this is the number $N_{\mathcal{A}}(z_1, \ldots, z_n)$, defined in (8.13) — by definition, this number must be smaller than the shatter coefficient $\mathcal{S}(\mathcal{A}, n)$, for any choice of $z_1, \ldots, z_n$. Notice that $\tilde{\nu}_n(A)$, conditioned on $Z_1 = z_1, \ldots, Z_n = z_n$, is still a random variable, through the random signs $\sigma_1, \ldots, \sigma_n$. Since this random variable is a function of the vector $(I_{z_i \in A}, \ldots, I_{z_i \in A})$, the number of values it can take as A ranges over $\mathcal{A}$ is also bounded by $\mathcal{S}(\mathcal{A}, n)$. Therefore, $\sup_{A \in \mathcal{A}} |\tilde{\nu}_n(A)|$ turns out to be a maximum of at most $\mathcal{S}(\mathcal{A}, n)$ values, so that one can employ the Union Bound (A.10) as follows:

$$
\begin{aligned}
P\left(\sup_{A \in \mathcal{A}} |\tilde{\nu}_n(A)| > \frac{\tau}{4} \,\bigg|\, Z_1, \ldots, Z_n\right) &= P\left(\bigcup_{A \in \mathcal{A}} \left\{|\tilde{\nu}_n(A)| > \frac{\tau}{4}\right\} \,\bigg|\, Z_1, \ldots, Z_n\right) \\
&\leq \sum_{A \in \mathcal{A}} P\left(|\tilde{\nu}_n(A)| > \frac{\tau}{4} \,\bigg|\, Z_1, \ldots, Z_n\right) \leq \mathcal{S}(\mathcal{A}, n) \sup_{A \in \mathcal{A}} P\left(|\tilde{\nu}_n(A)| > \frac{\tau}{4} \,\bigg|\, Z_1, \ldots, Z_n\right),
\end{aligned}
\tag{A.171}
$$

with the understanding that the union, sum, and suprema are finite. Now we apply Hoeffding's Inequality (Theorem A.14) to bound the probability $P\left(|\tilde{\nu}_n(A)| > \frac{\tau}{4} \,\big|\, Z_1, \ldots, Z_n\right)$. Conditioned on $Z_1 = z_1, \ldots, Z_n = z_n$, $\tilde{\nu}_n(A) = \sum_{i=1}^n \sigma_i I_A(z_i \in A)$ is a sum of independent zero-mean random variables, which are bounded in the interval $[-1, 1]$ (they are not identically-distributed, but this is not necessary for application of Theorem A.14). Hoeffding's Inequality then yields:

$$
P\left(|\tilde{\nu}_n(A)| > \frac{\tau}{4} \,\bigg|\, Z_1, \ldots, Z_n\right) \leq 2e^{-n\tau^2/32}, \quad \text{for all } \tau > 0.
\tag{A.172}
$$

Applying (A.171) and integrating on both sides with respect to $Z_1, \ldots, Z_n$ yields

$$
P\left(\sup_{A \in \mathcal{A}} |\tilde{\nu}_n(A)| > \frac{\tau}{4}\right) \leq 2\mathcal{S}(\mathcal{A}, n)e^{-n\tau^2/32}, \quad \text{for all } \tau > 0.
\tag{A.173}
$$

Now, if $n < 2\tau^{-2}$, the inequality in (A.170) is trivial. If $n \geq 2\tau^{-2}$, we can apply Lemma A.2 and get the desired result. $\diamond$

If $\mathcal{S}(\mathcal{A}, n)$ grows polynomially with n (this is the case if the VC dimension of $\mathcal{A}$ is finite), then, by an application of Theorem A.8, (A.170) yields the uniform LLN: $\sup_{A \in \mathcal{A}} |\nu_n(A) - \nu(A)| \xrightarrow{a.s.} 0$.

Specializing Theorem A.20 to the classification case yields the required proof.

Proof of Theorem 8.2

Consider the probability space $(R^{d+1}, \mathcal{B}^{d+1}, P_{\mathbf{X}, Y})$, where $P_{\mathbf{X}, Y}$ is the joint feature-label probability measure constructed in Section 2.6.3. Let the i.i.d. training data be $S_n = \{(\mathbf{X}_1, Y_1), \ldots, (\mathbf{X}_n, Y_n)\}$. Given a family of classifiers $\mathcal{C}$, apply Theorem A.20 with $\nu = P_{\mathbf{X}, Y}$, $Z_i = (\mathbf{X}_i, Y_i) \sim P_{\mathbf{X}, Y}$, for $i = 1, \ldots, n$, and $\tilde{\mathcal{A}}_{\mathcal{C}}$ containing all set of the kind

$$
\tilde{A}_{\psi} = \{\psi(\mathbf{X}) \neq Y\} = \{\psi(\mathbf{X}) = 1, Y = 0\} \cup \{\psi(\mathbf{X}) = 0, Y = 1\},
\tag{A.174}
$$

for each $\psi \in \mathcal{C}$ (the sets $\tilde{A}_\psi$ are Borel since classifiers are measurable functions). Then $\nu(\tilde{A}_\psi) = \varepsilon[\psi]$, $\nu_n(\tilde{A}_\psi) = \hat{\varepsilon}[\psi]$, and $\sup_{\tilde{A}_\psi \in \tilde{\mathcal{A}}_\mathcal{C}} |\nu_n(\tilde{A}_\psi) - \nu(\tilde{A}_\psi)| = \sup_{\psi \in \mathcal{C}} |\hat{\varepsilon}[\psi] - \varepsilon[\psi]|$. It remains to show that $\mathcal{S}(\tilde{\mathcal{A}}_\mathcal{C}, n) = \mathcal{S}(\mathcal{A}_\mathcal{C}, n)$, where $\mathcal{A}_\mathcal{C} = \{A_\psi \mid \psi \in \mathcal{C}\}$, and A_ψ is defined in (8.23). First note that there is a one-to-one correspondence between $\tilde{\mathcal{A}}_\mathcal{C}$ and $\mathcal{A}_\mathcal{C}$, since, for each $\psi \in \mathcal{C}$, we have $\tilde{A}_\psi = A_\psi \times \{0\} \cup A_\psi^c \times \{1\}$. Given a set of points $\{x_1, \ldots, x_n\}$, if k points are picked by A_ψ, then k points can be picked by $\tilde{A}_\psi$ in the set $\{(x_1, 1), \ldots, (x_n, 1)\}$; hence $\mathcal{S}(\mathcal{A}_\mathcal{C}, n) \leq \mathcal{S}(\tilde{\mathcal{A}}_\mathcal{C}, n)$. On the other hand, given a set of points $\{(x_1, 0), \ldots, (x_{n_0}, 0), (x_{n_0+1}, 1), \ldots, (x_{n_0+n_1}, 1)\}$, suppose that $\tilde{A}_\psi$ picks out the subset $\{(x_1, 0), \ldots, (x_l, 0), (x_{n_0+1}, 1), \ldots, (x_{n_0+m}, 1)\}$ (the sets can be unambiguously written this way, since order does not matter). Then A_ψ picks out the subset $\{(x_1, \ldots, x_l, x_{n_0+m+1}, x_{n_0+n_1}\}$, among the set of points $\{x_1, \ldots, x_{n_0+n_1}\}$, and the two subsets determine each other uniquely, so $\mathcal{S}(\tilde{\mathcal{A}}_\mathcal{C}, n) \leq \mathcal{S}(\mathcal{A}_\mathcal{C}, n)$ Thus, $\mathcal{S}(\tilde{\mathcal{A}}_\mathcal{C}, n) = \mathcal{S}(\mathcal{A}_\mathcal{C}, n)$. (Thus, the VC dimensions also agree: $V_{\tilde{\mathcal{A}}_\mathcal{C}} = V_{\mathcal{A}_\mathcal{C}}$.)

A7   Proof of Convergence of the EM Algorithm

Here we present a proof of convergence of the general Expectation-Maximization algorithm to a local maximum of the log-likelihood function.

Let $\mathbf{X}$, $\mathbf{Z}$, $\boldsymbol{\theta} \in \Theta$ be the observed data, the hidden variables, and the vector of model, respectively. meters. The EM method relies on a clever application of Jensen's inequality to obtain the following lower bound on the "incomplete" log-likelihood $L(\boldsymbol{\theta}) = \ln p_{\boldsymbol{\theta}}(\mathbf{X})$:

$$B(\boldsymbol{\theta}) = \sum_{\mathbf{Z}} q(\mathbf{Z}) \ln \frac{p_{\boldsymbol{\theta}}(\mathbf{Z}, \mathbf{X})}{q(\mathbf{Z})} \leq \ln \sum_{\mathbf{Z}} q(\mathbf{Z}) \frac{p_{\boldsymbol{\theta}}(\mathbf{Z}, \mathbf{X})}{q(\mathbf{Z})} = \ln \sum_{\mathbf{Z}} p_{\boldsymbol{\theta}}(\mathbf{Z}, \mathbf{X}) = L(\boldsymbol{\theta}) , \qquad (A.175)$$

for all $\boldsymbol{\theta} \in \boldsymbol{\theta}$, where $q(\mathbf{Z})$ is an arbitrary probability distribution to be specified shortly. The inequality follows directly from concavity of the logarithm function and Jensen's inequality.

One would like to maximize the lower bound function $B(\boldsymbol{\theta})$ so that it touches $L(\boldsymbol{\theta})$, at a value $\boldsymbol{\theta} = \boldsymbol{\theta}^{(m)}$. We show by inspection that the choice $q(\mathbf{Z}; \boldsymbol{\theta}^{(m)}) = p_{\boldsymbol{\theta}^{(m)}}(\mathbf{Z} \mid \mathbf{X})$ accomplishes this. First we replace this choice of $q(\mathbf{Z})$ in (A.175) to obtain:

$$B(\boldsymbol{\theta}, \boldsymbol{\theta}^{(m)}) = \sum_{\mathbf{Z}} p_{\boldsymbol{\theta}^{(m)}}(\mathbf{Z} \mid \mathbf{X}) \ln \frac{p_{\boldsymbol{\theta}}(\mathbf{Z}, \mathbf{X})}{p_{\boldsymbol{\theta}^{(m)}}(\mathbf{Z} \mid \mathbf{X})} . \qquad (A.176)$$

Now we verify that indeed this lower bound touches the log-likelihood at $\boldsymbol{\theta} = \boldsymbol{\theta}^{(m)}$:

$$\begin{aligned} B(\boldsymbol{\theta}^{(m)}, \boldsymbol{\theta}^{(m)}) &= \sum_{\mathbf{Z}} p_{\boldsymbol{\theta}^{(m)}}(\mathbf{Z} \mid \mathbf{X}) \ln \frac{p_{\boldsymbol{\theta}^{(m)}}(\mathbf{Z}, \mathbf{X})}{p_{\boldsymbol{\theta}^{(m)}}(\mathbf{Z} \mid \mathbf{X})} = \sum_{\mathbf{Z}} p_{\boldsymbol{\theta}^{(m)}}(\mathbf{Z} \mid \mathbf{X}) \ln p_{\boldsymbol{\theta}^{(m)}}(\mathbf{X}) \\ &= \ln p_{\boldsymbol{\theta}^{(m)}}(\mathbf{X}) \sum_{\mathbf{Z}} p_{\boldsymbol{\theta}^{(m)}}(\mathbf{Z} \mid \mathbf{X}) = L(\boldsymbol{\theta}^{(m)}) . \end{aligned} \qquad (A.177)$$

The main idea behind EM is that choosing a value of $\boldsymbol{\theta} = \boldsymbol{\theta}^{(m+1)}$ that increases $B(\boldsymbol{\theta}, \boldsymbol{\theta}^{(m)})$ over its previous value $B(\boldsymbol{\theta}^{(m)}, \boldsymbol{\theta}^{(m)})$ will also increase $L(\boldsymbol{\theta})$ over its previous value $L(\boldsymbol{\theta}^{(m)})$. This can be proved as follows:

$$
\begin{aligned}
B(\boldsymbol{\theta}^{(m+1)}, \boldsymbol{\theta}^{(m)}) - B(\boldsymbol{\theta}^{(m)}, \boldsymbol{\theta}^{(m)}) &= \sum_{\mathbf{Z}} p_{\boldsymbol{\theta}^{(m)}}(\mathbf{Z} \mid \mathbf{X}) \ln \frac{p_{\boldsymbol{\theta}^{(m+1)}}(\mathbf{Z}, \mathbf{X})}{p_{\boldsymbol{\theta}^{(m)}}(\mathbf{Z}, \mathbf{X})} \\
&= \sum_{\mathbf{Z}} p_{\boldsymbol{\theta}^{(m)}}(\mathbf{Z} \mid \mathbf{X}) \ln \frac{p_{\boldsymbol{\theta}^{(m+1)}}(\mathbf{Z} \mid \mathbf{X})}{p_{\boldsymbol{\theta}^{(m)}}(\mathbf{Z} \mid \mathbf{X})} + \sum_{\mathbf{Z}} p_{\boldsymbol{\theta}^{(m)}}(\mathbf{Z} \mid \mathbf{X}) \ln \frac{p_{\boldsymbol{\theta}^{(n+1)}}(\mathbf{X})}{p_{\boldsymbol{\theta}^{(m)}}(\mathbf{X})} \\
&= - D(p_{\boldsymbol{\theta}^{(m)}}(\mathbf{Z} \mid \mathbf{X}) \,\|\, p_{\boldsymbol{\theta}^{(m+1)}}(\mathbf{Z} \mid \mathbf{X})) + L(\boldsymbol{\theta}^{(m+1)}) - L(\boldsymbol{\theta}^{(m)}),
\end{aligned}
\tag{A.178}
$$

where $D(p \,\|\, q)$ is the *Kullback-Leibler distance* between two probability mass functions. The KL distance is always nonnegative [Kullback, 1968], with equality if and only if $p = q$ with probability 1. We conclude that

$$
B(\boldsymbol{\theta}^{(m+1)}, \boldsymbol{\theta}^{(m)}) - B(\boldsymbol{\theta}^{(m)}, \boldsymbol{\theta}^{(m)}) \leq L(\boldsymbol{\theta}^{(m+1)}) - L(\boldsymbol{\theta}^{(m)}),
\tag{A.179}
$$

and that setting

$$
\boldsymbol{\theta}^{(m+1)} = \arg \max_{\boldsymbol{\theta} \in \Theta} B(\boldsymbol{\theta}, \boldsymbol{\theta}^{(m)}),
\tag{A.180}
$$

will increase the log-likelihood $L(\boldsymbol{\theta})$, unless one is already at a local maximum of $L(\boldsymbol{\theta})$.[2] This fact is graphically represented in Figure A.4. This proves the eventual convergence of the EM procedure to a local maximum of $L(\boldsymbol{\theta})$. Now,

$$
B(\boldsymbol{\theta}, \boldsymbol{\theta}^{(m)}) = \sum_{\mathbf{Z}} p_{\boldsymbol{\theta}^{(m)}}(\mathbf{Z} \mid \mathbf{X}) \ln p_{\boldsymbol{\theta}}(\mathbf{Z}, \mathbf{X}) - \sum_{\mathbf{Z}} p_{\boldsymbol{\theta}^{(m)}}(\mathbf{Z} \mid \mathbf{X}) \ln p_{\boldsymbol{\theta}^{(m)}}(\mathbf{Z} \mid \mathbf{X}).
\tag{A.181}
$$

Since the second term in the previous equation does not depend on $\boldsymbol{\theta}$, the maximization in (A.180) can be accomplished by maximizing the first term only:

$$
Q(\boldsymbol{\theta}, \boldsymbol{\theta}^{(m)}) = \sum_{\mathbf{Z}} \ln p_{\boldsymbol{\theta}}(\mathbf{Z}, \mathbf{X})\, p_{\boldsymbol{\theta}^{(m)}}(\mathbf{Z} \mid \mathbf{X}) = E_{\boldsymbol{\theta}^{(m)}}[\ln p_{\boldsymbol{\theta}}(\mathbf{Z}, \mathbf{X}) \mid \mathbf{X}].
\tag{A.182}
$$

The unknown hidden variable $\mathbf{Z}$ is "averaged out" by the expectation.

The resulting EM procedure consists of picking an initial guess $\boldsymbol{\theta} = \boldsymbol{\theta}^{(0)}$ and iterating two steps:

- **E-Step:** Compute $Q(\boldsymbol{\theta}, \boldsymbol{\theta}^{(m)})$

- **M-Step:** Find $\boldsymbol{\theta}^{(m+1)} = \arg \max_{\boldsymbol{\theta}} Q(\boldsymbol{\theta}, \boldsymbol{\theta}^{(m)})$

for $n = 0, 1, \ldots$ until the improvement in the log-likelihood $|\ln L(\boldsymbol{\theta}^{(m+1)}) - \ln L(\boldsymbol{\theta}^{(m)})|$ falls below a pre-specified positive value.

---

[2]In fact, just selecting $\boldsymbol{\theta}^{(m+1)}$ such that $B(\boldsymbol{\theta}^{(m+1)}, \boldsymbol{\theta}^{(m)}) - B(\boldsymbol{\theta}^{(m)}, \boldsymbol{\theta}^{(m)}) > 0$ will do — this is called "Generalized Expectation Maximization"

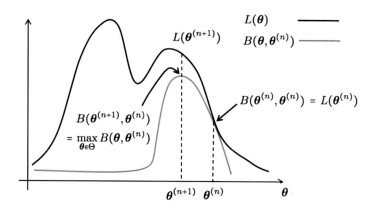

Figure A.4: The lower bound $B(\boldsymbol{\theta}, \boldsymbol{\theta}^{(m)})$ touches the log-likelihood $L(\boldsymbol{\theta})$ at $\boldsymbol{\theta} = \boldsymbol{\theta}^{(m)}$. Maximizing $B(\boldsymbol{\theta}, \boldsymbol{\theta}^{(m)})$ with respect to $\boldsymbol{\theta}$ to obtain $\boldsymbol{\theta}^{(m+1)}$ increases $L(\boldsymbol{\theta})$. Repeating the process leads to eventual convergence to a local maximum of $L(\boldsymbol{\theta})$. (Adapted from Figure 1 of Minka [1998].)

.

A8   Data Sets Used in the Book

In this section we describe the synthetic and real data sets that are used throughout the book. The real data sets can be downloaded from the book website.

A8.1   Synthetic Data

We employ a general multivariate Gaussian model to generate synthetic data, which consists of blocked covariance matrices of the form

$$
\Sigma_{d \times d} = \begin{bmatrix} \Sigma_{l_1 \times l_1} & 0 & \cdots & 0 \\ 0 & \Sigma_{l_2 \times l_2} & \cdots & 0 \\ \vdots & \vdots & \ddots & \vdots \\ 0 & 0 & \cdots & \Sigma_{l_k \times l_k} \end{bmatrix} \tag{A.183}
$$

where $l_1 + \cdots l_k = d$. The features are thus clustered into k independent groups. If $k = d$, then all features are independent. The individual covariance matrices $\Sigma_{l_i \times l_i}$ could be arbitrary, but here we will consider a simple parametric form

$$
\Sigma_{l_i \times l_i}(\sigma_i^2, \rho_i) = \sigma_i^2 \begin{bmatrix} 1 & \rho_i & \cdots & \rho_i \\ \rho_i & 1 & \cdots & \rho_i \\ \vdots & \vdots & \ddots & \vdots \\ \rho_i & \rho_i & \cdots & 1 \end{bmatrix} \tag{A.184}
$$

for $i = 1, \ldots, k$, where $-1 < \rho_i < 1$. Hence, the features within each block have the same variance σ_i^2 and are all correlated with the same correlation coefficient ρ_i.

The class mean vectors $\boldsymbol{\mu}_0$ and $\boldsymbol{\mu}_1$ and prior probabilities $c_0 = P(Y = 0)$ and $c_1 = P(Y = 1)$ are arbitrary. Heteroskedastic Gaussian models result from specifying the class-conditional covariance matrices Σ_0 and Σ_1 separately. "Noisy features" can be obtained by matching mean components across classes and matching corresponding singleton blocks in the covariance matrices. Each noisy feature is an independent feature with the same mean and variance across the classes.

The python script `app_synth_data.py` generates sample data from this model.

A8.2   Dengue Fever Prognosis Data Set

This is gene-expression microarray data from a dengue fever diagnosis study performed in the Northeast of Brazil. The primary purpose of the study was to be able to predict the ultimate clinical outcome of dengue (whether the benign classical form or the dangerous hemorrhagic fever) from gene expression profiles of peripheral blood mononuclear cells (PBMCs) of patients in the early days of fever. The study is reported in Nascimento et al. [2009]. See also Example 1.1. The data consist of 26 training points measured on 1981 genes and three class labels, corresponding to: 8 classical dengue fever (DF) patients, 10 dengue hemorrhagic fever (DHF) patients, and 8 febrile non-dengue (ND) patients, as classified by an experienced clinician. This is a retrospective study, meaning that the patients were tracked and their outcomes verified by a clinician, but their status could not be determined clinically at the time the data was obtained, which was within one week of the start of symptoms.

A8.3   Breast Cancer Prognosis Data Set

This is gene-expression microarray data from the breast cancer prognosis study conducted in the Netherlands and reported in van de Vijver et al. [2002]. The data set consists of 295 training points of dimensionality 70 and two class labels. The feature vectors are normalized gene-expression profiles from cells harvested from 295 beast tumor samples in a retrospective study, meaning that patients were tracked over the years and their outcomes recorded. Using this clinical information, the authors labeled the tumor samples into two classes: the "good prognosis" group (label 1) were disease-free for at least five years after first treatment, whereas the "bad prognosis" group developed distant metastasis within the first five years. Of the 295 patients, 216 belong to the "good-prognosis" class, whereas the remaining 79 belong to the "poor- prognosis" class.

A8.4 Stacking Fault Energy Data Set

This data set contains the experimentally recorded values of the stacking fault energy (SFE) in austenitic stainless steel specimens with different chemical compositions; see Yonezawa et al. [2013]. The SFE is a microscopic property related to the resistance of austenitic steels. High-SFE steels are less likely to fracture under strain and may be desirable in certain applications. The data set contains 17 features corresponding to the atomic element content of 473 steel specimens and the continuous-valued measured SFE for each.

A8.5 Soft Magnetic Alloy Data Set

This is a data set on Fe-based nanocrystalline soft magnetic alloys, which is part of on-going work [Wang et al., 2020]. This data set records the atomic composition and processing parameters along with several different electromagnetic properties for a large number of magnetic alloys. We will be particularly interested in the magnetic coercivity as the property to be predicted. Larger values of coercivity mean that the magnetized material has a wider hysteresis curve and can withstand larger magnetic external fields without losing its own magnetization. By contrast, small values of coercivity mean that a material can lose its magnetization quickly. Large-coercivity materials are therefore ideal to make permanent magnets, for example.

A8.6 Ultrahigh Carbon Steel Data Set

This is the Carnegie Mellon University Ultrahigh Carbon Steel (CMU-UHCS) dataset [Hecht et al., 2017; DeCost et al., 2017]. This data set consists of 961 high-resolution 645×484 images of steel samples subjected to a variety of heat treatments. The images are *micrographs* obtained by scanning electron microscopy (SEM) at several different magnifications. There are a total of seven different labels, corresponding to different phases of steel resulting from different thermal processing (number of images in parenthesis): spheroidite (374), carbide network (212), pearlite (124), pearlite + spheroidite (107), spheroidite+Widmanstätten (81), martensite (36), and pearlite+Widmanstätten (27). The main goal is to be able to predict the label of a new steel sample given its micrograph.

List of Symbols

$\mathbf{X} = (X_1, \ldots, X_d) \in R^d$ feature vector

$Y \in \{0, 1\}$ target

$c_i = P(Y = i), \, i = 0, 1$ class prior probabilities

$P_{\mathbf{X}Y}$ feature-target distribution

$p(\mathbf{x})$ feature vector density (if it exists)

$p_i(\mathbf{x}) = p(\mathbf{x} \mid Y = i), \, i = 0, 1$ class-conditional densities (if they exist)

$\eta(\mathbf{x}) = P(Y = 1 \mid \mathbf{X} = \mathbf{x})$ posterior-probability function

$\psi : R^d \to \{0, 1\}$ classifier

$\varepsilon = \varepsilon[\psi] = P(\psi(\mathbf{X}) \neq Y)$ classifier error rate

ψ^*, ε^* Bayes classifier and Bayes error

$\varepsilon^i = P(\psi(\mathbf{X}) = 1 - i \mid Y = i), \, i = 0, 1$ population-specific true error rates

$D_n : R^d \to R$ sample-based discriminant

$\mu_i, \, i = 0, 1$ class means

$\sigma_i^2, \, i = 0, 1$ class variances

$\Sigma_i, \, i = 0, 1$ class covariance matrices

$\Phi(x) = (1/2\pi) \int_{-\infty}^{x} e^{-u^2} du$ cdf of a $N(0, 1)$ Gaussian random variable

$S_n = \{(\mathbf{X}_1, Y_1), \ldots, (\mathbf{X}_n, Y_n)\}$ sample training data

$n, n_0, n_1 = n - n_0$ total and population-specific sample sizes

$\Psi_n : S_n \mapsto \psi_n$ classification rule

$\psi_n : R^d \to \{0, 1\}$ classifier designed from training data

$\varepsilon_n = \varepsilon[\psi_n] = P(\psi_n(\mathbf{X}) \neq Y \mid S_n)$ error rate of sample-based classifier

© Springer Nature Switzerland AG 2020
U. Braga-Neto, *Fundamentals of Pattern Recognition and Machine Learning*,
https://doi.org/10.1007/978-3-030-27656-0

$\boldsymbol{\theta}, \hat{\boldsymbol{\theta}}$ parameter vector and estimator

$k(\mathbf{x}, \mathbf{x}')$ kernel

$\lambda_i,\, i = 1, \ldots, d$ Lagrange multipliers

L_P, L_D primal and dual Lagrangians

$\mathcal{C}$ set of classification rules

$V_{\mathcal{C}},\, \mathcal{S}(\mathcal{C}, n)$ VC dimension and shatter coefficients

$p_i,\, q_i,\, U_i,\, V_i,\, i = 1, \ldots, b$ population-specific bin probabilities and counts

$\Xi_n : (\,\Psi_n, S_n, \xi) \mapsto \hat{\varepsilon}_n$ error estimation rule

$\hat{\varepsilon}_n$ error estimator for mixture sample

$\mathrm{Bias}(\hat{\varepsilon}_n),\, \mathrm{Var}_{\mathrm{dev}}(\hat{\varepsilon}_n),\, \mathrm{RMS}(\hat{\varepsilon}_n)$ bias, deviation variance, root mean square error

$S_m = \{(\mathbf{X}_i^t, Y_i^t); i = 1, \ldots, m\}$ independent test sample

$\hat{\varepsilon}_{n,m}$ test-set error estimator

$\hat{\varepsilon}_n^{\,r}$ resubstitution error estimator

$\hat{\varepsilon}_n^{\,\mathrm{cv}(k)}$ k-fold cross-validation error estimator

$\hat{\varepsilon}_n^{\,l}$ leave-one-out error estimator

$L[f]$ regression error of f

$\mathcal{F}$ σ-algebra

$\mathcal{B}$ Borel σ-algebra

μ, ν measures

λ Lebesgue measure

$X_n \xrightarrow{a.s.} X$ almost-sure convergence (with probability 1) of X_n to X

$X_n \xrightarrow{L^p} X$ L^p convergence of X_n to X

$X_n \xrightarrow{P} X$ convergence of X_n to X in probability

$X_n \xrightarrow{D} X$ convergence of X_n to X in distribution

Bibliography

Afsari, B., Braga-Neto, U., and Geman, D. (2014). Rank discriminants for predicting phenotypes from rna expression. *Annals of Applied Statistics*, 8(3):1469–1491.

Aitchison, J. and Dunsmore, I. (1975). *Statistical Predication Analysis*. Cambridge University Press, Cambridge, UK.

Alberts, B., Bray, D., Lewis, J., Raff, M., Roberts, K., and Watson, J. (2002). *Molecular Biology of the Cell*. Garland, 4th edition.

Alvarez, S., Diaz-Uriarte, R., Osorio, A., Barroso, A., Melchor, L., Paz, M., Honrado, E., Rodriguez, R., Urioste, M., Valle, L., Diez, O., Cigudosa, J., Dopazo, J., Esteller, M., and Benitez, J. (2005). A predictor based on the somatic genomic changes of the brca1/brca2 breast cancer tumors identifies the non-brca1/brca2 tumors with brca1 promoter hypermethylation. *Clin Cancer Res*, 11(3):1146–1153.

Ambroise, C. and McLachlan, G. (2002). Selection bias in gene extraction on the basis of microarray gene expression data. *Proc. Natl. Acad. Sci.*, 99(10):6562–6566.

Anderson, T. (1951). Classification by multivariate analysis. *Psychometrika*, 16:31–50.

Anderson, T. (1973). An asymptotic expansion of the distribution of the studentized classification statistic W. *The Annals of Statistics*, 1:964–972.

Anderson, W. (1984). *An Introduction to Multivariate Statistical Analysis*. Wiley, New York, 2nd edition.

Bartlett, P., Boucheron, S., and Lugosi, G. (2002). Model selection and error estimation. *Machine Learning*, 48:85–113.

Bertsekas, D. (1995). *Nonlinear Programming*. Athena Scientific.

Billingsley, P. (1995). *Probability and Measure*. John Wiley, New York City, New York, third edition.

Bishop, C. (2006). *Pattern Recognition and Machine Learning*. Springer, New York.

Boser, B., Guyon, M., and Vapnik, V. (1992). A training algorithm for optimal margin classifiers. In *Proceedings of the Workshop on Computational Learning Theory*.

Bowker, A. (1961). A representation of hotelling's t^2 and anderson's classification statistic w in terms of simple statistics. In Solomon, H., editor, *Studies in Item Analysis and Prediction*, pages 285–292. Stanford University Press.

Bowker, A. and Sitgreaves, R. (1961). An asymptotic expansion for the distribution function of the w-classification statistic. In Solomon, H., editor, *Studies in Item Analysis and Prediction*, pages 292–310. Stanford University Press.

Boyd, S. and Vandenberghe, L. (2004). *Convex optimization*. Cambridge university press.

Braga-Neto, U. (2007). Fads and fallacies in the name of small-sample microarray classification. *IEEE Signal Processing Magazine*, 24(1):91–99.

Braga-Neto, U., Arslan, E., Banerjee, U., and Bahadorinejad, A. (2018). Bayesian classification of genomic big data. In Sedjic, E. and Falk, T., editors, *Signal Processing and Machine Learning for Biomedical Big Data*. Chapman and Hall/CRC Press.

Braga-Neto, U. and Dougherty, E. (2004). Bolstered error estimation. *Pattern Recognition*, 37(6):1267–1281.

Braga-Neto, U. and Dougherty, E. (2015). *Error Estimation for Pattern Recognition*. Wiley, New York.

Breiman, L. (1996). Bagging predictors. *Machine Learning*, 24(2):123–140.

Breiman, L. (2001). Random forests. *Machine Learning*, 45(1):5–32.

Breiman, L., Friedman, J., Olshen, R., and Stone, C. (1984). *Classification and Regression Trees*. Wadsworth.

Bryson, A. and Ho, Y.-C. (1969). *Applied Optimal Control: Optimization, Estimation, and Control*. Blaisdell Publishing Company.

Buduma, N. and Locascio, N. (2017). *Fundamentals of deep learning: Designing next-generation machine intelligence algorithms*. O'Reilly Media, Inc.

Burges, C. J. (1998). A tutorial on support vector machines for pattern recognition. *Data mining and knowledge discovery*, 2(2):121–167.

Casella, G. and Berger, R. (2002). *Statistical Inference*. Duxbury, Pacific Grove, CA, 2nd edition.

Castro, R. (2020). Statistical Learning Theory Lecture Notes. Accessed: Jun 12, 2020. https://www.win.tue.nl/~rmcastro/2DI70/files/2DI70_Lecture_Notes.pdf.

Chapelle, O., Scholkopf, B., and Zien, A., editors (2010). *Semi-Supervised Learning*. MIT Press.

Cherkassky, V. and Ma, Y. (2003). Comparison of model selection for regression. *Neural computation*, 15(7):1691–1714.

Cherkassky, V., Shao, X., Mulier, F. M., and Vapnik, V. N. (1999). Model complexity control for regression using vc generalization bounds. *IEEE transactions on Neural Networks*, 10(5):1075–1089.

Chernick, M. (1999). *Bootstrap Methods: A Practitioner's Guide*. John Wiley & Sons, New York.

Chung, K. L. (1974). *A Course in Probability Theory, Second Edition*. Academic Press, New York City, New York.

Cover, T. (1969). Learning in pattern recognition. In Watanabe, S., editor, *Methodologies of Pattern Recognition*, pages 111–132. Academic Press, New York, NY.

Cover, T. and Hart, P. (1967). Nearest-neighbor pattern classification. *IEEE Trans. on Information Theory*, 13:21–27.

Cover, T. and van Campenhout, J. (1977). On the possible orderings in the measurement selection problem. *IEEE Trans. on Systems, Man, and Cybernetics*, 7:657–661.

Cover, T. M. (1974). The best two independent measurements are not the two best. *IEEE Transactions on Systems, Man, and Cybernetics*, SMC-4(1):116–117.

Cramér, H. (1999). *Mathematical methods of statistics*, volume 43. Princeton university press.

Cressie, N. (1991). *Statistics for Spatial Data*. John Wiley, New York City, New York.

Cybenko, G. (1989). Approximation by superpositions of a sigmoidal function. *Mathematics of control, signals and systems*, 2(4):303–314.

Dalton, L. and Dougherty, E. (2011a). Application of the bayesian mmse error estimator for classification error to gene-expression microarray data. *IEEE Transactions on Signal Processing*, 27(13):1822–1831.

Dalton, L. and Dougherty, E. (2011b). Bayesian minimum mean-square error estimation for classification error part I: Definition and the bayesian mmse error estimator for discrete classification. *IEEE Transactions on Signal Processing*, 59(1):115–129.

Dalton, L. and Dougherty, E. (2011c). Bayesian minimum mean-square error estimation for classification error part II: Linear classification of gaussian models. *IEEE Transactions on Signal Processing*, 59(1):130–144.

Dalton, L. and Dougherty, E. (2012a). Exact mse performance of the bayesian mmse estimator for classification error part i: Representation. *IEEE Transactions on Signal Processing*, 60(5):2575–2587.

Dalton, L. and Dougherty, E. (2012b). Exact mse performance of the bayesian mmse estimator for classification error part ii: Performance analysis and applications. *IEEE Transactions on Signal Processing*, 60(5):2588–2603.

Dalton, L. and Dougherty, E. (2013). Optimal classifiers with minimum expected error within a bayesian framework – part I: Discrete and gaussian models. *Pattern Recognition*, 46(5):1301–1314.

Davis, J. and Goadrich, M. (2006). The relationship between precision-recall and roc curves. In *Proceedings of the 23rd international conference on Machine learning*, pages 233–240.

De Leeuw, J. and Mair, P. (2009). Multidimensional scaling using majorization: Smacof in r. *Journal of Statistical Software*, 31(3).

DeCost, B. L., Francis, T., and Holm, E. A. (2017). Exploring the microstructure manifold: image texture representations applied to ultrahigh carbon steel microstructures. *Acta Materialia*, 133:30–40.

Dempster, A. D., Laird, N. M., and Rubin, D. B. (1977). Maximum likelihood from incomplete data via the EM algorithm (with Discussion). *Journal of the Royal Statistical Society, Series B*, 39:1–38.

Deng, J., Dong, W., Socher, R., Li, L.-J., Li, K., and Fei-Fei, L. (2009). Imagenet: A large-scale hierarchical image database. In *2009 IEEE conference on computer vision and pattern recognition*, pages 248–255. Ieee.

Devroye, L., Gyorfi, L., and Lugosi, G. (1996). *A Probabilistic Theory of Pattern Recognition*. Springer, New York.

Devroye, L. and Wagner, T. (1976). Nonparametric discrimination and density estimation. Technical Report 183, Electronics Research Center, University of Texas, Austin, TX.

Dougherty, E. R. and Brun, M. (2004). A probabilistic theory of clustering. *Pattern Recognition*, 37(5):917–925.

Duda, R., Hart, P., and Stork, G. (2001). *Pattern Classification*. John Wiley & Sons, New York, 2nd edition.

Dudley, R. M. (1978). Central limit theorems for empirical measures. *The Annals of Probability*, pages 899–929.

Dudoit, S. and Fridlyand, J. (2002). A prediction-based resampling method for estimating the number of clusters in a dataset. *Genome biology*, 3(7):research0036–1.

Dudoit, S., Fridlyand, J., and Speed, T. (2002). Comparison of discrimination methods for the classification of tumors using gene expression data. *Journal of the American Statistical Association*, 97(457):77–87.

Efron, B. (1979). Bootstrap methods: Another look at the jacknife. *Annals of Statistics*, 7:1–26.

Efron, B. (1983). Estimating the error rate of a prediction rule: Improvement on cross-validation. *Journal of the American Statistical Association*, 78(382):316–331.

Efron, B. and Tibshirani, R. (1997). Improvements on cross-validation: The .632+ bootstrap method. *Journal of the American Statistical Association*, 92(438):548–560.

Elashoff, J. D., Elashoff, R., and COLDMAN, G. (1967). On the choice of variables in classification problems with dichotomous variables. *Biometrika*, 54(3-4):668–670.

Esfahani, S. and Dougherty, E. (2014). Effect of separate sampling on classification accuracy. *Bioinformatics*, 30(2):242–250.

Evans, M., Hastings, N., and Peacock, B. (2000). *Statistical Distributions.* Wiley, New York, 3rd edition.

Fei-Fei, L., Deng, J., Russakovski, O., Berg, A., and Li, K. (2010). ImageNet Summary and Statistics. http://www.image-net.org/about-stats. Accessed: Jan 2, 2020.

Fisher, R. (1935). The fiducial argument in statistical inference. *Ann. Eugen.*, 6:391–398.

Fisher, R. (1936). The use of multiple measurements in taxonomic problems. *Ann. Eugen.*, 7(2):179–188.

Fix, E. and Hodges, J. (1951). Nonparametric discrimination: Consistency properties. Technical Report 4, USAF School of Aviation Medicine, Randolph Field, TX. Project Number 21-49-004.

Foley, D. (1972). Considerations of sample and feature size. *IEEE Transactions on Information Theory*, IT-18(5):618–626.

Freund, Y. (1990). Boosting a weak learning algorithm by majority. In *Proc. Third Annual Workshop on Computational Learning Theory*, pages 202–216.

Fukushima, K. (1980). Neocognitron: A self-organizing neural network model for a mechanism of pattern recognition unaffected by shift in position. *Biological cybernetics*, 36(4):193–202.

Funahashi, K.-I. (1989). On the approximate realization of continuous mappings by neural networks. *Neural networks*, 2(3):183–192.

Galton, F. (1886). Regression towards mediocrity in hereditary stature. *The Journal of the Anthropological Institute of Great Britain and Ireland*, 15:246–263.

Geisser, S. (1964). Posterior odds for multivariate normal classification. *Journal of the Royal Statistical Society: Series B*, 26(1):69–76.

Geman, D., d'Avignon, C., Naiman, D., Winslow, R., and Zeboulon, A. (2004). Gene expression comparisons for class prediction in cancer studies. In *Proceedings of the 36th Symposium on the Interface: Computing Science and Statistics*, Baltimore, MD.

Girosi, F. and Poggio, T. (1989). Representation properties of networks: Kolmogorov's theorem is irrelevant. *Neural Computation*, 1(4):465–469.

Glick, N. (1973). Sample-based multinomial classification. *Biometrics*, 29(2):241–256.

Glick, N. (1978). Additive estimators for probabilities of correct classification. *Pattern Recognition*, 10:211–222.

Groenen, P. J., van de Velden, M., et al. (2016). Multidimensional scaling by majorization: A review. *Journal of Statistical Software*, 73(8):1–26.

Hamamoto, Y., Uchimura, S., Matsunra, Y., Kanaoka, T., and Tomita, S. (1990). Evaluation of the branch and bound algorithm for feature selection. *Pattern Recognition Letters*, 11:453–456.

Hanczar, B., Hua, J., and Dougherty, E. (2007). Decorrelation of the true and estimated classifier errors in high-dimensional settings. *EURASIP Journal on Bioinformatics and Systems Biology*, 2007. Article ID 38473, 12 pages.

Hand, D. (1986). Recent advances in error rate estimation. *Pattern Recognition Letters*, 4:335–346.

Harter, H. (1951). On the distribution of wald's classification statistics. *Ann. Math. Statist.*, 22:58–67.

Hassan, M. (2018). VGG16 convolutional network for classification and detection. https://neurohive.io/en/popular-networks/vgg16/. Accessed: Jan 1, 2020.

Hastie, T., Tibshirani, R., and Friedman, J. (2001). *The Elements of Statistical Learning*. Springer.

Hecht, M. D., Picard, Y. N., and Webler, B. A. (2017). Coarsening of inter-and intra-granular proeutectoid cementite in an initially pearlitic 2c-4cr ultrahigh carbon steel. *Metallurgical and Materials Transactions A*, 48(5):2320–2335.

Hills, M. (1966). Allocation rules and their error rates. *Journal of the Royal Statistical Society. Series B (Methodological)*, 28(1):1–31.

Hirst, D. (1996). Error-rate estimation in multiple-group linear discriminant analysis. *Technometrics*, 38(4):389–399.

Hoeffding, W. (1963). Probability inequalities for sums of bounded random variables. *Journal of the American Statistical Association*, 58:13–30.

Horn, R. and Johnson, C. (1990). *Matrix Analysis*. Cambridge University Press, New York, NY.

Hornik, K., Stinchcombe, M., and White, H. (1989). Multilayer feedforward networks are universal approximators. *Neural networks*, 2(5):359–366.

Hua, J., Tembe, W., and Dougherty, E. (2009). Performance of feature-selection methods in the classification of high-dimension data. *Pattern Recognition*, 42:409–424.

Hughes, G. (1968). On the mean accuracy of statistical pattern recognizers. *IEEE Transactions on Information Theory*, IT-14(1):55–63.

Izmirlian, G. (2004). Application of the random forest classification algorithm to a SELDI-TOF proteomics study in the setting of a cancer prevention trial. *Ann. NY. Acad. Sci.*, 1020:154–174.

Jain, A. and Zongker, D. (1997). Feature selection: Evaluation, application, and small sample performance. *IEEE Trans. on Pattern Analysis and Machine Intelligence*, 19(2):153–158.

Jain, A. K., Dubes, R. C., et al. (1988). *Algorithms for clustering data*, volume 6. Prentice hall Englewood Cliffs, NJ.

James, G., Witten, D., Hastie, T., and Tibshirani, R. (2013). *An introduction to statistical learning*, volume 112. Springer.

Jazwinski, A. H. (2007). *Stochastic processes and filtering theory*. Courier Corporation.

Jeffreys, H. (1961). *Theory of Probability*. Oxford University Press, Oxford, UK, 3rd edition.

Jiang, X. and Braga-Neto, U. (2014). A naive-bayes approach to bolstered error estimation in high-dimensional spaces. Proceedings of the IEEE International Workshop on Genomic Signal Processing and Statistics (GENSIPS'2014), Atlanta, GA.

John, G. H., Kohavi, R., and Pfleger, K. (1994). Irrelevant features and the subset selection problem. In *Machine Learning Proceedings 1994*, pages 121–129. Elsevier.

John, S. (1961). Errors in discrimination. *Ann. Math. Statist.*, 32:1125–1144.

Kaariainen, M. (2005). Generalization error bounds using unlabeled data. In *Proceedings of COLT'05*.

Kaariainen, M. and Langford, J. (2005). A comparison of tight generalization bounds. In *Proceedings of the 22nd International Conference on Machine Learning*. Bonn, Germany.

Kabe, D. (1963). Some results on the distribution of two random matrices used in classification procedures. *Ann. Math. Statist.*, 34:181–185.

Kaufman, L. and Rousseeuw, P. J. (1990). *Finding groups in data: an introduction to cluster analysis*, volume 344. John Wiley & Sons.

Kim, S., Dougherty, E., Barrera, J., Chen, Y., Bittner, M., and Trent, J. (2002). Strong feature sets from small samples. *Computational Biology*, 9:127–146.

Knights, D., Costello, E. K., and Knight, R. (2011). Supervised classification of human microbiota. *FEMS microbiology reviews*, 35(2):343–359.

Kohane, I., Kho, A., and Butte, A. (2003). *Microarrays for an Integrative Genomics*. MIT Press, Cambridge, MA.

Kohavi, R. (1995). A study of cross-validation and bootstrap for accuracy estimation and model selection. In *Proc. of Fourteenth International Joint Conference on Artificial Intelligence (IJCAI)*, pages 1137–1143, Montreal, CA.

Kohavi, R. and John, G. (1997). Wrappers for feature subset selection. *Artificial Intelligence*, 97(1–2):273–324.

Kolmogorov, A. (1933). *Grundbegriffe der Wahrscheinlichkeitsrechnung*. Springer.

Krizhevsky, A., Sutskever, I., and Hinton, G. E. (2012). Imagenet classification with deep convolutional neural networks. In *Advances in neural information processing systems*, pages 1097–1105.

Kudo, M. and Sklansky, J. (2000). Comparison of algorithms that select features for pattern classifiers. *Pattern Recognition*, 33:25–41.

Kullback, S. (1968). *Information Theory and Statistics*. Dover, New York.

Lachenbruch, P. (1965). *Estimation of error rates in discriminant analysis*. PhD thesis, University of California at Los Angeles, Los Angeles, CA.

Lachenbruch, P. and Mickey, M. (1968). Estimation of error rates in discriminant analysis. *Technometrics*, 10:1–11.

LeCun, Y., Bottou, L., Bengio, Y., Haffner, P., et al. (1998). Gradient-based learning applied to document recognition. *Proceedings of the IEEE*, 86(11):2278–2324.

Linnaeus, C. (1758). *Systema naturae*. Impensis Laurentii Salvii, 10th edition.

Lloyd, S. (1982). Least squares quantization in pcm. *IEEE transactions on information theory*, 28(2):129–137.

Lockhart, D., Dong, H., Byrne, M., Follettie, M., Gallo, M., Chee, M., Mittmann, M., Wang, C., Kobayashi, M., Horton, H., and Brown, E. (1996). Expression monitoring by hybridization to high-density oligonucleotide arrays. *Nature Biotechnology*, 14(13):1675–1680.

Loève, M. (1977). *Probability Theory I*. Springer.

Lorentz, G. G. (1976). The 13th problem of hilbert. In *Proceedings of Symposia in Pure Mathematics*, volume 28, pages 419–430. American Mathematical Society.

Lu, Z., Pu, H., Wang, F., Hu, Z., and Wang, L. (2017). The expressive power of neural networks: A view from the width. In *Advances in neural information processing systems*, pages 6231–6239.

Lugosi, G. and Pawlak, M. (1994). On the posterior-probability estimate of the error rate of nonparametric classification rules. *IEEE Transactions on Information Theory*, 40(2):475–481.

Mallows, C. L. (1973). Some comments on c p. *Technometrics*, 15(4):661–675.

Marguerat, S. and Bahler, J. (2010). Rna-seq: from technology to biology. *Cellular and molecular life science*, 67(4):569–579.

Martins, D., Braga-Neto, U., Hashimoto, R., Bittner, M., and Dougherty, E. (2008). Intrinsically multivariate predictive genes. *IEEE Journal of Selected Topics in Signal Processing*, 2(3):424–439.

McCulloch, W. and Pitts, W. (1943). A logical calculus of the ideas immanent in nervous activity. *Bulletin of Mathematical Biophysics*, 5:115–133.

McFarland, H. and Richards, D. (2001). Exact misclassification probabilities for plug-in normal quadratic discriminant functions. i. the equal-means case. *Journal of Multivariate Analysis*, 77:21–53.

McFarland, H. and Richards, D. (2002). Exact misclassification probabilities for plug-in normal quadratic discriminant functions. ii. the heterogeneous case. *Journal of Multivariate Analysis*, 82:299–330.

McLachlan, G. (1976). The bias of the apparent error in discriminant analysis. *Biometrika*, 63(2):239–244.

McLachlan, G. (1987). Error rate estimation in discriminant analysis: recent advances. In Gupta, A., editor, *Advances in Multivariate Analysis*. D. Reidel, Dordrecht.

McLachlan, G. (1992). *Discriminant Analysis and Statistical Pattern Recognition*. Wiley, New York.

McLachlan, G. and Krishnan, T. (1997). *The EM Algorithm and Extensions*. Wiley-Interscience, New York.

Minka, T. (1998). Expectation maximization as lower bound maximization. Technical report, Microsoft Research. Tutorial published on the web at http://www-white.media.mit.edu/tpminka/papers/em.html.

Moran, M. (1975). On the expectation of errors of allocation associated with a linear discriminant function. *Biometrika*, 62(1):141–148.

Murphy, K. (2012a). *Machine Learning: A Probabilistic Perspective*. MIT Press.

Murphy, K. P. (2012b). *Machine learning: a probabilistic perspective*. MIT press.

Narendra, P. and Fukunaga, K. (1977). A branch and bound algorithm for feature subset selection. *IEEE Trans. on Computers*, 26(9):917–922.

Nascimento, E., Abath, F., Calzavara, C., Gomes, A., Acioli, B., Brito, C., Cordeiro, M., Silva, A., Andrade, C. M. R., Gil, L., and Junior, U. B.-N. E. M. (2009). Gene expression profiling during early acute febrile stage of dengue infection can predict the disease outcome. *PLoS ONE*, 4(11):e7892. doi:10.1371/journal.pone.0007892.

Nilsson, R., Peña, J. M., Björkegren, J., and Tegnér, J. (2007). Consistent feature selection for pattern recognition in polynomial time. *Journal of Machine Learning Research*, 8(Mar):589–612.

Nocedal, J. and Wright, S. (2006). *Numerical optimization*. Springer Science & Business Media.

Nualart, D. (2004). Kolmogorov and probability theory. *Arbor*, 178(704):607–619.

Okamoto, M. (1963). An asymptotic expansion for the distribution of the linear discriminant function. *Ann. Math. Statist.*, 34:1286–1301. Correction: Ann. Math. Statist., 39:1358–1359, 1968.

Pollard, D. (1984). *Convergence of Stochastic Processes*. Springer, New York.

Poor, V. and Looze, D. (1981). Minimax state estimation for linear stochastic systems with noise uncertainty. *IEEE Transactions on Automatic Control*, AC-26(4):902–906.

Rajan, K., editor (2013). *Informatics for Materials Science and Engineering*. Butterworth-Heinemann, Waltham, MA.

Rasmussen, C. E. and Williams, C. K. (2006). *Gaussian processes for machine learning*. MIT Press, Cambridge, MA.

Raudys, S. (1972). On the amount of a priori information in designing the classification algorithm. *Technical Cybernetics*, 4:168–174. in Russian.

Raudys, S. (1978). Comparison of the estimates of the probability of misclassification. In *Proc. 4th Int. Conf. Pattern Recognition*, pages 280–282, Kyoto, Japan.

Raudys, S. and Jain, A. (1991). Small sample size effects in statistical pattern recognition: Recommendations for practitioners. *IEEE Transactions on Pattern Analysis and Machine Intelligence*, 13(3):4–37.

Raudys, S. and Young, D. (2004). Results in statistical discriminant analysis: a review of the former soviet union literature. *Journal of Multivariate Analysis*, 89:1–35.

Rissanen, J. (1989). *Stochastic complexity in statistical inquiry*. World Scientific.

Robert, C. (2007). *The Bayesian Choice: From Decision-Theoretic Foundations to Computational Implementation*. Springer, 2nd edition.

Rogers, W. and Wagner, T. (1978). A finite sample distribution-free performance bound for local discrimination rules. *Annals of Statistics*, 6:506–514.

Rosenblatt, F. (1957). The perceptron – a perceiving and recognizing automaton. Technical Report 85-460-1, Cornell Aeronautical Laboratory, Buffalo, NY.

Rosenthal, J. (2006). *A First Look At Rigorous Probability Theory*. World Scientific Publishing, Singapore, 2nd edition.

Ross, S. (1994). *A first course in probability*. Macmillan, New York, 4th edition.

Ross, S. (1995). *Stochastic Processes*. Wiley, New York, 2nd edition.

Rumelhart, D. E., Hinton, G. E., and Williams, R. J. (1985). Learning internal representations by error propagation. Technical report, California Univ San Diego La Jolla Inst for Cognitive Science.

Sayre, J. (1980). The distributions of the actual error rates in linear discriminant analysis. *Journal of the American Statistical Association*, 75(369):201–205.

Schafer, J. and Strimmer, K. (2005). A shrinkage approach to large-scale covariance matrix estimation and implications for functional genomics. *Statistical Applications in Genetics and Molecular Biology*, 4(1):32.

Schena, M., Shalon, D., Davis, R., and Brown, P. (1995). Quantitative monitoring of gene expression patterns via a complementary DNA microarray. *Science*, 270:467–470.

Schiavo, R. and Hand, D. (2000). Ten more years of error rate research. *International Statistical Review*, 68(3):295–310.

Schroeder, M. (2009). *Fractals, chaos, power laws: Minutes from an infinite paradise*. Dover.

Sima, C., Attoor, S., Braga-Neto, U., Lowey, J., Suh, E., and Dougherty, E. (2005a). Impact of error estimation on feature-selection algorithms. *Pattern Recognition*, 38(12):2472–2482.

Sima, C., Braga-Neto, U., and Dougherty, E. (2005b). Bolstered error estimation provides superior feature-set ranking for small samples. *Bioinformatics*, 21(7):1046–1054.

Sima, C. and Dougherty, E. (2006). Optimal convex error estimators for classification. *Pattern Recognition*, 39(6):1763–1780.

Sima, C., Vu, T., Braga-Neto, U., and Dougherty, E. (2014). High-dimensional bolstered error estimation. *Bioinformatics*, 27(21):3056–3064.

Simonyan, K. and Zisserman, A. (2014). Very deep convolutional networks for large-scale image recognition. *arXiv preprint arXiv:1409.1556*.

Sitgreaves, R. (1951). On the distribution of two random matrices used in classification procedures. *Ann. Math. Statist.*, 23:263–270.

Sitgreaves, R. (1961). Some results on the distribution of the W-classification. In Solomon, H., editor, *Studies in Item Analysis and Prediction*, pages 241–251. Stanford University Press.

Smith, C. (1947). Some examples of discrimination. *Annals of Eugenics*, 18:272–282.

Snapinn, S. and Knoke, J. (1985). An evaluation of smoothed classification error-rate estimators. *Technometrics*, 27(2):199–206.

Snapinn, S. and Knoke, J. (1989). Estimation of error rates in discriminant analysis with selection of variables. *Biometrics*, 45:289–299.

Stark, H. and Woods, J. W. (1986). *Probability, random processes, and estimation theory for engineers*. Prentice-Hall, Inc.

Stein, M. L. (2012). *Interpolation of spatial data: some theory for kriging.* Springer Science & Business Media.

Stigler, S. M. (1981). Gauss and the invention of least squares. *The Annals of Statistics*, 9:465–474.

Stone, C. (1977). Consistent nonparametric regression. *Annals of Statistics*, 5:595–645.

Stone, M. (1974). Cross-validatory choice and assessment of statistical predictions. *Journal of the Royal Statistical Society. Series B (Methodological)*, 36:111–147.

Sutton, R. S. and Barto, A. G. (1998). *Introduction to reinforcement learning.* MIT press Cambridge.

Tamayo, P., Slonim, D., Mesirov, J., Zhu, Q., Kitareewan, S., Dmitrovsky, E., Lander, E. S., and Golub, T. R. (1999). Interpreting patterns of gene expression with self-organizing maps: methods and application to hematopoietic differentiation. *Proceedings of the National Academy of Sciences*, 96(6):2907–2912.

Tan, A. C., Naiman, D. Q., Xu, L., Winslow, R. L., and Geman, D. (2005). Simple decision rules for classifying human cancers from gene expression profiles. *Bioinformatics*, 21(20):3896–3904.

Tanaseichuk, O., Borneman, J., and Jiang, T. (2013). Phylogeny-based classification of microbial communities. *Bioinformatics*, 30(4):449–456.

Teichroew, D. and Sitgreaves, R. (1961). Computation of an empirical sampling distribution for the *w*-classification statistic. In Solomon, H., editor, *Studies in Item Analysis and Prediction*, pages 285–292. Stanford University Press.

Tibshirani, R. (1996). Regression shrinkage and selection via the lasso. *Journal of the Royal Statistical Society: Series B (Methodological)*, 58(1):267–288.

Tibshirani, R., Hastie, T., Narasimhan, B., and Chu, G. (2002). Diagnosis of multiple cancer types by shrunken centroids of gene expression. *PNAS*, 99:6567–6572.

Toussaint, G. (1971). Note on optimal selection of independent binary-valued features for pattern recognition. *IEEE Transactions on Information Theory*, 17(5):618.

Toussaint, G. (1974). Bibliography on estimation of misclassification. *IEEE Transactions on Information Theory*, IT-20(4):472–479.

Toussaint, G. and Donaldson, R. (1970). Algorithms for recognizing contour-traced hand-printed characters. *IEEE Transactions on Computers*, 19:541–546.

Toussaint, G. and Sharpe, P. (1974). An efficient method for estimating the probability of misclassification applied to a problem in medical diagnosis. *IEEE Transactions on Information Theory*, IT-20(4):472–479.

Tutz, G. (1985). Smoothed additive estimators for non-error rates in multiple discriminant analysis. *Pattern Recognition*, 18(2):151–159.

van de Vijver, M., He, Y., van't Veer, L., Dai, H., Hart, A., Voskuil, D., Schreiber, G., Peterse, J., Roberts, C., Marton, M., Parrish, M., Astma, D., Witteveen, A., Glas, A., Delahaye, L., van der Velde, T., Bartelink, H., Rodenhuis, S., Rutgers, E., Friend, S., and Bernards, R. (2002). A gene-expression signature as a predictor of survival in breast cancer. *The New England Journal of Medicine*, 347(25):1999–2009.

Vapnik, V. (1998). *Statistical Learning Theory*. Wiley, New York.

Vitushkin, A. (1954). On hilberts thirteenth problem. *Dokl. Akad. Nauk SSSR*, 95(4):701–704.

Vu, T., Braga-Neto, U., and Dougherty, E. (2008). Preliminary study on bolstered error estimation in high-dimensional spaces. In *Proceedings of GENSIPS'2008 - IEEE International Workshop on Genomic Signal Processing and Statistics*. Phoenix, AZ.

Vu, T., Sima, C., Braga-Neto, U., and Dougherty, E. (2014). Unbiased bootstrap error estimation for linear discrimination analysis. *EURASIP Journal on Bioinformatics and Systems Biology*, 2014:15.

Wald, A. (1944). On a statistical problem arising in the classification of an individual into one of two groups. *Ann. Math. Statist.*, 15:145–162.

Wang, Y., Tian, Y., Kirk, T., Laris, O., Ross Jr, J. H., Noebe, R. D., Keylin, V., and Arróyave, R. (2020). Accelerated design of fe-based soft magnetic materials using machine learning and stochastic optimization. *Acta Materialia*, 194:144–155.

Ward Jr, J. H. (1963). Hierarchical grouping to optimize an objective function. *Journal of the American statistical association*, 58(301):236–244.

Webb, A. (2002). *Statistical Pattern Recognition*. John Wiley & Sons, New York, 2nd edition.

Werbos, P. (1974). *Beyond Regression: New Tools for Prediction and Analysis in the Behaviorial Sciences*. PhD thesis, Harvard University, Cambridge, MA.

Wolpert, D. (2001). The supervised learning no-free-lunch theorems. In *World Conference on Soft Computing*.

Wyman, F., Young, D., and Turner, D. (1990). A comparison of asymptotic error rate expansions for the sample linear discriminant function. *Pattern Recognition*, 23(7):775–783.

Xiao, Y., Hua, J., and Dougherthy, E. (2007). Quantification of the impact of feature selection on cross-validation error estimation precision. *EURASIP J. Bioinformatics and Systems Biology*.

Xie, S. and Braga-Neto, U. M. (2019). On the bias of precision estimation under separate sampling. *Cancer informatics*, 18:1–9.

Xu, Q., Hua, J., Braga-Neto, U., Xiong, Z., Suh, E., and Dougherty, E. (2006). Confidence intervals for the true classification error conditioned on the estimated error. *Technology in Cancer Research and Treatment*, 5(6):579–590.

Yonezawa, T., Suzuki, K., Ooki, S., and Hashimoto, A. (2013). The effect of chemical composition and heat treatment conditions on stacking fault energy for fe-cr-ni austenitic stainless steel. *Metallurgical and Materials Transactions A*, 44A:5884–5896.

Zhou, X. and Mao, K. (2006). The ties problem resulting from counting-based error estimators and its impact on gene selection algorithms. *Bioinformatics*, 22:2507–2515.

Zollanvari, A., Braga-Neto, U., and Dougherty, E. (2009a). On the sampling distribution of resubstitution and leave-one-out error estimators for linear classifiers. *Pattern Recognition*, 42(11):2705–2723.

Zollanvari, A., Braga-Neto, U., and Dougherty, E. (2010). Joint sampling distribution between actual and estimated classification errors for linear discriminant analysis. *IEEE Transactions on Information Theory*, 56(2):784–804.

Zollanvari, A., Braga-Neto, U., and Dougherty, E. (2011). Analytic study of performance of error estimators for linear discriminant analysis. *IEEE Transactions on Signal Processing*, 59(9):1–18.

Zollanvari, A., Braga-Neto, U., and Dougherty, E. (2012). Exact representation of the second-order moments for resubstitution and leave-one-out error estimation for linear discriminant analysis in the univariate heteroskedastic gaussian model. *Pattern Recognition*, 45(2):908–917.

Zollanvari, A., Cunningham, M. J., Braga-Neto, U., and Dougherty, E. R. (2009b). Analysis and modeling of time-course gene-expression profiles from nanomaterial-exposed primary human epidermal keratinocytes. *BMC Bioinformatics*, 10(11):S10.

Zollanvari, A. and Dougherty, E. (2014). Moments and root-mean-square error of the bayesian mmse estimator of classification error in the gaussian model. *Pattern Recognition*, 47(6):2178–2192.

Zolman, J. (1993). *Biostatistics: Experimental Design and Statistical Inference*. Oxford University Press, New York, NY.

Zou, H. and Hastie, T. (2005). Regularization and variable selection via the elastic net. *Journal of the royal statistical society: series B (statistical methodology)*, 67(2):301–320.

Index

Akaike information criterion (AIC), 280

apparent error, 152

approximation error, 186

bagging, 61, 139

balanced sampling, 70

bandwidth, 90, 96

Bayes classifier, 20

Bayes decision rule, 37

Bayes error, 24

Bayes Theorem, 292

Bayesian information criterion (BIC), 280

best linear unbiased estimator (BLUE), 263

binary tree, 136

bolstered empirical distribution, 165

boosting, 63

bootstrap, 61, 63

    balanced, 164

    complete, 164

    sample, 163

Borel σ-algebra, 288, 290

Borel set, 16, 38, 288, 289, 294, 296

Borel-Cantelli Lemma

    Second, 291

Borel-Cantelli Lemma First, 291

Borel-measurable function, 17, 26, 52, 227, 255, 289, 299, 301, 305

Borel-measurable space, 289

Bounded Convergence Theorem, 311

branch-and-bound algorithm, 224

categorical feature, 136

Central Limit Theorem, 312

Chernoff error, 35

class-conditional densities, 12, 16

class-specific errors, 18

classification, 4

Classification and Regression Tree (CART), 137

classification error, 4, 18, 54, 206

classification rule

    Bayesian parametric, 81

    consistent, 6, 55

    covariance plug-in, 71

    cubic histogram, 91

    discrete histogram, 53

    ensemble, 60

    histogram, 91

    k-top scoring pair, 142

    kernel, 95

    nearest-centroid, 52

    nearest-neighbor, 52, 93

    parametric plug-in, 67

    random, 61

    smart, 64

    strongly consistent, 55

    super, 64

    symmetric, 161

    top-scoring median, 142

    top-scoring-pair, 141

    universally consistent, 6, 55

classifier, 4, 17

cluster membership, 236

cluster responsibility, 236

Printed in the United States
by Baker & Taylor Publisher Services